Solutions to Problems of Controlling Long Waves with the Help of Micro-Structure Tools

Authored By

Vladimir V. Arabadzhi

Senior Researcher Department of Hydroacoustics and Hydrodynamics
Institute of Applied Physics
(Russian Academy of Science)
Nizhny Novgorod
Russia

CONTENT

Foreword

After radar and sonar were founded, problems of the "visibility" reduction for physical bodies in air (electromagnetic waves) or in water (acoustical waves) have immediately become serious in physics and technics. Facilities of the "visibility" reduction are oriented physically for the reduction of radiation field and reduction of scattering field of physical bodies. Now many problems of this circle have become classical, and many researchers thought that these problems cannot have solutions for a practical scenario. New technologies (sensors, actuators) of high spatial-temporal resolution plus new computers with fast and accurate calculations in combination with untraditional versions of solutions of above boundary problems, allow to obtain the success in some cases. Traditional thick and weakly absorbing coatings are unacceptable today because of their greater size and weight. Interference coatings have a thickness about ¼ wavelength, but they have narrow band of working directions and frequencies. The next attempt was the active damping (with radiation of cancellation anti-wave), suggested and developed by G. Maljuzhinets, M. Jessel, G. Mangiante. This idea permits to design coating of thickness much smaller than the wavelength. The works by B. Widrow on adaptive filtering have been applied by C. Fuller, C. Hansen, S. Elliott successfully for active noise control with adaptation ("training" during real time) to a prior unknown boundary value problem (which can also change). Training of active control system requires enough big time, because it must have total acoustical information (vast amount) on the boundary value problem. In real conditions (elastic shell in various depths and temperatures, with vibroacoustical characterictics, which were investigated by M.C. Junger and D. Feit) this vast volume of information deviates more quickly than the process of training. In recent times the idea of cloacking becomes very popular, suggested by J. Pendry. This presents the solution of above general problem for some body with cavity. Incident wave does not penetrate the cavity and, on the other hand, can not be scattered by the external surface of a body. Therefore any body can be spaced into this cavity, and will be invisible for outside observer. This result is achieved, due to special microdistribution of parameters of body material (called "metamaterials"). The cancellation of radiation and scattering field of shell in liquid is connected with following serious problems: (a) wideband acoustical fields to be damped; (b) neutral floatability of a shell in liquid; (c) the absence of dynamical support (consequence of (b)); (d) the absence of inertial coordinate system to measure shell's surface displacements (consequence of (b)); (e) the uncertainty of many parameters of shell and incident wave; and (f) problem of compactness of active control system. The book by V. Arabadzhi presents a rare attempt to solve the problem as a whole. Author uses stable technological tendencies in miniaturization and acceleration of sensors, actuators, computer components. On the other hand the lengths of waves to be damped, were constant due to the constant condition of their far propagation. Taking into account the technological progress, the author also used several untraditional approaches to the problem: conversion of the waves to be damped, into the waves of spatial and temporal frequencies which are "invisible" for the used receivers. I hope this book will be useful for a lot of readers with various interests in physics of waves. This book can help them to find new approaches and solutions of a wide area of problems.

Mal'tsev, A.A.

Nizhny Novgorod State University,
Russia

Preface

Basic wave problems annotated in this book are the control of radiation, scattering and absorption, "black body," and "transparent body." Possible technical applications include but not limited to the absorption of sound by linear multipoles and unidirectional multipoles from Huygens sources; parametric coating of small waves for the absorption of long electromagnetic waves; active system for conversion of any physical body in liquid to an acoustically transparent one; parametric muffler that is small size in comparison with the length of suppressed wave. We propose superwideband (temporal, with initial conditions) representation of the boundary-value wave problem. Active control systems considered here use special control algorithms. These algorithms form the control signals that are much smaller than the minimum time scale of the waves to be damped. The control algorithms include logical nonlinear or modulation operations that do not require interpretation of the control system found in traditional combination of linear electric circuits where all parameters are constant in time. The proposed active control systems need accurate information on the geometry, but not full information on the boundary-value problem.

Traditional self-restrictions. There are some self-restrictions in the statement of the active wave damping problem. The most self-restriction of such a type is the operating of complex amplitude at the frequency of the wave to be damped. Traditional control of complex amplitudes applied to problems of reflective damping for instance leads to some principle contradiction i.e. having the complete spectral information on the incident wave when we see that this wave is over already. Traditional approaches require accumulating vast amounts of information on the wave field and the boundary problem such as factorization of the wave field in the incident and the reflected one, spectral estimates of fields, and identification of Green's functions. This requires time demanding training (tuning) compared with the drift time scale of the boundary problem parameters and the maximum time scale of the damped field. Note that optimum version of such a system (if we had chance to identify them precisely) would need some combination of linear electric circuits between sensors and emitters (actuators) with parameters constant in time.

Some opinions have appeared concerning the universal solvability of active damping problems by means the adaptive 'Widrow'like structures if at most we had sufficient number of adaptive filters of sufficient time-length. Such an intelligent system has tried in some form to build a copy of the boundary problem. A very useful property of such a system was the modeling of chain factors including acoustical, electric, mechanical, etc. into one total linear electric circuit with parameters constant in time. But principal problems still existed in tuning time and accuracy for the large dimension when a number of emitters and sensors were used for damping the system; and in finite drift time scale of the boundary problem. The modeling noise (tuning noise, finite accuracy) was a natural paying price for adaptivity in these structures.

In addition we review the problem of acoustically transparent body of neutral floatability in liquid. To solve this problem we computer simulate the complicated vibroacoustical characteristics of the protected body very accurately, i.e. know all dynamic parameters of the body. To control the radiation and scattering fields of the protected body we need to either support it by mechanical attachment (vibrostat), or control the body's oscillations based on full information about its dynamical behavior. In many cases of practical interest the protected body has neutral floatability (vibrostat is absent). This body is rigid if surrounded by air, but soft for acoustical pressure in a liquid. Vibroacoustical parameters of the body are characterized simultaneously by a vast volume of information, and fast temporal changes. The temporal changes of temperature and hydrostatic pressure outside the protected body does not permit identification of a body's parameters with sufficient speed for effective control.

We propose a super wideband (temporal, with initial conditions) representation of the boundary value wave problem, and fast control of parameters in real-time that permits us to avoid previous difficulties.

Vladimir V. Arabadzhi

Department of Hydroacoustics and Hydrophysics,
Institute of Applied Physics (RAS)
Russia

CHAPTER 1

Introduction

Abstract: In this chapter we formulate the basic principles of the approach used in this book. In the beginning we present a brief description of known traditional passive and active approaches to the problem of "invisibility" of physical bodies in the wave fields of various physical genesis. All these approaches can be reduced to a combination of linear electric circuits with constant parameters. Due to the physical conditions of far propagation, the wavelengths of incident waves retain of the constant order for many years. We consider the contemporary achievements of electronic actuators and sensors: growth of rapidity and decreasing of geometrical sizes of actuators and sensors together with the growth of speed and accuracy of computer calculations present the stable technological tendency. What benefit we could get from this progress in solution of above mentioned problems? Below we consider several simplest solutions of one-dimensional boundary value problems. The suggested control algorithms use fast parametric and fast logical operations. Therefore, these algorithms can not be reduced to a combination of linear electric circuits with constant parameters (as in traditional approaches above). Control algorithms suggested are characterized by spatial locality and temporal locality of control. This means, for instance, that we need not know the period and wavelength of incident wave for effective control.

Any physical body manifest itself *via* heat, radiowave and acoustic trace. Our goal is the cancellation of microwave and acoustical signatures of physical bodies, that is reduction of the their "visibility", disguise. For instance the presence of any body in a liquid is acoustically proclaimed by its radiation and scattering fields. The absence of these fields means that the body is invisible (for example "acoustically invisible"). Invisible bodies are desirable in many cases, and this has led to numerous works devoted to the control of acoustical (and electromagnetic) radiation and scattering fields created by bodies [1-3].

This problem is often divided into three physical sub-problems [4]: radiation suppression, suppression of backscattering, suppression of forwardscattering, absorption maximization. A body, which does not produce backscattering and forwardscattering is called a "transparent body" (see Chapter 4). We note again that an absorbing coating reduces only backscattering and creates a "black body" (see Chapter 3), but does not reduce forwardscattering, and in principle cannot create a "transparent body".

It is always desirable to achieve effective suppression of the "visibility" if:

1) the amount of available information about the boundary problem is minimal;
2) the amount of information about the incident waves is minimal;
3) the size of the suppressing sound system is minimum (for example, thickness of the protective coating, *i.e.*, $D_V / h >> 1$, $\lambda / h >> 1$, where D_V is the body's size, λ is the length of the wave to be suppressed, h is the coating thickness);
4) frequency range of the waves to be damped is wide, *i.e.* $\omega_{max} / \omega_{min} >> 1$ (where ω_{max} and $\omega_{max} / \omega_{min} >> 1$ are the maximum and minimum frequencies of the waves to be suppressed).

Below we present a short review of known tools (methods) for the visibility reduction of physical bodies.

1.1. PASSIVE METHODS (PASSIVE COATINGS)

Passive methods to reduce the scattering field and radiating field requires equipping the protected body with some coating which uses wave absorption or interference, or a complicated coating based on metamaterials, but always without the use of any outside source of energy.

1.1.1. Absorbing Coatings

Nonresonant absorption by coatings can be effective if: (a) the external surface of the coating does (almost) not reflect the incident wave; (b) the incident wave, passing through the coating and reflected from the internal surface of the coating, becomes very weak (due to absorption) when returning to the external surface of coating. The main reasons to choose such passive coatings are their simplicity, their wideband frequency range of incident waves and the wideband range of angles of incoming incident waves of arbitrary structures. Besides this, such coatings also suppress the radiation of the protected body (**Fig. 1-1-a**). The main drawback of passive coatings is their large thickness $h << \lambda$ (equivalent to several incident wavelengths $h >> \lambda$). This precludes the use of passive coatings in low frequency applications. Besides this, good absorbing coatings can only effectively suppress backscattering, but cannot suppress forward scattering.

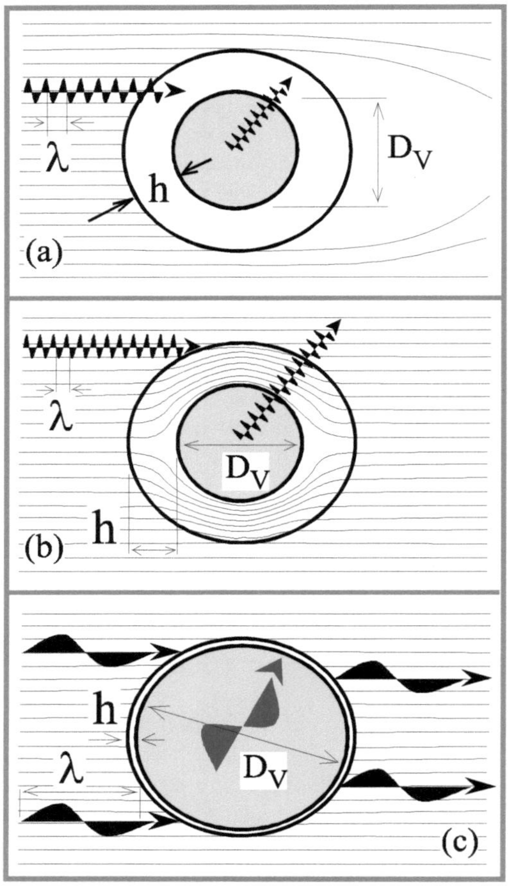

Figure 1-1: Traditional thick passive coating for absorption ($h >> \lambda$) suppresses the body's radiation and its back scattering, but does not suppress forward scattering (a); (cloaking) thick ($h >> \lambda$) passive coating, constructed from metamaterials, makes the body transparent, but does not suppress the body's radiation (b); thin ($h << \lambda$) active coating, which suppresses both scattering to the back and forward scattering and radiation of the protected body (c). The horizontal parallel lines on the left side of (a-c) represent the power stream lines.

Passive absorbing coatings can also be constructed from a large ensemble of passive oscillators with resonant absorption (see Chapter 2). In principle, the thickness of such a coating can be much smaller than the length λ of the incident wave at a wide frequency band. However to absorb incident waves with oscillators, we need to compensate for the dynamical interaction of oscillators, for instance, with special electroacoustical (for instance) chains. In this case, the parameters of each oscillator must depend on the arrival direction of each incident wave. This fact presents a great difficulty for effective absorption over a wide range of arrival directions, so a coating becomes a very complicated oscillatory system, which is very sensitive to any small perturbation of its parameters such as temperature, humidity, and material aging. We may here consider scattering of a plane wave from a body of an isotropic shape, for example, a sphere covered by an interference coating (see below). Maximization of the absorbed power of this wave uniquely determines optimum dynamic couplings (elements with constant parameters) between absorbing oscillators. This means that this coating is inefficient, in particular, for the absorption of an incident wave arriving from the opposite direction.

1.1.2. Coatings Based on Interference

Interference coatings are homogeneous in the tangential direction and discrete in the longitudinal direction. Widely used to suppress backscattering, these structures are based on mutual compensation of waves reflected from the exterior and interior surfaces of a coating. The same effect is exploited in bloomed optics. Dissipation localized on the aforementioned surfaces causes an inevitable reflection. A method is based on the exact quasi-resonance relationship between the wavelength (frequency) of an incident wave and the coating thickness when a plane wave propagates normally to a plane coating. Thus, the thickness of an interference coating should be no less than a quarter-wavelength to provide for the phase shift equal to π. Any departure from the problem formulated above under the assumption that a plane monochromatic wave is normally incident on 1D coating results in an increase in the backscattered field. In particular, when the parameters of such a coating on a 3D convex body are tangentially homogeneous, the desired interference relationships between the waves reflected from coating layers are violated. Therefore, an object to be masked by the aforementioned coating is sometimes shaped as close to a plane as possible (*i.e.*, shaped as a locally 1D body). For example, in a spherical interference coating (in the case when the sphere's dimension considerably exceeds the wavelength), only the waves reflected by a section parallel to the incident wavefront are perfectly compensated for at a given frequency. As the distance from this section of the sphere increases, the contributions of the remaining coating areas to the backscattered field grow. This growth is limited by the grazing incidence.

1.1.3. Coatings Based on Metamaterials

Pendry and others [1-3], [5-9] have provided several very attractive analytical solutions of the problem of transformation of any opaque body into a transparent one. These solutions are obtained with a ray-like representation of the wave fields both for electromagnetic and acoustical waves and are represented by passive coating. This coating contains metamaterials (where, for instance, the phase and group velocities of a wave have mutually opposite signs). Due to the exotic characteristics of metamaterials, the incident wave avoids the protected body without scattering behind the body (see **Fig. 1-1-b**). For the solution by Pendry and others in its acoustical version [8-10] we need the anisotropic mass density, which is a difficult problem in practice. The drawbacks of this approach are: (a) the energy concentration of rays (and field intensity) in the tangential area of a coating with thickness h (**Fig. 1-1-b**) becomes too large if $D_V >> h$; (b) the thickness h of this coating must be much greater than the length λ of an incident wave (like in the passive absorptive coatings mentioned above) because the wave field is represented by rays; (c) this coating does not suppress any radiation from the protected body; (d) any practical realization of these coatings can achieve the desired exotic parameters only for sufficiently narrowband incident waves, due to the effects of the wave dispersion in the coating.

Recently the work devoted to passive coating with distributed mechanical connections was published [11]. This coating reduces scattering and has less thickness than a classic thick passive absorbing coating. However, this coating requires (assumes) the mechanical support vibrostat This causes the inapplicability of coating for the bodies of neutral floatability in liquid.

1.2. ACTIVE METHODS (ACTIVE COATINGS)

Active methods of sound control present an alternative to the passive ones and require an outside source of energy. The main difference between active and passive methods is the radiation of special waves, which have the same propagation direction and magnitude as the controlled waves, but the opposite wave sign. In the monochromatic case this means radiation of anti-phase waves. Active methods can be attractive for the reason that active coatings can be much thinner than the length of the wave to be damped (**Fig. 1-1-c**). To realize this coating we need very small sensors and actuators, electrically connected with the control center (computer). From a rigorous mathematical point of view, the methods for active control of waves represent a large branch of automatic control theory or an optimal control having a with criteria of minimum errors or fastest descent,...etc, see for reference the books by I. Lasiecka and others [12-14]).

1.2.1. Malyuzhinets' Method

The possibility of controlling both radiation and scattering fields created by an arbitrary physical body (**Fig. 1-1-c**) was shown [15-19] by G.D. Malyuzhinets, which becomes possible due to the use of Huygens' surfaces, characterized by unidirectional radiation (or receipt, see Chapter 5). Practical implementation of this method requires the synthesis of Huygens' radiating (and receiving) surfaces (with discrete elements) and development of the algorithm, which connects them. The Malyuzhinets solution has been formulated for monochromatic fields. Practical solutions require causal algorithmic realization of this system in real time, wideband frequency range, and a spatially discrete form, which lead to great difficulties in a literal realization of Malyuzhinets' theory. To realize Huygens' surfaces, we need to use spatially discrete antenna arrays. The latter (and any devices required to attach them) must be acoustically transparent, *i.e.* the sizes of their elements and the spatial period of their spacing must be much smaller than the length of the waves to be suppressed. This is the reason why the magnitudes of the array's elements must be much greater than the magnitude of the waves to be cancelled. On the other hand, to retain the unidirectional characteristics of the Huygens surface, the field of the discrete version of the Huygens surface (with addition of the reactive field) must be very small (this permits exclusion of their interaction with the protected body). This interaction could break the unidirectional characteristics of Huygens' surfaces. This fact requires a permissible, but not very small distance between the array and the surface of the protected body. This creates a large technical difficulty: there is no way of placing a Huygens' surface immediately on the surface of the protected body. Malyuzhinets' theory has been tested experimentally and numerically for acoustical and electromagnetic waves [20].

1.2.2. Adaptive Methods

Some opinions have appeared concerning the universal solvability of active damping problems by means of the adaptive. "Widrow like" structures [21] we had sufficient number of adaptive filters of sufficient time-length. Such an intelligent system tried in some form to build a copy of the boundary problem. A very useful property of such a system was the modeling of chain factors including acoustical, electric, mechanical, *etc.* into one total linear electric circuit with parameters constant in time. But principal problems still existed in tuning time and accuracy for the large dimension when a number of emitters and sensors were used for damping the system; and in finite drift time scale of the boundary problem. The modeling noise (tuning noise, finite accuracy) was a natural paying price for adaptivity in these structures.

The conventional approach to the problem of active wave damping includes measurement of the wave field, identification (by measurement or using a priori data) of the complete spatio-temporal structure of the damping wave and the boundary condition, and generation of a cancellation wave, which is the inverse of the quenched wave. Moreover, the following ideas are commonly accepted in regard to the active wave damping problem:

1) The wave damping system can be efficient only if complete information on the boundary-value problem, including Green's function, and information on the damped waves (obtained by division of the field into incident and reflected, spectral estimation, *etc.*) is available. This condition is based on the classical "sensor-linear processor-actuator" scheme. The modern active wave damping system, based on adaptive algorithms and having an array of pickups, a matrix of adaptive linear processors with correlation feedbacks, and an array of actuators, permits one to take into account and renew information on the parameters of the boundary-value problem in real time. In the stationary (nearly optimal) state, such an adaptive quenching system reduces to a combination of linear electric circuits with constant parameters. The conventional methods of adaptation represent correlation tuning of either complex amplitudes (in quasimonochromatic problems where the amplitude and phase are of independent significance and can be tuned separately) or weight coefficients of multigate delay lines (in wideband problems).

One of the widely known self-limitations is operation by complex amplitudes of the actuators at the damped wave frequency. This is due to the traditional use of hydroacoustic piezoceramics at resonance frequencies. However, according to some papers on sound control [22-27], the effect of the intrinsic temporal scales of piezoceramics is suppressed efficiently by the feedback. Earlier, such a mode was used in piezoceramic devices for automatic adjustment of laser mirrors in the presence of slow vibrations. Traditional control of complex amplitudes applied to the problems of reflection suppression leads to some principle contradiction *i.e.* obtaining complete spectral information on the incident wave when we see that this wave is over already.

Solution of any problem of active wave control requires

a) a sufficient number of identification nodes (for example, weight coefficients [21]) in the boundary value problem;

b) a sufficient length of digital representation of the measured wave-field values;

c) a sufficient rate of calculation;

d) a sufficient accuracy of physical measurement of the wave field and of the actuators.

Thus, all the active control systems, usually considered, carry the total information copy of the wave field and the boundary problem (factorization of the wave field, the spectral estimates of fields, and the identification of Green's functions) as the necessary condition of the effective operation. However (in classical adaptive acive control systems [28-42]), identification of the boundary-value problem (or Green's functions) reduces to estimation of no less than $n_r \times n_t$ numbers, where $n_r = (2D_V / \lambda)^2$ is the number of spatial identification nodes, $n_t = \tau_G / \tau$ is the number of temporal identification nodes, D_V is the characteristic linear scale of the three-dimensional object being protected, λ and τ are the characteristic minimum spatial and temporal scales of the quenched wave, respectively ($\lambda = c\tau$ and c is the velocity of the wave propagation), and τ_G is the duration of Green's function, which is due to either finite wave dimensions of the boundary-value problem or multiple reflections, or wave dispersion (the latter takes place, when the waves are produced by a stone thrown into a deep basin). For the maximum achievable finite accuracy of measuring, calculating, and actuating devices, the relative root-mean-square error $\sqrt{n_r^2 n_t \tau / \tau_{aver}}$ of assigning the resultant amplitude of the quenched wave can only be decreased by increasing the time τ_{aver} of estimation (*i.e.*, the temporal averaging of test signals) of identification nodes. However, the drift of parameters in the boundary-value problem (*i.e.*, the drift of true values of the parameters in the identification nodes) can be greater than the apparent small error of parameter estimation, reached by increasing τ_{aver}, under the assumption of time constancy of the parameters. This makes the identification procedure increasingly ineffective with an increase in n_t and n_r, respectively.

1.2.3. Bobrovnitskii's Active Method

Yu.I.Bobrovnitskii [43] has proposed a solution of the problem of an acoustically transparent (nonscattering) body as some expression of external active forces, which are applied to the protected surface, *via* full acoustical pressure, measured immediately on the protected surface, or *via* full oscillatory normal velocity, also measured immediately on this surface. The relation between measured and controlled values gives the active force for the compensation of a scattering field, *i.e.* it converts the body into a transparent one. However, under the conditions of a body's neutral floatability in a liquid, active outside forces need mechanical support (*i.e.* a protected body), and these active forces induce additional oscillations of the protected body. These oscillations can be of very complicated shapes (in the case of a finite relation between elasticity of the body and compressibility of the external fluid) and of very simple shapes (in the case of oscillations of an ideally rigid body in a very compressible fluid). The scattering field cannot be suppressed without knowing these oscillations of the protected body, so the complicated vibroacoustical characteristics of the protected body must be simulated (by computer) very accurately, *i.e.* find out all the dynamic parameters of the body. To create the arbitrary desired force applied to the body's boundary the body's oscillations need to be controlled, based on full information about its dynamical behavior. In a lot of cases of practical interest, the protected body has neutral floatability. This body is rigid, if it is surrounded by air, but this body is soft for acoustical pressure in a liquid. Vibroacoustical parameters of the body are characterized simultaneously:

1) by a vast volume of information;
2) by fast temporal changes of this information, due to temporal changes of temperature and hydrostatic pressure outside the protected body, aging of materials.

This does not permit identification of a body's parameters with sufficient quickness for effective control [21].

1.3. SUPERWIDEBAND BOUNDARY VALUE PROBLEMS

Below we will consider solutions of several problems connected with wave suppression by active methods. The traditional active control approach includes the following steps: (1) measurement of the wave field; (2) extraction (identification) of total space-time structure of the wave to be damped from data measured; (3) generation of the damping wave *i.e.* the one inverted with respect to the wave to be damped.

We propose a super wideband (temporal, with initial conditions) representation of the boundary value wave problem, and fast control of parameters in real-time that permits us to avoid previous difficulties. In other words, we use very high spatial and temporal resolution microstructure tools for controlling long (low frequency) waves. Below we consider solutions using the space-time control of the distribution of parameters of the boundary-value problem in the direction normal to a layer (coating) as alternatives to the above approaches. This book is devoted to the search of alternative approaches, which may assist to produce cancellation wave without knowing the structure of the wave to be suppressed or make active wave control without cancellation wave at all.

At present, owing to rapid progress in miniaturization (for example, spatial step of chips today presents 32 nanometers and shrinks in 1.5 times per year) and temporal resolution (fastness) [44-45], which are two inseparable fields of microelectronics development, much attention is focused on ultrawideband temporal representations of electromagnetic and acoustic processes in various dynamical systems. On the other hand, the length of the wave to be suppressed is the same as tens years ago due to the constancy of natural physical conditions of its far propagation (in problems of "visibility suppression"). Therefore it seems perfereable to use microelectronics in these problems of long waves to be suppressed.

Representation of processes in a frequency range that extends from zero to the microwave (or ultrasound for instance, having new piezoelectric materials [26-27]) band is opening new avenues in computer technology [44-45], antenna engineering [46], radar (smart arrays and coatings [47]), and hydro-acoustical signatures of physical bodies. The development of these new areas of science and technology is stimulated by the necessity of enhancing the reliability of aerial target recognition [47] by certain radar (or sonar) systems and reducing the target detectability [48] by other radar (or sonar) systems. The latter problem, which is sometimes referred to as the problem of scattering (scattering cross section) control [49] can be solved by changing the direction of reflected-wave propagation; transforming the reflected - wave frequency; transforming the reflected-wave polarization for angles, frequencies, and polarizations that are inaccessible for reception; or providing for absorption of an incident wave in a coating.

Four algorithms (approaches, devices) are proposed below (section 1.6) for the space one-dimensional active damping problem. These control systems and algorithms ensure the solution operating in the space-time region of the scales to be much less than the minimum space-time scales of the wave to be suppressed, *i.e.* this control is local in space and time.

1.4. STATEMENT OF THE PROBLEM

We consider (see **Fig.1-2**) the problem of damping the wave, reflected from the boundary x_0 of a semi-infinite ($x_0 \le x < \infty$) elastic rod of cross-section area S the impedance $Z = S\sqrt{\rho E}$, wave number $k = \omega\sqrt{\rho / E}$, mass density ρ , and elastic modulus E . This problem is described by the simplest wave equation

$$(\partial^2 u / \partial x^2) - c^{-2}(\partial^2 u / \partial t^2) = F_I \qquad \textbf{(1-1)}$$

(without dispersion), where $u(x,t)$ is the longitudinal shift of the rod particles from the equilibrium state, $c = \sqrt{E / \rho}$ is the sound speed in the rod, $F_I(x,t)$ is the source (concentrated far from x_0) of the waves incident from the right to the boundary x_0 with the boundary condition of the form

$$\left[\overline{p}(\partial / \partial t)u + \overline{\overline{p}}(\partial / \partial x)u = B\right]_{x=x_0} \qquad \textbf{(1-2)}$$

where $\overline{p} = \overline{p}(t)$ and $\overline{\overline{p}} = \overline{\overline{p}}(t)$ are boundary parameters, $B = B(t)$ (boundary source) are the control functions with characteristic temporal scale τ_c and temporal period T_c . The solution of the problem (1-1), (1-2) near the boundary x_0 , where $F = 0$, is described by the sum of waves of mutually opposite directions $u(x,t) = \upsilon(x - ct) + w(x + ct)$. The $w(x + ct)$ means the incident wave with characteristic temporal scale

$$\sim \tau_w >> T_c >> \tau_c \qquad \textbf{(1-3)}$$

and spatial scale $\sim \ell_w = c\tau_w$.

The problem of damping consists in minimizing some criterion of the component $u(x-ct)$ by the control functions $\overline{p}$, $\overline{\overline{p}}$, B without knowing anything about the incident wave, besides slowness of the value

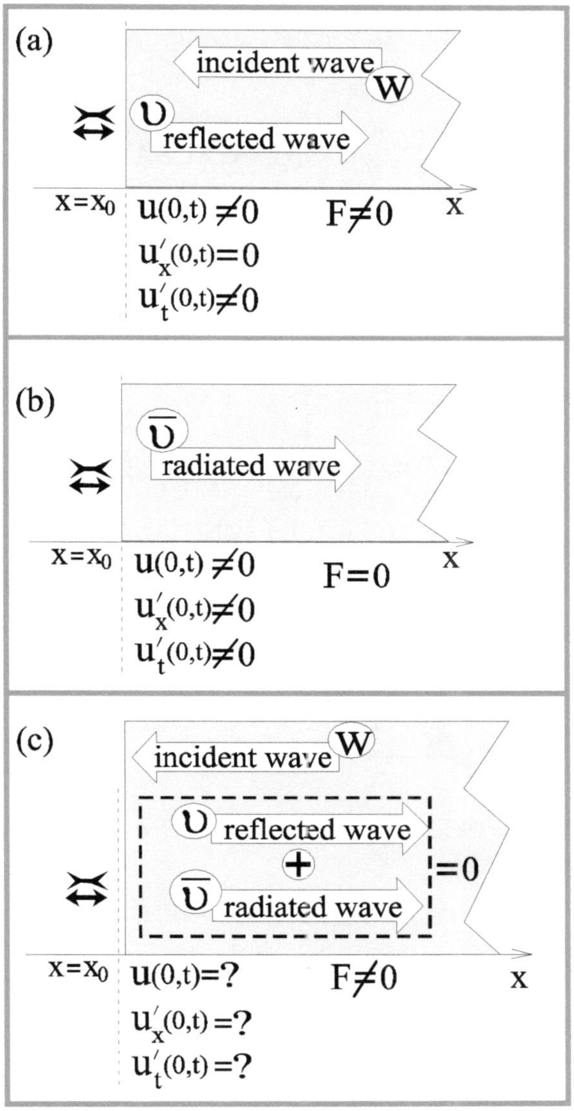

Figure 1-2: Geometry of the one-dimensional boundary-value problem: (a) free end of elastic rod, incident wave of particles longitudinal shifts $w(x+ct)$ produces the reflected wave $\upsilon(x-ct)$; (b) incident wave is absent in $w=0$; the end x_0 is driven (not free) by some force and produces the radiated wave $\overline{\upsilon}(x-ct)\neq 0$; (c) linear combination of problems (a) and (b) above and their solution $u(x,t)=\upsilon(x-ct)+\overline{\upsilon}(x-ct)+w(x+ct)$.

$$\max\left|(\partial/\partial t)w(x+ct)\right| << a/\tau_c \tag{1-4}$$

and smoothness of the value

$$\max\left|(\partial/\partial x)w(x+ct)\right| << a/(c\tau_c), \tag{1-5}$$

where $a \sim |w|$ is a characteristic scale of longitudinal displacements amplitude in the incident wave. More precisely and accurately we will minimize the average value as

$$(aT_w)^{-1}\left|\int_t^{t+T_C}(\upsilon+\bar{\upsilon})dt\right| << 1 \text{ or } (aT_w)^{-1}\left|\int_t^{t+T_C}(\upsilon+\bar{\upsilon})dt\right| \to 0. \quad \textbf{(1-6)}$$

We assume that on very high frequencies (of order $\geq 2\pi / T_c$), the dissipation (with distance $r << c\tau_w$ of relaxation in $\sim e$ times) exists in the waveguiding media.

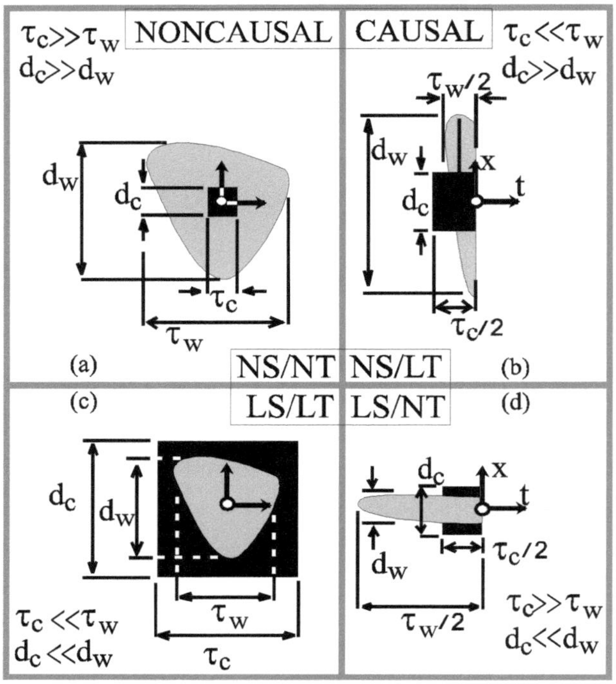

Figure 1-3: Classification of algorithms on the base of their spatio-temporal requirements on the data on the wave to be damped, and about boundary problem, and causality of algorithms.

1.5. CLASSIFICATION OF THE CONTROL ALGORITHMS

In principle, a stable active system represents artificial constructions (with sensors and actuators) and can be formally replaced by corresponding combination of functions $\bar{p}$, $\bar{\bar{p}}$, B. This construction can help to achieve the following goals: cancellation of radiated or and reflected waves, and absorption of incident waves.

All algorithms and active control systems below can be separated in the following conditions: (a) so called feedback systems (F-systems) refer to the operation in time and space with right part $B(t)$ of boundary conditions (sources); (b) so called modulated systems (M-systems) refer to the parametric temporal operations with coefficients $\bar{p}(t)$, $\bar{\bar{p}}(t)$ of boundary conditions. There are however no reasons to combine these feedback and modulation principles.

Below we will use full classification of control algorithms, including temporal locality (LT or NT), spatial locality (LS or NS) and causality terms. **Figs. 1-3** illustrate this classification as "x" denoting the spatial dimension, and "t" denoting temporal dimension. Black area denotes the ones for input information measured to access the control action in white point (center $x=t=0$ of each picture (a)-(d)). Gray area denotes the smallness along "x", with characteristic scale d_c and fastness along "t", with characteristic scale τ_c of actuators of active control system. Symbols τ_w and d_w denote characteristic temporal scale and spatial scale respectively of area (black) of data collection measuring the control action in $x=t=0$. Now we can classify control algorithms as following: $\tau_c >> \tau_w$, $d_c >> d_w$ (**Fig. 1-3-a**) algorithm nonlocal temporally (NT), nonlocal spatially (NS), noncausal (NC), *i.e.* current action depends of the future measurements; $\tau_c << \tau_w$, $d_c >> d_w$ (**Fig. 1-3-b**) algorithm local temporally (LT), nonlocal spatially (NS), causal (CA), *i.e.* current action depends of the measurements in past; $\tau_c << \tau_w$, $d_c << d_w$ (**Fig. 1-3-c**) algorithm local temporally (LT), nonlocal spatially (NL), noncausal (NC); $\tau_c >> \tau_w$, $d_c << d_w$ (**Fig. 1-3-d**) algorithm nonlocal temporally (NT), local spatially (LS), causal (CA). Therefore the full classification of algorithms will have the form:

{LS or NS / LT or NT / CA or NC / F or M}.

1.6. EXAMPLES OF LOCAL ALGORITHMS OF ACTIVE WAVE CONTROL

Below we consider several examples of boundary value problems, controlled by {LS/LT/CA/M)} and {LS/LT/CA/F} algorithms, where we may not complete the information on the wave to be damped and on the boundary value problem in acoustics.

1.6.1. Half-Return Algorithm {LS/LT/C/F}.

At the first step we assume the absence of incident wave, *i.e.* $w = 0$. The control functions for such an algorithm of the half return have the form [50]:

$$\overline{p} = 0,\ B(t) = \overline{\overline{p}}(t_n)\sum_n \Pi[(t - t_n)/\tau_c] \tag{1-7}$$

where $t_n = nT_c$, $n = 1,2,\ldots$, $\Pi(\xi) = 0$, if $\xi < 0$, $\Pi(\xi) = 1$, if $0 < \xi \le 1$, $\Pi(\xi) = 0$, if $0 < \xi \le 1$. Let the aim shift boundary x_0 from position $u(x_0, t_{n-1})$ to position $u(x_0, t_n)$ during time interval τ_c and save the shift for the infinite time $t \ge t_n + \tau_c$. To achieve this result we must blow the free end of the elastic rod by the longitudinal force pulse $B(t) = \overline{\overline{p}}(t_n)\Pi[(t - t_n)/\tau_c]$ (rectangular, for instance) for the duration τ_c and with amplitude $\overline{\overline{p}}(t_n) = [u(x_0, t_n) - u(x_0, t_{n-1})]Z/\tau_c$. For the free ($\left[(\partial/\partial x)u = 0\right]_{x = x_0}$, without control) boundary x_0 of the rod, the shift

$$u(x_0, t) = 2w(x_0 + ct) \tag{1-8}$$

is twice as much as in the incident wave $w(x_0 + ct)$. We note that the remarkable property of the free boundary of semi infinite rod with entirely real impedance is the following: the boundary "remembers" the shift caused by the shock and adds it to the displacement $w(x_0 + ct)$. Further we return the boundary *by half of its free shift* periodically with the period T_c ($T_c << \tau_w$) the boundary $u(x_0, t)$, by the longitudinal blows

$$\overline{\overline{p}}(t_n) = [u(x_0, t_n) - u(x_0, t_{n-1})]Z/(2\tau_c). \tag{1-9}$$

As a result we get the trajectory $u(x_0, t) = w(x_0 + ct)$ (on average on several periods T_c, see **Fig. 1–4–a**). It is seen that the control action $\overline{\overline{p}}(t_n)$ applied to the boundary x_0 is switched off as the free shift returns by half. At $\overline{\overline{a}} \to \infty$ this action approaches the shock action. The algorithm (1-9) is characterized by the absorption by the boundary x_0 of low frequency incident waves (of the order of $\sim \tau_w^{-1}$) and the powerful high-frequency radiation at the frequencies of the order of $\sim T_c^{-1}$, which however decrease exponentially with the moving away from the boundary with the decrement of viscous damping proportional to the frequency squared. The flux of the power of the high-frequency radiation of the boundary x_0 exceeds the power flux in the incident wave on average by two times, and at the moment of blow – by $\sim T_c^2 \tau_w^{-1} \tau_c^{-1}$ times correspondingly (τ_c is the characteristic duration of the returning pulses $\overline{\overline{p}}(t_n)$). Note that the sequence of shocks (δ–pulses of boundary velocity in ideal representation) is very much analogous to the sequence of discrete samples, through which continuous signals with finite spectrum can be restored (according to the known Shannon's theorem in the information theory) assisting these pulses in passing through a low-frequency filter with rectangular frequency characteristic. In the considered case the role of such a filter is played by viscous damping of high-frequency waves in a rod. **Fig. 1-4-a** shows the rod particle shift in the incident wave (the broken line) and the shift of the boundary x_0 at the control (1-9).

1.6.2. Algorithm of Maximum of the Instant Power Absorbed {LS/LT/C/F}.

The algorithm of the maximum of the instantaneous absorbed power [50] is based on the step-wise adjustment of the boundary velocity $V(t_n)=V_n$ ($t_n = nT_c$, $n=1,2,...n$, T_c is the temporal step-interval) and on measuring the corresponding increment of the absorbed power W which is square function $W = V(2E - ZV)$ of the boundary velocity $V(t)$ with the unique maximum at $V(t) = E(t)/Z$ (Z is the wave impedance (real value) of the rod), $E(t)$ is the stress in the incident wave at the fixed boundary x_0 *i.e.* double pressure in the incident wave). This algorithm has the following control functions:

$$\overline{p}=1,\ \overline{\overline{p}}=0,\ V_n = V_{n-1} + \overline{V}\,\mathrm{sgn}(\overline{W}_{n-1}), \qquad \textbf{(1-10)}$$

where $\overline{W}_{n-1}$ is the increase of the absorbed power of the form $\overline{W}_{n-1} = F_{n-1}V_{n-1} - F_{n-2}V_{n-2}$, F_{n-1}, and V_{n-1} are the measured values of stress at the boundary and its velocities correspondingly at the moment t_{n-1}, $F_{n-1} = \Psi'_x(x_0, t_{n-1})$, $V_{n-1} = \Psi'_t(x_0, t_{n-1})$. $\overline{V} > 0$ is the step of the velocity tuning of the boundary x_0. If at the previous step the velocity increase causes the decrease ($\overline{W} < 0$) of the absorbed power, at the next step the velocity increase will change its sign and will not change it in the opposite case. The absorption maximum corresponds to the reflection and radiation minimum and the weak high-frequency boundary radiation is defined by the scale of the velocity-tuning step. The algorithm (1-10) effectively traces the incident wave if the following conditions are satisfied: $\overline{V} << c_1$, $(\overline{V}/T) >> c_2$, where c_1 and c_2 are the maximum velocities and the particle acceleration in the incident wave respectively. **Fig. 1-4-b** shows the velocity of the rod particle shift in the incident wave (the broken line) and the velocity of the boundary x_0 at the control (1-10).

1.6.3. Modulation Algorithm {LS/LT/C/M}.

The modulation algorithm [50] is defined by the control of the form:

$$\overline{p}(t) = (1/2)[1+\mathrm{sgn}(\cos(2\pi t/T))],\ \overline{\overline{p}}(t) = (1/2)[1-\mathrm{sgn}(\cos(2\pi t/T))],\ \gamma(t)=0 \qquad \textbf{(1-11)}$$

and alternates with period T soft (free) and rigid states of the boundary x_0, which neither absorbs nor radiates energy (always we have $F=0$ (*i.e.* $W=FV=0$) or $V=0$ (*i.e.* $W=FV=0$ too), but transforms on the reflected wave to visco-damping high frequency waves. The algorithm (1-11) makes the up-conversion of low frequency incident wave into the high frequency technological waves (which are relaxed dissipatively in the waveguiding media). This algorithm was tested in the experiment with the surface gravitational waves in liquid (Section 3.2.1). **Fig. 1-4-c** shows the velocity of the rod particle shift in the incident wave (gray area) and the velocity (black area) of the boundary x_0 at the control (1-11). Note that the algorithms (1-9), (1-10), (1-11) can use the random flux of pulses $\{t_n\}$ with the average interpulse interval $< t_n - t_{n-1} >= T$ instead of the periodic succession of the controlling actions with the period T.

1.6.4. On the One Alternative Approach in Mufflers Design {LS/LT/CA/M}.

There are many works devoted to passive and active mufflers [51-57] with constant parameters known. Below we suggest an alternative approach to the design of mufflers. The problem of the muffler of exhaust gas noise is to miss the flow of the gas, but block sound waves in the flow of the gas. Here, one can use the following criterion Ξ of muffler's quality (high Ξ means high quality):

$$\Xi = 1/(\alpha+\beta+\gamma) \qquad \textbf{(1-12)}$$

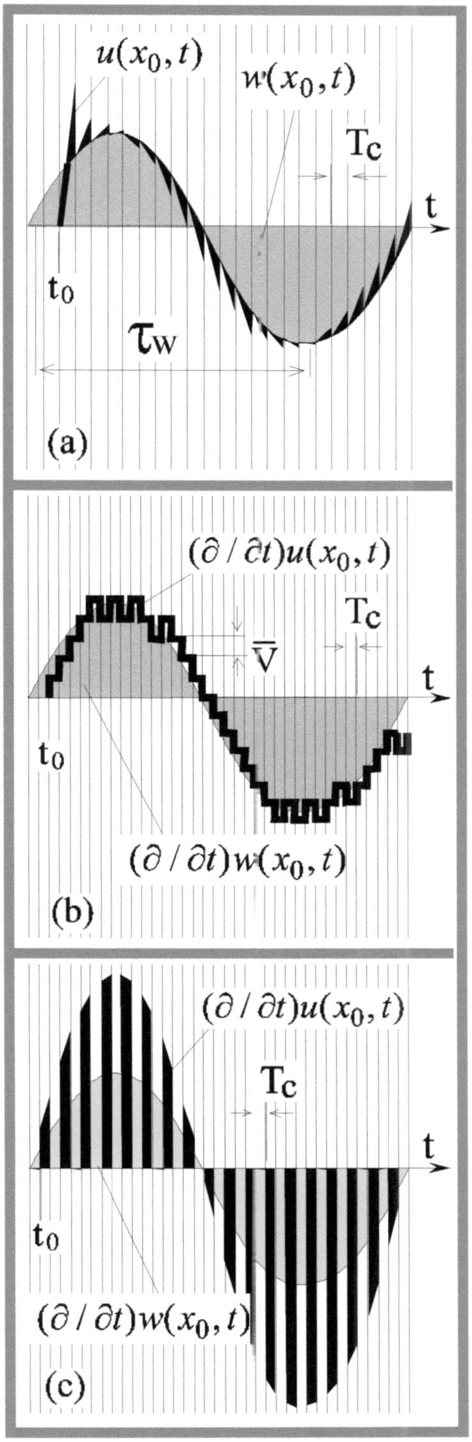

Figure 1-4: Oscillograms of the boundary x_0 controlled by the LSLT algorithms: (a) coordinate (black graph) of the boundary under the half-return algorithm (1-9) (gray graph $w(x_0,t)$ denotes oscillogram of particles shifts in incident wave); (b) velocity $(\partial/\partial t)u(x_0,t)$ (black graph) of the boundary under the algorithm (section (1-10) of maximum instant absorbing power (gray graph $(\partial/\partial t)w(x_0,t)$ denotes oscillogram of particles velocities in incident wave); (c) velocity $(\partial/\partial t)u(x_0,t)$ (black graph) of the boundary under the modulation algorithm (1-11) (gray graph denotes oscillogram of particles velocities $(\partial/\partial t)w(x_0,t)$ in incident wave).

where $\alpha = W_{out} / W_{in}$ characterizes the effectiveness of noise suppression, $\beta = W_{loss} / W_0$ characterizes the energy losses caused by passing of gas *via* muffler, $\gamma = q / \lambda$, W_{in} −low frequency noise power in the input of muffler, W_{out} −low frequency noise power in the output of muffler, W_{loss} −power of losses inside muffler, W_0 −power of ideal gas flow in the tube without noise, Q −characteristic geometrical size of muffler, and λ −characteristic wavelength of noise in the input of muffler. We can easily achieve the shrinking of one component: α or β or γ. However to get high Ξ is not easily.

Let's consider the following construction of muffler (see **Fig. 1-5**). Let's suppose, we can instantly switch gas flow (from one output tube 1 of engine with internal cross section S_0) between input ends of two tubes (2 and 3) of length L and with the same internal cross section S_0 (like superfast fauset). We can also connect the output ends of these pairs of tubes (2 and 3) alternately and instantly to the output tube 4 of muffler with the same internal cross section S_0. In reality the switch produces its operations during the final time $\tau_c << L / V_w$, where V_w is the speed of sound waves in the tubes 1, 2, 3, 4. Of course, we assume the one-dimensional sound propagation in the tubes, provided by the condition

$$\omega < 1.84 \times V_w / \sqrt{S_0 / \pi} \qquad \textbf{(1-13)}$$

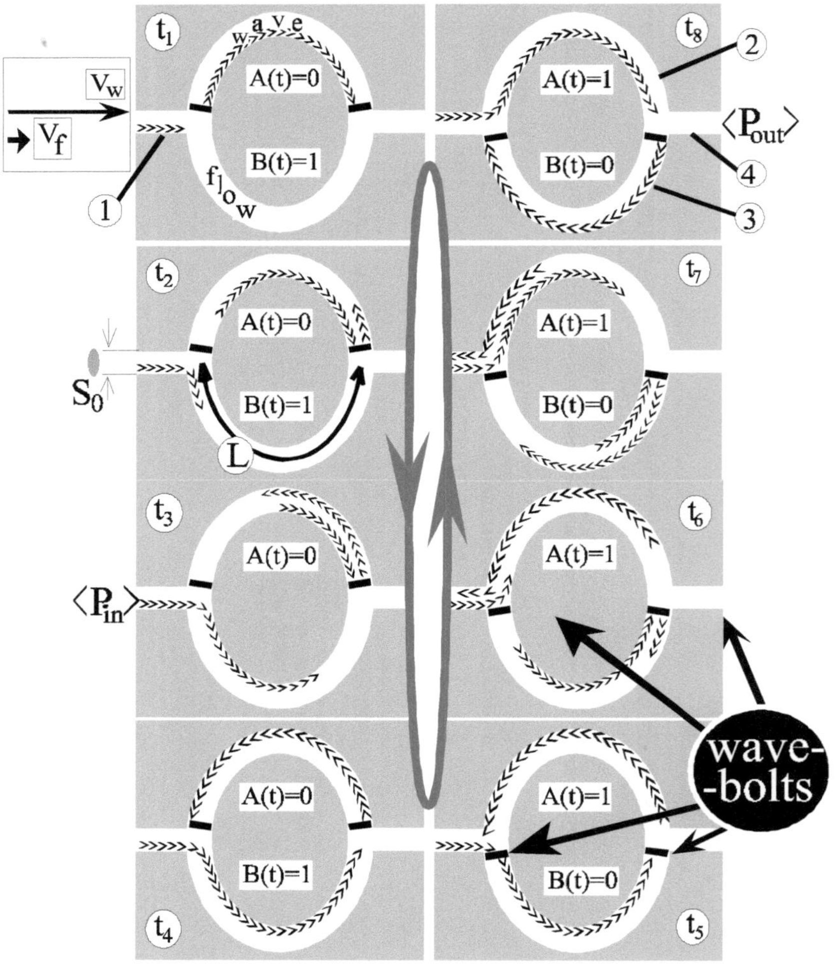

Figure 1-5: The cyclic temporal-spatial diagram of the muffler action. The way for the gas transportation is always free. Arrows mean the directions of wave propagation Instant photos at the moments: $t_1 < t_2 < t_3 < t_4 < t_5 < t_6 < t_7 < t_8$, where $t_5 = t_4 + \tau_S$, $t_1 = t_8 + \tau_S$, $\tau_S << L / c$. Wave-bolts are shown (visible black) in the closed state.

(for round pipes with internal cross section S_0) for the frequencies $\omega = \omega_{tech} = \pi V_w / L$, where $\omega_{tech} = 2\pi / T$ –technological frequency (analogously the above algorithms) and $T = 2L / V_w$. Note, that rotation frequency ($\omega_{tech} / 2\pi$) of winchester disk in ordinary computer is about ~ 250 cycles per second. Temporal diagram of suggested parametric muffler is presented in **Fig. 1-6**. Closed state means $A(t) = 0$ or (and) $B(t) = 0$, opened state would mean $A(t) = 1$ or (and) $B(t) = 1$. By these switches we alternate the propagation of the sound waves *via* pipe 2 and *via* pipe 3 with temporal period $T = 2L / V_w$, which is much smaller than the characteristic period of sound wave in pipes, *i.e.* $T << \lambda / V_w$. Transitive processes of switching take the time of about $\tau_c << T$.

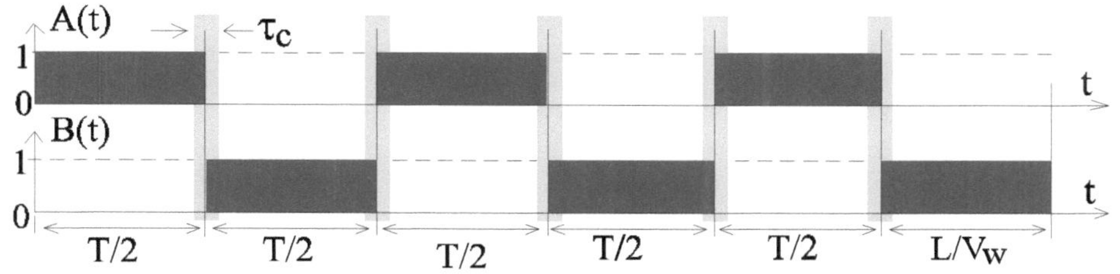

Figure 1-6: Temporal diagrams of the switches functions $A(t)$ (on the ends of pipe 2) and $B(t)$ (on the ends of pipe 3) o the parametric muffler.

Within each half-period L / V_w the piece of wave between two closed switches $A(t) = 0$ (or $B(t) = 0$) on the ends of pipe 2 (or pipe 3) has the time to change fully the direction of its propagation (i. e. to the back). In the interim, the input wave freely propagates the distance L in pipe 3 (or pipe 2). On the other hand, the gas flow with velocity

$$V_f << V_w$$

has free way for motion all the time: *via* pipe 2 or *via* pipe 3. Further we alternate this process from one pipe (2 or 3) to another pipe (3 of 2) with temporal period T. Therefore we obtain the parametric rejecting acoustical filter passing the gas flow only (with input flow velocity υ_{in} and output flow velocity $\upsilon_{out} = \upsilon_{in} = V_f$).

Using the concrete possible construction of muffler (see **Fig. 1-7**) we can determine the technological frequency as

$$\omega_{tech} = N_P / \Omega_0 \qquad \textbf{(1-14)}$$

where N_P is the number of petals of the wave-bolt (**Fig. 1-7-c**, $N_P = 3$), and Ω_0 is the angular frequency of the wave-bolt rotation. Besides this we can determine the transition time scale as $\tau_c = \theta_0 / \Omega_0$, when $\theta_0 < \overline{\theta}_0$, where θ_0 is the angular size of internal cross section of pipes 2 and 3 (see **Fig. 1-7-b**), $\overline{\theta}_0 = 2\pi / N_P$ is the angular size of the wave-bolt's petal.

Of course even in the absence of noise in gas flow, this muffler produces high frequency technological radiation (like in above sections 1.6.1-1.6.3) on the frequencies $\omega = n\omega_{tech}$ (where $n = 1, 2, 3, \ldots$). High frequency radiation causes the main part W_{rad} of energy losses ($W_{loss} \approx W_{rad}$) inside parametric muffler. One can estimate the relative power $\beta \approx W_{loss} / W_0$ of technological radiation using the temporal mean pressure $< P_{in} >$ and $< P_{out} >$ on the input and output of the muffler

$$\beta \approx 8\rho V_w V_f / < P_{in} >, \qquad \textbf{(1-15)}$$

where $W_0 \approx < P_{in} > V_f S_0$. We must provide $\beta << 1$. Effectiveness $\alpha = W_{out} / W_{in}$ of low frequency noise suppression (for the muffler version in **Fig. 1-7**) can be estimated as

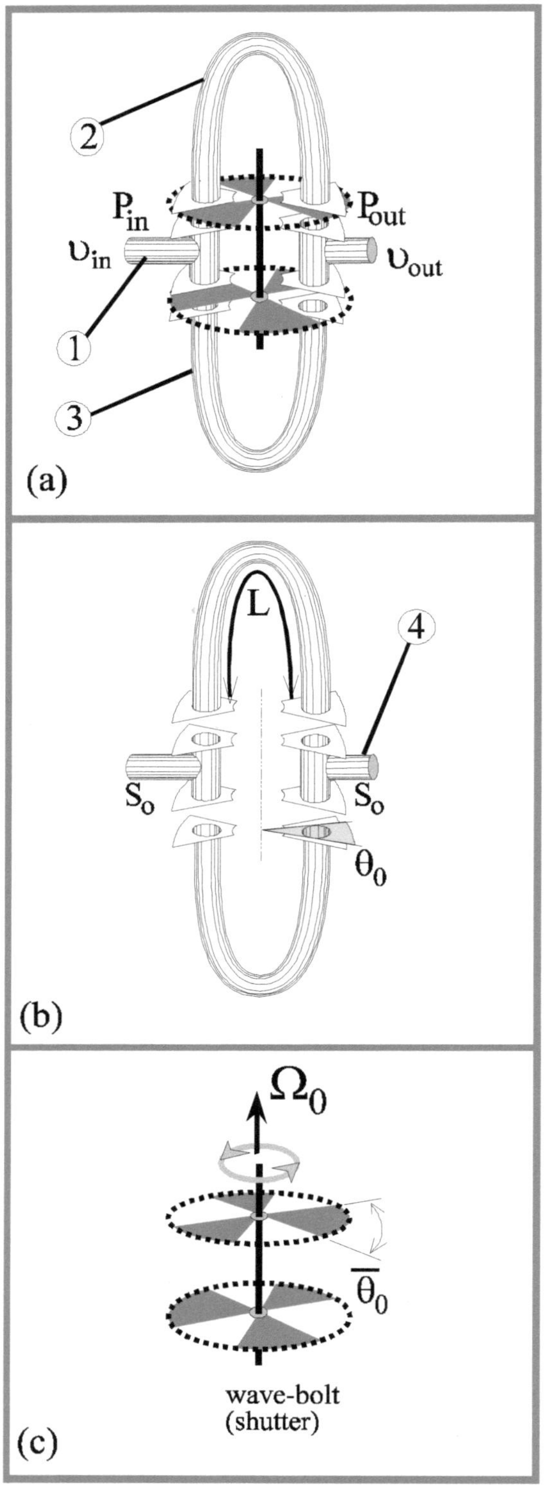

Figure 1-7: On the construction of possible implementation of parametric muffler. $\overline{\theta}_0$ –angular size of the wave-bolt's petal, θ_0 –angular size of the internal cross section of the pipes 2 and 3, Ω_0 –angular frequency of the wave-bolt rotation.

$$\alpha = (W_{out} / W_{in}) \approx (\theta_0 / \overline{\theta}_0)^2 , \tag{1-16}$$

trying to provide $\alpha << 1$. Smallness γ of geometrical dimensions of muffler can be estimated as

$$\gamma \approx (2L/\lambda)^2 << 1. \qquad \textbf{(1-17)}$$

We can choose so high Ω_0 that the high frequency radiation can be absorbed by ordinary dissipative coating of thickness $\sim V_w/(N_P\Omega_0) << \lambda$. So, taking into account (1-15), (1-16), (1-17) one can achieve $\Xi >> 1$ by the above parametric muffler. This muffler presents the active system of type {LS/LT/CA/M} (see **Fig. 1-3**).

1.7. CONCLUSION

Above we briefly considered several simple examples of spatially and temporally local algorithms (see section 1.5.) of wave control. These algorithms are characterized by the following particularities:

1) algorithms are local temporally and spatially, and they do not require information on the suppressed wave (excluding demands of smoothness with spatial scale $\sim c\tau_w$ and slowness with temporal scale $\sim \tau_w$ of this wave);

2) algorithms have characteristic periods $T >> \tau_w$ and control temporal scales $\tau_c << T$;

3) alongside with the suppression of the wave to be damped, these algorithms produce the so called "technological" high frequency radiation, concentrated near the frequencies $2\pi/T$, $4\pi/T$, $6\pi/T$,..., $2\pi/\tau_c$.

4) above algorithms include nonlinear, logical or parametric operations, and can not be exchanged equivalently by any combination of linear electric chains with temporally constant parameters.

More thorough investigation of one- and three- dimensional boundary value problems for water surface waves, acoustical waves, electromagnetic waves has been made in [50], [58-67], [90-92].

CHAPTER 2

Absorption Maximization

Abstract: The effect of resonant absorption of long waves by the oscillator of little sizes is investigated analytically and numerically. This effect means that absorption cross-section of the oscillator (monopole, dipole...) is defined by wavelength absorbed only, and does not depend on wave geometrical dimensions (much smaller, than the wavelength absorbed) of the oscillator. The expression of optimum amplitudes of excitation of the group of degrees of freedom (or oscillators) in the boundary problem of general type is obtained in the form of generalized velocities and generalized forces. Using linear microstructures (formed by monopoles, which are located on the axis periodically) we investigated the possibility to achieve maximum absorption cross-section of the acoustic waves by these microstructures of small wave dimensions. We consider the examples of linear microstructures, which provide unlimited logarithmic, linear and square growing of the total absorption cross-section, with growing of the quantity of elements (monopoles) in the linear microstructure with wave dimensions remaining small. The examples of cooperative and the individual strategies of absorbing oscillators are also compared.

2.1. SUPPRESSION AND ABSORPTION

An absorption (**Fig. 2-1-b**) (together with suppression, presented in **Fig. 2-1-a**) presents one of the main physical mechanisms of passive and active wave damping (in particular, acoustical waves [68–75]). Due to the specific relation between antennas, characteristics of reception, radiation and scattering, their absorption cross-section (ACS) is defined by the wavelength and by constructive coefficient of antenna, when the wave dimensions of such an antenna are approaching zero. First formulation of the idea of construction of multipole electromagnetic antenna is presented in the famous book by S. Shelkunoff and G. Frees [72]. Today we have much more wide possibilities to design antennas with superdirectivity. For instance, the acoustical superlenses using methamaterials [76] can scan the researched objects by their own near (reactive) field containing very high spatial frequencies. For instance, the new acoustic superlens (designed on the base of methamaterials [76]) which can scan the objects by own near field (or reactive field), where very high spatial frequencies are presented. This causes the superresolution of these lences. The aim of this chapter – to investigate the possibilities of absorption of maximum power from incident wave by the system of emitters with minimum total wave dimensions and minimum number of elements in the absorbing system [67].

2.2. MULTIPOLAR SOURCE (MS)

The basic element of linear multipole source (MS) is the pulsing sphere of radius a with fixed center (i.e. monopole). Sound field of pulsing sphere is given by the following equations

$$p = i(\omega\rho V_0 a^2)r^{-1}\exp(-ikr)\,,\quad \upsilon = (V_0 a^2)r^{-2}\exp(-ikr) + i(V_0 a^2 k)r^{-1}\exp(-ikr)\,, \tag{2-1}$$

where p – is the sound pressure, υ – is the radial velocity at the distance r from the sphere center, V_0 – is the complex amplitude of the velocity of sphere surface pulsations, ω – is the angular frequency, k – is the wave number. Neglecting near field of velocity, we obtain

$$\upsilon = i(V_0 a^2 k)r^{-1}\exp(-ikr)\,. \tag{2-2}$$

Pressure and velocity fields of incident wave can be accepted as following

$$p_e = \rho c V_e \exp(+ikr\cos\vartheta) \text{ and } \upsilon_e = V_e \cos\vartheta \exp(+ikr\cos\vartheta)\,, \tag{2-3}$$

respectively and the absorption cross section (ACS) $\sigma_{M,N}$ of monopole (when $N=0$) or multipole (when $N \ge 1$) we define by the expression

$$\sigma_{M,N} = W_{M,N} / S_e\,, \tag{2-4}$$

Vladimir V. Arabadzhi

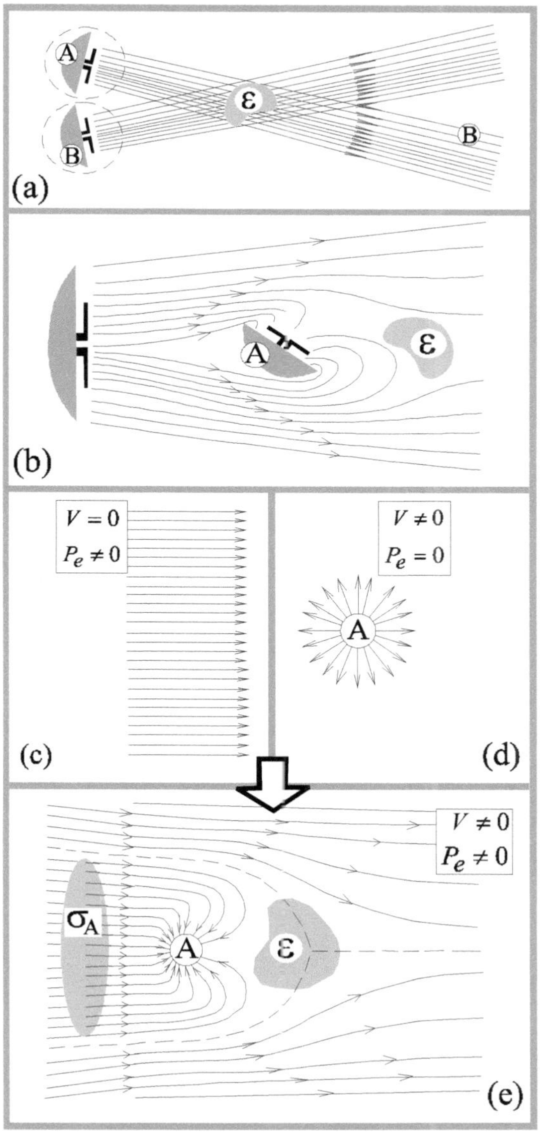

Figure 2-1: Variants of active wave damping: (a) mutual suppression of waves of antennas " A " and " B " without absorption; (b) absorption of waves of antenna " B " by antenna " A ". Conversion of power stream lines of a acoustic wave field: (c) power stream lines of planar wave with sound pressure $P_e \neq 0$ (monopole is switched off with volume velocity $V = 0$); (d) power stream lines of a monopole source with volume velocity $V \neq 0$ (planar wave is switched off with $P_e = 0$); (e) power stream lines of a monopole source " A " in interaction with plane wave ($P_e \neq 0$, $V \neq 0$, σ_A –absorption cross-section of a monopole " A ", ε –area of damped intensity of wave field).

where $W_{M,N}$ – is the power absorbed by structure, $S_e = \rho c|V_e|^2 / 2$ – is the power stream density in the incident wave, V_e – is the complex amplitude of the oscillatory velocity of incident wave and ϑ – is the angle between the direction of incident wave propagation and axis " x ". Acoustical monopole has maximum (resonant) absorption cross-section (ACS) $\sigma_0 = \lambda^2 / 4\pi$. The structure of linear MS (**Fig. 2-2**) presents the chain of $n = N_M = 2^N$ monopoles (pulsing spheres with radius a) with a total length

$$L_{M,N} = 2(2^N - 1)h. \qquad \textbf{(2-5)}$$

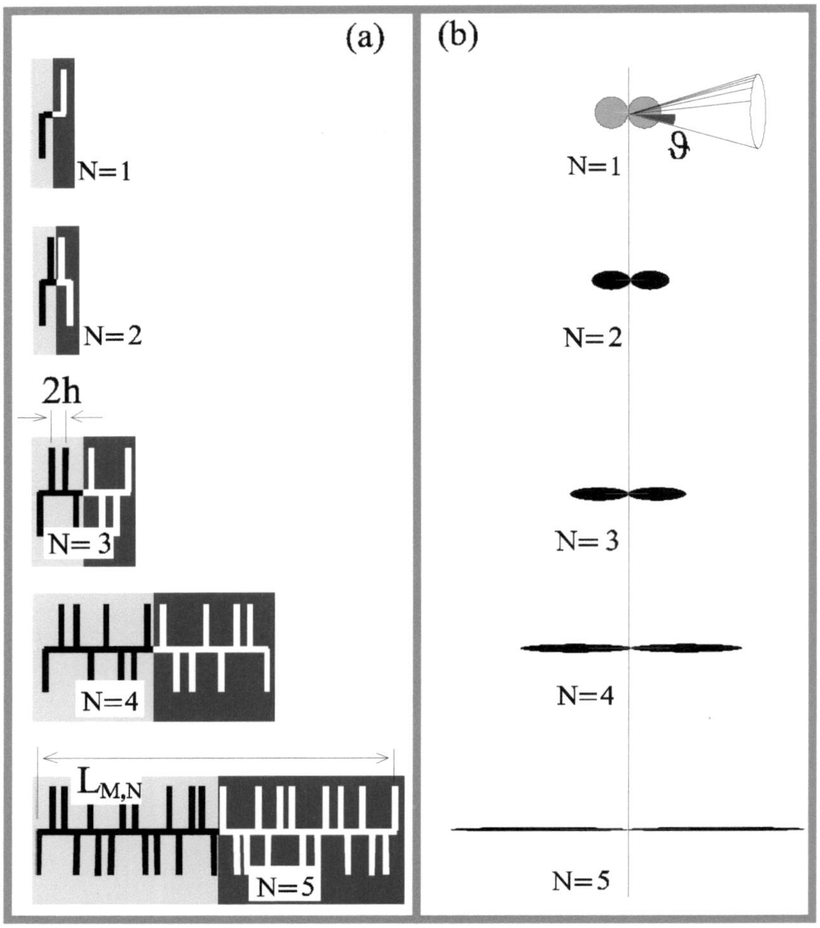

Figure 2-2: The structure of MS when the multipolarity order N grows (a), and also the corresponding directivity patterns (in power), here the ϑ means the angle of observation (b).

Monopoles are spaced equidistantly on the axis "x" with spatial period $h >> a$: (when $N=1$) 2 monopoles in the points $x=-h,+h$ with amplitudes $(-1,+1)V_0$; (when $N=2$) 4 monopoles in the points $x=-3h,-h,+h,+3h$ with amplitudes $(+1,-1,-1,+1)V_0$; (when $N=3$) 8 monopoles in the points $x=-7h,-5h,-3h,-h,+h,+3h,+5h,+7h$ with amplitudes $(-1,+1,+1,-1,+1,-1,-1,+1)V_0,\ldots$etc (**Fig. 2-2**). In finite-differential representation spatial distribution of amplitudes in linear multipole of N-th order corresponds to the N-th spatial derivative of the incident wave field along the axis "x".

When $kh << 1$ we note that thge wave fields of pressure $p_{M,N}$ and velocity $\upsilon_{M,N}$ of multipole are connected with corresponding fields $p(r)$ and $\upsilon(r)$ of monopole by the following equations

$$p_{M,N} = i^N 2^{(3N/2)+(N^2/2)} (kh)^N (\cos\vartheta)^N p(r)\,,\ \upsilon_{M,N} = i^N 2^{(3N/2)+(N^2/2)} (kh)^N (\cos\vartheta)^N \upsilon(r)\,, \qquad \textbf{(2-6)}$$

with radiation resistance of the form $\operatorname{Re} Z_{M,N} = 4\pi a^4 k\omega\rho (kh)^{2N} 2^{N(N+3)} (2N+1)^{-1}$, where ϑ–is the angle of observation relating to the axis "x", r– is the distance between observation point and point $x=0$ on the multipole axis, when

$$ka << L_M k << 1\,. \qquad \textbf{(2-7)}$$

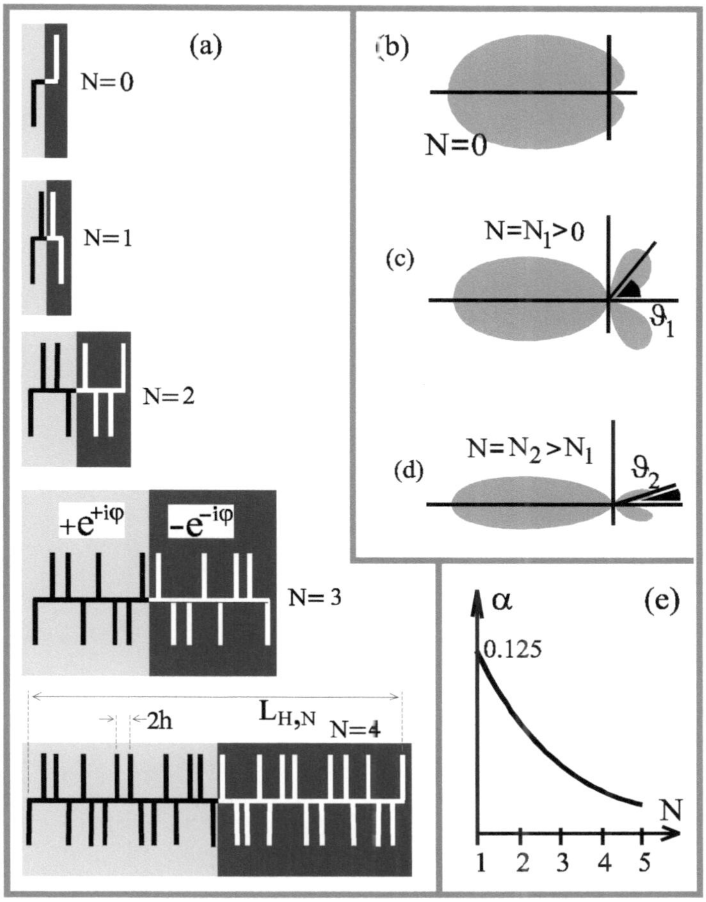

Figure 2-3: Structure of the MSUR in multipolarity order N is growing (a). MSUR's directivity patterns of MSUR (in power) (b–d) and ratio (e) between module amplitude of back petal and module amplitude of main petal as a function of N.

After consideration of power stream *via* sphere, overlapping MS, we observe [10], that maximum ACS

$$\sigma_{M,N} = (2N+1)(\lambda^2 / 4\pi) \tag{2-8}$$

is achieved by MS with complex amplitude $V_0 = V_{M,N}$ of the module

$$\left|V_{M,N}\right| = (2N+1)(ka)^{-2}(kh)^{-N}\,2^{-\left(1+\frac{3}{2}N+\frac{1}{2}N^2\right)}\left|V_e\right| \tag{2-9}$$

and argument (phase)

$$\arg(V_{M,N}) = \arg(V_e) - N(\pi/2)\,. \tag{2-10}$$

2.3. MULTIPOLAR SOURCE OF UNIDIRECTIONAL RADIATION (MSUR)

Here, we will present one multipole source of unidirectional radiation (MSUR) using two MS with the multipolarity order N and total length

$$L_{H,N} = 2L_{M,N} + 2h\,. \tag{2-11}$$

MSUR presents $n = N_H = 2^{N+1}$ pulsing spheres spaced equidistantly on the axis "x" with spatial period $2h$ (**Fig. 2-3-a**). Complex amplitudes V_0 of the left multipole of MSUR are converted into $V_0 \exp(ikL_{H,N}/2)$ and right multipole of MSUR are converted into $-V_0 \exp(-ikL_{H,N}/2)$. So the pressure field $p_{H,N}(\mathbf{r})$ of MSUR in the point $\mathbf{r}$ is connected with the pressure field $p_{M,N}(\mathbf{r})$ of multipole by the equation

$$p_{H,N}(\mathbf{r}) = +p_{M,N}[\mathbf{r} - \mathbf{x}_0(L_{H,N}/2)]\exp(+ikL_H/2) - p_{M,N}[\mathbf{r} + \mathbf{x}_0(L_{H,N}/2)]\exp(-ikL_H/2)\,, \quad \textbf{(2-12)}$$

where $\mathbf{x}_0$ – is the vector of axis "x". When

$$ka << L_H k << 1 \quad \textbf{(2-13)}$$

we see that the pressure and velocity fields $p_{H,N}$, $\upsilon_{H,N}$ of MSUR can be represented by the following expressions:

$$p_{H,N}(r,\vartheta) \approx (2i)\sin[(k\overline{L}_{H,N}/2)(1-\cos\vartheta)]p_{M,N}(r,\vartheta)$$

and

$$\upsilon_{H,N}(r,\vartheta) \approx (2i)\sin[(k\overline{L}_{H,N}/2)(1-\cos\vartheta)]\upsilon_{M,N}(r,\vartheta)\,, \quad \textbf{(2-14)}$$

with the radiation resistance of the form

$$\operatorname{Re} Z_{H,N} = 2^6 \pi (kh)^{2(N+1)} 2^{N(N+5)} (N+1)(2N+1)^{-1}(2N+3)^{-1} a^4 k\omega\rho\,,$$

where $\overline{L}_{H,N} = L_{M,N} + 2h$.

When $N = 0$ MSUR has the only maximum $\vartheta = 0$ and the only zero $\vartheta = \pi$ (**Fig. 2-3-b**) of directivity pattern in power. When $N \geq 1$, zeroes of radiation are spaced in directions $\vartheta = \pi$ and $\vartheta = \pm\pi/2$. When $N \geq 1$, maximums of MSUR field are spaced in directions $\vartheta = 0$ and $\vartheta = \pm\vartheta_B = \pm\arccos[N/(N+1)]$ (**Figs. 2-3-c, 2-3-d**). We estimate the ratio (**Fig. 2-3-e**) between the amplitude of back petal to the amplitude of main (forward) petal of MSUS pattern of directivity as $\sim 2^{-1} N^N (N+1)^{-(N+1)}$ ($N = 1,2,3,\ldots$, do not depend on L_H, a, k, if (2-13) has been satisfied). Back petals of MSUR directivity pattern decreased smoothly with the growth of N (proportionally to α) and deviated backwards ($\vartheta_1 < \vartheta_2$ when $0 < N_1 < N_2$). Both the main petals and the back petals narrowed. Power $W_{FH,N}$, radiated by MSUR to the forward half-space and power $W_{BH,N}$, radiated by MSUR to the back space are connected by the following ratio

$$W_{FH,N} / W_{BH,N} = 8N^2 + 16N + 7\,. \quad \textbf{(2-15)}$$

After consideration of power stream *via* sphere, overlapping MS, we observe [10], that maximum ACS

$$\sigma_{H,N} = (2N+1)(2N+3)(N+1)^{-1}(\lambda^2/4\pi) \quad \textbf{(2-16)}$$

is achieved by MS with complex amplitude $V_0 = V_{H,N}$ of the module

$$\left|V_{H,N}\right| = (2N+1)(2N+3)(N+1)^{-1}(ka)^{-2}(kh)^{-(N+1)} 2^{-\left(3+\frac{5}{2}N+\frac{1}{2}N^2\right)} \left|V_e\right| \quad \textbf{(2–17)}$$

and argument (phase)

$$\arg(V_{H,N}) = \arg(V_e) - (-1)^N (\pi / 2)\,. \qquad \textbf{(2–18)}$$

It is notable, that on the border of ACS area (in the maximum of directivity pattern of MS or MSUR) the field of optimally absorbing structure exceeds (in module) the field of incident wave, maintaining the same order.

We must note that MSUR absorption appears only in the case when power stream of MSUR scattered field has the same direction with the incident wave. For example, MS's petal, as against the incident wave, only constitutes the scattering field (back scattering). But another petal constitutes both scattering field (forward scattering) and absorption of incident wave.

Note, that in the case of big wave dimensions of acoustical antennas, we must increase the structure of antenna in cross-section direction, to increase ACS. But in the case of small wave dimensions of antenna, we must increase and complicate the longitudinal structure of acoustical antenna. The ratio between real part $|\mathrm{Re}\, Z_{ex}|$ and image part $|\mathrm{Im}\, Z_{ex}|$ of impedance of any wave source is called "ringingness", which for the MS (or MSUR) has an approximate value $|\mathrm{Im}\, Z_{ex}| / |\mathrm{Re}\, Z_{ex}| \approx 2^{-N^2} (2^N - 1)^{2N} (ka)^{-1} (kL_{M,N})^{-2N} >> 1$.

We see from (2-8) and (2-16) that ACS (both for MS and MSUR) grows proportionally to the multipolarity order N . Unlike MS, the MSUR practically does not produce the back scattering (see (2-15)). Such an MSUR acts like an acoustically "black" body [71] with finite cross-sectional linear dimension $\sim \lambda\sqrt{N}$, in spite of its small wave dimension $kL_H << 1$.

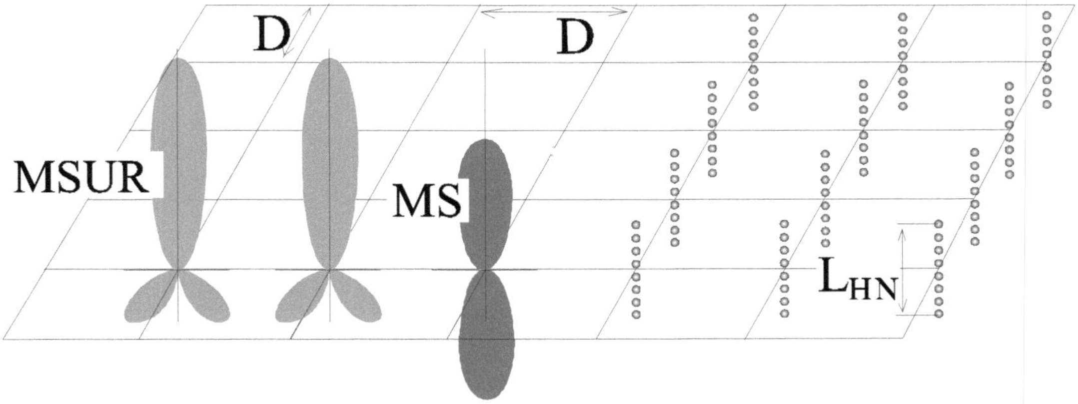

Figure 2-4: The planar array of uninteracting MSs (or MSURs) and their directivity patterns in power.

2.4. MULTIPOLAR ARRAYS

The above MS and MSUR provided the growth $\sigma_A \sim N$ of ACS proportionally as the order N of multipolarity: or as the logarithm of the quantity n of elements in MS (MSUR), *i.e.* $\sigma_A \sim \ln(n)$ (**Fig. 2-8-b**). Now we consider [67] the lattice (with period D, **Fig. 2-4**) formed by $\overline{n}$ MS or MSUR, and they are directed to each other by zeroes of their directivity patterns. The quantity of monopoles of this lattice is $n = \overline{n}\,\overline{\overline{n}}$, where $\overline{\overline{n}} = 2^N$ is the quantity of elements in MS (for instance), and N is its multipolarity order. In this case total ASC is equal to the sum of $\overline{n}$ ASC of lonely MS (MSUR), *i.e.*

$$\sigma_A = (n / \overline{\overline{n}})\sigma_{M,N} \ \textbf{(or } \sigma_A = (n / \overline{\overline{n}})\sigma_{H,N}\textbf{)}. \qquad \textbf{(2-19)}$$

Consideration of the matrix

$$\hat{\mathbf{Z}} = \begin{bmatrix} Z_{A,A} & Z_{A,B} \\ Z_{B,A} & Z_{B,B} \end{bmatrix} \quad \textbf{(2-20)}$$

of impedances of system of two identical MS (A and B) gives us the following conditions of independent absorption of incident wave by two MS:

$$\left|Z_{A,B}\right| / \left|Z_{A,A}\right| \approx 2^{N^2} (2^N - 1)^{-2N} (kD)^{-N} (kL_{M,N})^{3N} (a / D) << 1 \quad \textbf{(2-21)}$$

and

$$a << h < L_{M,N} << D << \lambda \quad \textbf{(2-22)}$$

(the same for MSUR). The lattice (**Fig. 2-4**), formed by MS (MSUR), provides ACS as the following $\sigma_A \sim n$. A decrease of parameters $(kD)^{-N}$, $\left|Z_{A,B}\right| / \left|Z_{A,A}\right| \approx 2^{N^2} (2^N - 1)^{-2N} (kD)^{-N} (kL_{M,N})^{3N} (a / D) << 1$, (a / D) provides linear growth (**Fig. 2-8-b**) of total ACS of lattice, formed from MA (or MSUR), without breaking the condition of smallness of lattice total wave dimensions.

Note that, emitters of the wave of another genesis can do not permit to form the plane lattice (as shown in **Fig. 2-4**) of uninteracting actuators. For example, the elementary electromagnetic emitters have toroid (see below **Table 2-3**) directivity pattern (in power) unlike conical (see **Fig. 2-2-b**) directivity patterns of linear acoustical MS and MSUR. Linear growth of ACS is possible even in the electromagnetic case, but only when we place emitters on the line (unlike plane).

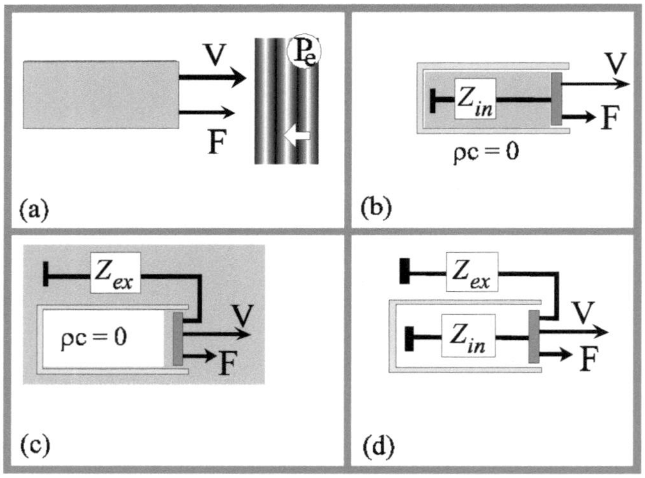

Figure 2-5: On the definition of acoustical emitter (a), its internal impedance (b) (vacuum outside, $Z_{ex} = 0$), external impedance (c) of emitter (interior vacuum, $Z_{in} = c\rho = 0$), and their resonant interaction (d).

2.5. GENERALIZED FORCES AND VELOCITIES

As the absorption can exist only on a moving boundary (in mechanical terminology) we will consider any boundary problem of absorption of some incident wave as a problem of scattering on a moving boundary. Further, due to the conception of linear superposition, we will divide the previous boundary problem into a pair of ones: (1)–boundary problem of radiation by the moving boundary in the absence of incident wave and (2)–boundary problem of an incident wave scattering on the fixed boundary.

The above considered wave sources, were acoustical. They were characterized, in spite of their complicated structure by the only degree of freedom – complex amplitude $V_0 = V$ of oscillatory velocity of monopoles forming these sources. The construction of piston acoustical emitter is presented in **Fig. 2-5**. To get the general approach to the problem of absorption of waves of various physical geneses by the emitters of various constructions, we formulate this problem in the titles of the complex amplitudes of generalized forces and generalized velocities. The concrete physical sense of these forces on velocities can be very different. But generalized forces and generalized velocities together must give an adequate description of boundary value problem, and the product of generalized velocity and generalized force must have the dimensionality of power [60]. The generalized emitter is characterized by internal (mechanical for instance) impedance Z_{in} and by external (sound field for instance) impedance Z_{ex}. Optimum complex amplitude of oscillatory velocity [60] of generalized microstructure (with one degree of freedom) is defined by the expression

$$V = V_{\text{opt}} = (1/2)E/\text{Re}(Z_{ex}), \text{ or } V_{\text{opt}} = (1/2)(E+ZV)/\text{Re}(Z_{ex}), \quad \textbf{(2-23)}$$

where E is the complex amplitude of the projection of force of incident wave action on the fixed ("frozen", when $V = 0$, and piston is stopped) degree of freedom. The velocity (2-23) provides maximum

$$W_{abs} = (1/8)|E|^2/\text{Re}(Z_{ex}) \quad \textbf{(2-24)}$$

power absorbed (and simultaneously the same value of scattered power). The amplitude (2-23) can be achieved both by active tools (servo-drive) and by passive tools, i. e. satisfaction of two impedance conditions:

$$\text{Re}(Z_{ex}) = \text{Re}(Z_{in}), \ \text{Im}(Z_{ex}) = -\text{Im}(Z_{in}), \quad \textbf{(2-25)}$$

where Z_{ex} is the impedance of absorbing microstructure in wave guiding media, Z_{in} is the impedance of the internal load of absorbing microstructure (see **Fig. 2-5**), F, V –are the measured complex amplitudes of the generalized force of action on emitter and the generalized velocity respectively. So the algorithm (2-3) can be classified as {LS/LT/CA/F} (see **Fig. 1-3**).

2.6. EXAMPLES OF WAVES OF DIFFERENT PHYSICAL GENESIS

Similarly, in analytical mechanics we will use generalized forces and generalized velocities. Further, we illustrate the universality of equations (2-23)-(2-25) in the examples of the waves of various physical genesis: acoustical waves (section 2.6.1.), electromagnetic waves (section 2.6.3.), and bending waves in thin elastic plates (section 2.6.4.), and surface gravitational waves in liquids (section 2 6.2.).

2.6.1. Sound Waves

Sound waves in compressible fluid with the particle velocity $\mathbf{V}(\mathbf{r},\omega)$ and the pressure $p(\mathbf{r},\omega)$ are defined by the wave potential $\phi(\mathbf{r},\omega)$ as $\mathbf{V} = -\nabla\phi$ and $p = i\omega\rho\phi$ (where ρ is medium mass density), which satisfies the wave equation $\Delta_n\phi + (\omega/c)^2\phi = 0$ (c is the sound speed, $n = 2,3$ the Laplace operator dimensionality) and the boundary condition on the surface $\overline{S}$ of emitter: $[(\partial/\partial\mathbf{n})\phi]_{\overline{S}} = \chi_m V$, where $[(\partial/\partial\mathbf{n})\phi]_{S_e} = \chi_m V$ is the unity normal to $\overline{S}$, V is the complex amplitude of the emitter excitation, $\chi_m(\mathbf{r})$ is the dimensionless normalized multiplier, describing the normal velocities distribution on the $\overline{S}$ corresponding to the m-th multipolarity order ($m = 0,1,\ldots$). **Table 2-1** presents: radiation resistance $\text{Re}\,Z_{ex}(\omega)$ of each emitter, the force and $E(\omega)$ of the action of the incident wave on the surface $\overline{S}$ of each emitter, maximum ACS σ_A of each emitter, and power stream density S_e of the incident wave with the complex amplitude $P_i(\omega)$ of the sound pressure. Lines (f) and (g) in **Table 2-1** contain module $|\omega|$, to provide $\text{Re}\,Z_{ex}(\omega) > 0$ for any signs of frequencies ω.

Table 2-1: Acoustical emitters

Emitter	Radiation resistance	Force	ACS
(a) 3D one-side oscillating piston	$\mathrm{Re}\,Z_{ex} = (\pi\rho a^4/4c)\omega^2$	$E = \pi a^2 P_e$	$\sigma^A = \lambda^2/4\pi$
(b) 3D one-side piston in screen	$\mathrm{Re}\,Z_{ex} = (\pi\rho a^4/2c)\omega^2$	$E = 2\pi a^2 P_e$	$\sigma^A = \lambda^2/2\pi$
(c) 3D pulsing sphere	$\mathrm{Re}\,Z_{ex} = (4\pi a^4\rho/c)\omega^2$	$E = 4\pi a^2 P_e$	$\sigma^A = \lambda^2/4\pi$
(d) 3D oscillating sphere	$\mathrm{Re}\,Z_{ex} = 3^{-1}\pi a^6\rho\omega^4 c^{-3}$	$E = \pi a^3 k P_e$	$\sigma^A = 3\lambda^2/4\pi$
(e) 3D, 4-pole source	$\mathrm{Re}\,Z_{ex} = (\pi\rho a^4 d^4/c^5)\omega^6$	$E = (4\pi a^2 d^2 k^2)P_e$	$\sigma^A = \lambda^2/2\pi$
(f) 2D pulsing cylinder	$\mathrm{Re}\,Z_{ex} = (\pi^2\rho a^2)\lvert\omega\rvert$	$E = (2\pi a)P_e$	$\sigma^A = \lambda/2\pi$
(g) 2D oscillating cylinder	$\mathrm{Re}\,Z_{ex} = (\pi^2 a^2\rho/c^3)\lvert\omega\rvert^3$	$E = (2\pi a^2 k)P_e$	$\sigma^A = \lambda/\pi$

(h) $S_e = \lvert P_e\rvert^2/2\rho c\,,\quad \omega = ck$

2.6.2. Gravity Waves on the Surface of Infinitely Deep Water

Gravitational liquid surface waves have the same dependence $\mathbf{V}(\mathbf{r},\omega)$, $p(\mathbf{r},\omega)$ of the potential $\phi(\mathbf{r},\omega)$, as for acoustical waves mentioned above, satisfying Laplace equation $\Delta_n\phi = 0$ and the boundary condition on the free liquid surface S_f : $[\phi''_{tt} - g\phi'_z]_{S_f} = 0$. For simplicity, we assume the liquid of infinite depth. **Table 2-2** presents: the power stream density S_e (per unit length of the front) of the incident wave with the potential $\phi = \omega k^{-1} A_i \exp(-kz - ikx)$ (A_i is the amplitude of the vertical shift of liquid surface S_f, x is the horizontal coordinate, and z is the vertical coordinate); the dispersion equation $\omega = \omega(k)$; the radiation resistance $\mathrm{Re}\,Z_{ex}(\omega)$, the force of the incident wave effect on the fixed emitter E; the maximum ACS σ_A as applied to monopole and dipole, three-dimensional and two-dimensional emitters. It can be seen from **Table 2-2** that ACS σ_A of the emitter of small wave dimension $ka \ll 1$ does not depend on size– a, or on the depth h of the emitter location and, as

Table 2-2: Emitters of surface gravitational waves in liquid

3D pulsing sphere (a)	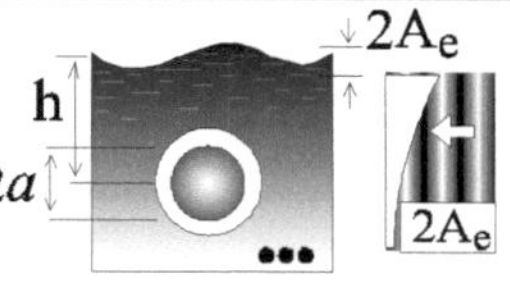	$\mathrm{Re}\,Z_{ex} = 8\pi^2 a^4 g^{-1} \rho\, \omega^3 \exp(-2kh)$ $E = 4\pi\rho g a^2 A_e \exp(-kh)$ $\sigma^A = \lambda / 2\pi$
3D oscillating sphere (b)	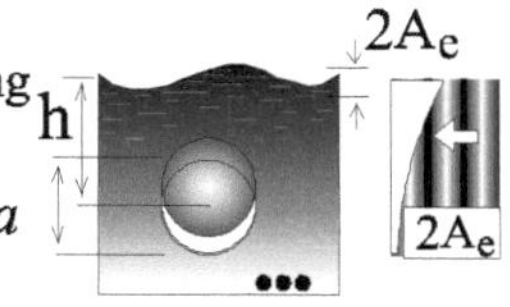	$\mathrm{Re}\,Z_{ex} = 3^{-2} 2^3 \pi^2 a^6 g^{-3} \rho \lvert\omega\rvert^7 \exp(-2kh)$ $E = (4\pi a^3 \rho g k / 3) A_e \exp(-2kh)$ $\sigma^A = \lambda / 2\pi$
2D pulsing cylinder (c)	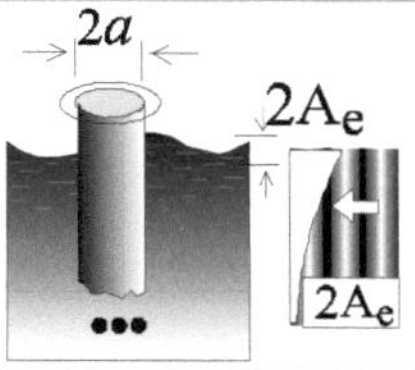	$\mathrm{Re}\,Z_{ex} = 16\pi^2 \rho a^2 g \lvert\omega\rvert^{-1} \exp(-2kh)$ $E = 2\pi a \rho g^2 \omega^{-2} A_e$ $\sigma^A = \lambda / 2\pi$
(d)	$S_e = (\rho g^2 / 4\lvert\omega\rvert) \lvert A_e \rvert^2, \ \omega = \sqrt{gk}$	power radiation directivity pattern 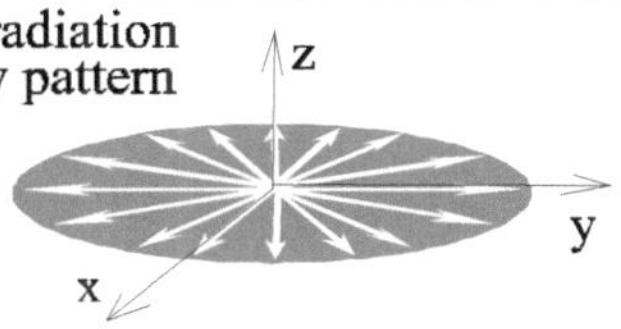

Table 2-3: Electromagnetic emitters

3D electric dipole (a)	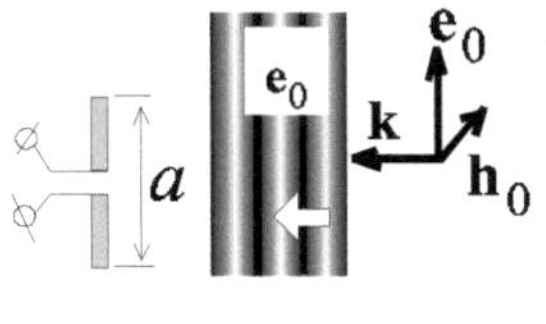	$\mathrm{Re}\,Z_{ex} = [(\mu / \varepsilon)^{1/2} a^2 / 6\pi c^2]\omega^2$ $\lvert E \rvert = a \lvert \mathbf{e}_0 \rvert$ $\sigma^A = 3\lambda^2 / 8\pi$	
3D magnetic dipole (b)	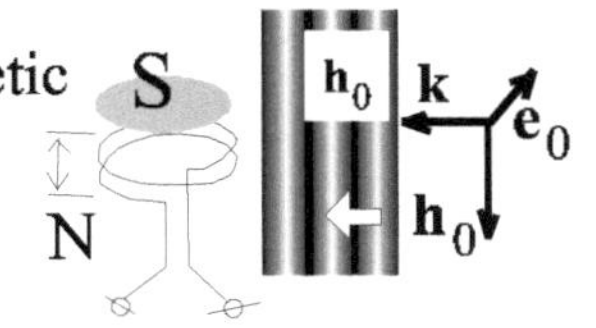	$\mathrm{Re}\,Z_{ex} = [(\varepsilon / \mu)^{1/2} S^2 N^2 / 6\pi c^2]\omega^2$ $\lvert E \rvert = \mu N S \omega \lvert \mathbf{h}_0 \rvert$ $\sigma^A = 3\lambda^2 / 8\pi$	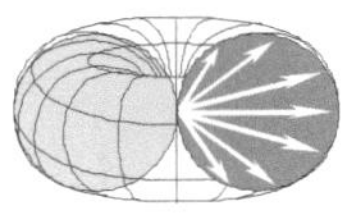
2D cylindrical current (c)	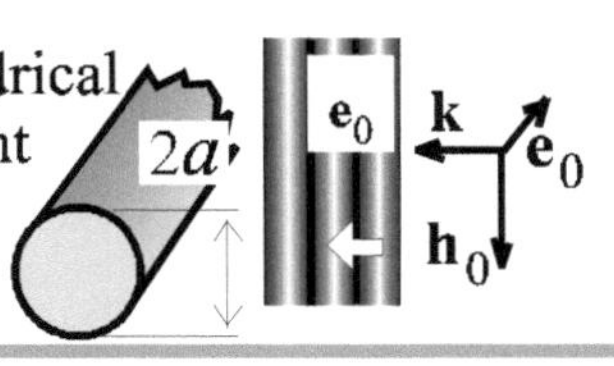	$\mathrm{Re}\,Z_{ex} = [\pi a / 2\varepsilon c^2]\lvert\omega\rvert$ $\lvert E \rvert = 2\pi \lvert \mathbf{e}_0 \rvert$ $\sigma^A = \lambda / 2\pi$	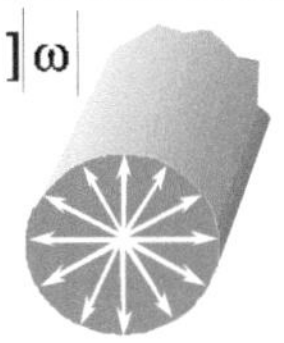
(d)	$S_e = (\mu / \varepsilon)^{1/2} \lvert \mathbf{h}_o \rvert^2 / 2 = (\varepsilon / \mu)^{1/2} \lvert \mathbf{e}_0 \rvert^2 / 2, \ \omega = ck$		

it is easy to make sure, it does not depend on the liquid depth. The analogous qualitative conclusions are valid obviously also in general for any gravitational waves in liquid (including internal waves) of arbitrary depth. Above sources of water surface waves have one particularity–they all have the isotropic directivity pattern of radiation power (in horizontal plane), and as a result, the identical ASC (in accordance with section 2.7.3. below).

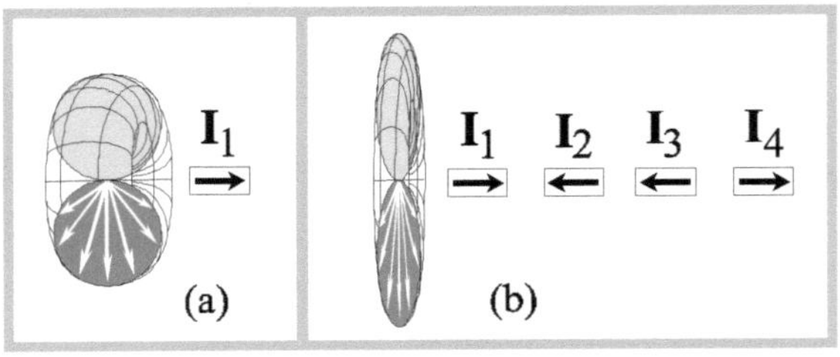

Figure 2-6: The compression of toroid directivity pattern of radiation (in power) of the elementary electric current $\mathbf{I}_1$ (a), when we built linear multipole antenna, with $N = 3$ from currents (b) $\mathbf{I}_1$, $\mathbf{I}_2 = -\mathbf{I}_1$, $\mathbf{I}_2 = -\mathbf{I}_1$, $\mathbf{I}_3 = \mathbf{I}_1$, $\mathbf{I}_3 = \mathbf{I}_1$ (both directivity patterns show only the shape, i. e. without calibration, axial radiation is zero).

2.6.3. Electromagnetic Waves

Here we consider the electromagnetic plane linearly polarized incident wave with amplitudes $\mathbf{e}(\omega)$ and $\mathbf{h}(\omega)$ of electric and magnetic fields, expressed *via* vector potential $\mathbf{f}(\mathbf{r},\omega)$: $\mathbf{h} = \mu^{-1} curl(\mathbf{f})$, $\mathbf{e} = -i\omega\mathbf{f}$ ($\mathbf{e}(\omega)$, $\mathbf{h}(\omega)$ are correspondingly the analogs of the pressure and velocity of particles in the acoustics, μ is the magnetic media or vacuum coefficient) satisfying the wave equation $\Delta_n\mathbf{f} + (\omega / c)\mathbf{f} = 0$ ($n = 2,3$ is the Laplacian dimensionality, c is the light speed) and the boundary condition on the emitter surface $\overline{S}$: $\mu^{-1}[\mathbf{n}, curl(\mathbf{f})] = \chi_m(\mathbf{r})V$ ($\chi_m(\mathbf{r})$ is the vector-analog of the multiplier $\chi(\mathbf{r})$ used above for acoustical case, V is the current density, the analog of the oscillating velocity of the emitter, $\mathbf{n}$ is the unity normal to the emitter surface $\overline{S}$).

Table 2-4: Emitters of bending waves in thin elastic plates

(a)	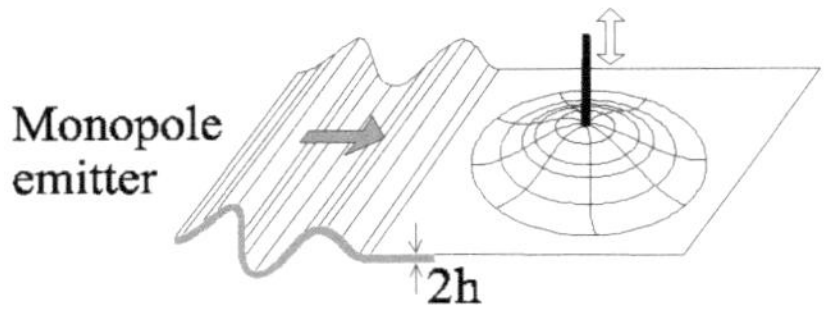 $\mathrm{Re}\,Z_{ex} = 2[\rho\Phi h^4 / 3(1-\nu^2)]^{1/2} = Z_0$ $E = \omega Z_0 A_e$ $\sigma^A = \lambda / 2\pi$ 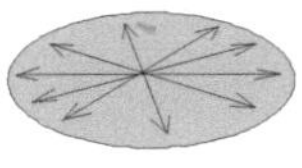
(b)	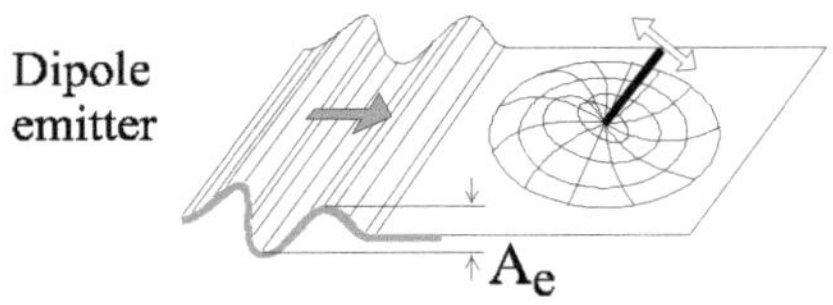 $\mathrm{Re}\,Z_{ex} = 3^{-3/2}2^{-1}(1-\nu^2)^{-1/2}h^2k^{-2}\Phi^{1/2}\rho^{1/2}$ $E = 2^{1/2}3^{-3/2}(1-\nu^2)^{-1}\Phi h^3 kA_e$ $\sigma^A = \lambda / \pi$ 
(c)	$S_e = 24^{-1}3^{-1/2}(1-\nu^2)^{-3/2}h^4k^5\Phi^{3/2}\rho^{-1/2}A_e^{\ 2}$ $\omega = k^2\gamma$ $\gamma = [h^2\Phi / 12\rho(1-\nu^2)]^{1/2}$

The power stream density S_e of the incident wave with the amplitudes $\mathbf{e}_0$, $\mathbf{h}_0$ of the electric and magnetic intensities, radiation resistance $\operatorname{Re} Z_{ex}(\omega)$, the force of the incident wave effect E (induced electromotive force), cross-sections σ_A for three-dimensional electric and magnetic (frame with current) dipoles and for two-dimensional cylinder current are given in **Table 2-3** (ε is the electric coefficient of media or vacuum). It is characteristic that the absorption cross-sections do not depend either on the electric dipole length– a , or on the square S and the number N of the magnetic dipole turns, or on the radius– a of the cylindrical current. At the right border of **Table 2-3**, the directivity patterns of radiation power of sources are presented: (a) electric dipole (toroid directivity pattern), (b) frame with current (toroid directivity pattern), and (c) cylindrical current (isotropic directivity pattern). Unlike the acoustical dipole (with conical directivity pattern in power, see **Fig. 2-2-b**), electromagnetic dipole is characterized by toroid directivity pattern (in power), which becomes more narrower with the growth of multipolarity order, but retains the toroid shape (see **Fig. 2-6**).

2.6.4. Bending Waves in a Thin Elastic Plates

Bending waves in a thin elastic plate with width $2h$ and density ρ , having the deviation $\phi(\mathbf{r},\omega)$ from the equilibrium are described by the equation $\Delta^2\phi + \varsigma^{-2}\phi''_{tt} = P(\mathbf{r},t)/(2\rho h\varsigma^2)$, where $\varsigma^2 = \Phi h^2 / 3\rho(1-\nu^2)$, Φ is the Young's modulus, ν is Poisson's coefficient, and is the external pressure, $\Delta^2 = (\partial^4/\partial x^4) + (\partial^4/\partial y^4)$. **Table 2-4** presents radiation resistance $\operatorname{Re} Z_{ex}(\omega)$, and complex amplitude E of the generalized force (point force and point twisting moment, with corresponding radiation directivity patterns). A_i is the amplitude of normal shift in the incident wave with frequency ω, the density of the power flux S_e and the dispersion equation $\omega = \omega(k)$. First line is occupied by monopole emitter, presented by normal (to the plate surface) optimum force E , applied to the point $\mathbf{r}_0$ (when $\operatorname{Re} Z_{ex}$ does not depend on frequency ω). Incident wave does the mechanical work on the source of force E . Second line is occupied by dipole source. This source presents thin weightless rigid rod, fastened (normally to the plane of plate) rigidly to the plate in the same point $\mathbf{r}_0$. External force moment is applied to the free end or rod (in this case $\operatorname{Re} Z_{ex} \sim \omega^2$).

2.7. MICROSTRUCTURE WITH AN ARBITRARY NUMBER OF THE INTERACTING FREEDOM DEGREES

Above we considered various emitters MS (MSUR, multipole lattices and others in **Tables 2-1, 2-2, 2-3, 2-4**) with only one degree of freedom – complex magnitude V of pulsation velocity of monopoles for instance (**Fig. 2-5**). Below we consider the system with n degrees of freedom, which can be described by symmetric impedance matrix $\hat{\mathbf{Z}}_{ex}(\omega) = [\hat{\mathbf{Z}}_{ex}]_{s,\ell}$ ($[\hat{\mathbf{Z}}_{ex}]_{s,\ell} = [\hat{\mathbf{Z}}_{ex}]_{\ell,s}$, $s,\ell = 1,2,...,n$, see **Fig. 2-7**) on the frequency ω . We define the element $Z_{s,\ell} = F_s / V_\ell$ as the ratio between force F_s and velocity V_ℓ F_s acts on the s –th degree of freedom *via* the pressure field, produced by the oscillations of ℓ –th degree of freedom (with velocity V_ℓ) , when all degrees of freedom (excluding ℓ –th) are "frozen" (i.e., when $V_s = 0$ for all $s \neq \ell$). Let us assume (analogously 2-23)), that the vector $\mathbf{V} = \mathbf{V}_{\text{opt}}$ of monopoles amplitudes, which provides maximum total absorbed power (MTAP), has the form [4]

$$\mathbf{V} = \mathbf{V}_{\text{opt}} = 2^{-1}\left[\operatorname{Re}\hat{\mathbf{Z}}_{ex}\right]^{-1}\mathbf{E}\,, \tag{2-26}$$

where −matrix (with elements $\left(\left[\operatorname{Re}\hat{\mathbf{Z}}_{ex}\right]^{-1}\right)_{s,\ell}$, which is inversed matrix $\operatorname{Re}\hat{\mathbf{Z}}_{ex}$. Let us show [67] that vector (2-26) is actually optimum. The power absorbed has the form

$$W(\mathbf{E};\mathbf{V}) = 2^{-1}\operatorname{Re}\left\{(\mathbf{V}^*)^T(\mathbf{E} - \hat{\mathbf{Z}}_{ex}V)\right\}\,, \tag{2-27}$$

where $\mathbf{E} = \{E_1, E_2, ..., E_n\}$ is the vector of amplitudes of incident wave action on the "frozen" (i.e., if $\mathbf{V} = 0$) oscillatory degrees of freedom of microstructure (spheres), and $\mathbf{V} = \{V_1, V_2, ..., V_n\}$, the vector of complex amplitudes of monopoles. We state optimum vector (2-26) and add it to the arbitrary vector ξ of velocities. The change of power W absorbed, caused by the arbitrary vector ξ of velocities, has the form

$$W(\mathbf{E};\mathbf{V}_{\text{opt}}+\xi)-W(\mathbf{E};\mathbf{V}_{opt})=-2^{-2}\,\text{Re}\left\{(\xi^*)^T[\text{Re}\,\hat{\mathbf{Z}}_{ex}]^{-1}\xi\right\}\le 0\,. \quad \textbf{(2-28)}$$

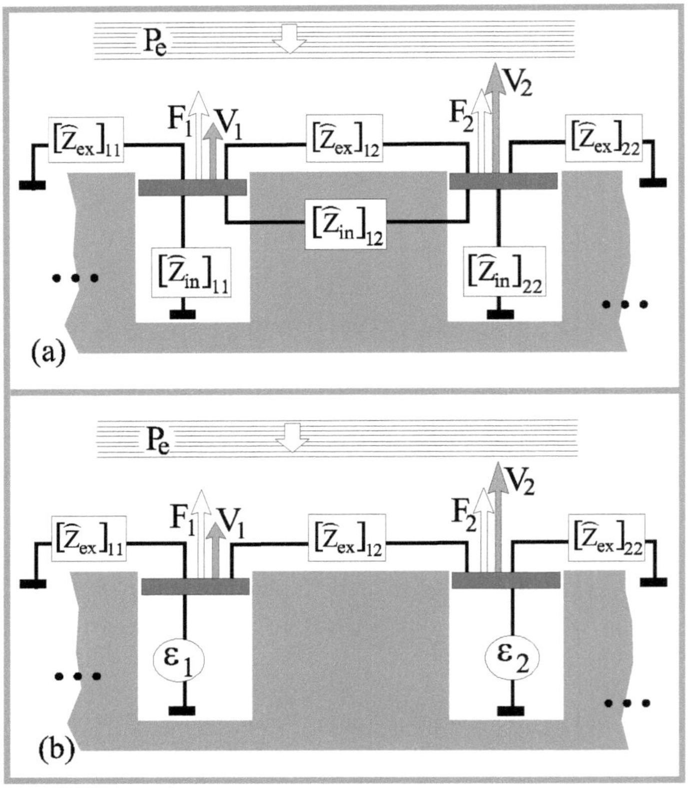

Figure 2-7: Two versions of absorbing system with two degrees of freedom ($n=2$, pistons in rigid screen): a) passive version with matrix $\hat{\mathbf{Z}}_{in}$ of load impedances, which satisfy condition (2-30); b) active version with internal sources of forces ε_1 and ε_2 , which provide the assigned optimum velocities V_1 and V_2 (components of vector $\mathbf{V}_{\text{opt}}$ of velocities (2-26)), independent on the forces F_1 и F_2 (components of the force vector $\mathbf{F}$ of the forces).

The value $\text{Re}\left\{(\xi^*)^T[\text{Re}\,\hat{\mathbf{Z}}_{ex}]^{-1}\xi\right\}\le 0$ has the constant sign and presents the power, radiated by microstructure with velocity vector ξ, in the absence of incident wave. This means that vector (2-26) actually provides the regime MTAP. From the vector (2-26) $\mathbf{V}_{\text{opt}}$, which provides maximum absorbed power

$$W_{\max}=2^{-3}\,\text{Re}\left\{(\mathbf{E}^*)^T[\text{Re}\,\hat{\mathbf{Z}}_{ex}]^{-1}\mathbf{E}\right\}, \quad \textbf{(2-29)}$$

(and simultaneously the same scattering power), we can design by active facilities or satisfying the resonant conditions

$$\text{Im}\,\hat{\mathbf{Z}}_{in}(\omega)=-\text{Im}\,\hat{\mathbf{Z}}_{ex}(\omega)\,,\ \ \text{Re}\,\hat{\mathbf{Z}}_{in}(\omega)=+\text{Re}\,\hat{\mathbf{Z}}_{ex}(\omega)\,, \quad \textbf{(2-30)}$$

where $\hat{\mathbf{Z}}_{in}$ is the matrix of monopoles loads ($\hat{\mathbf{Z}}_{in}$ and $\hat{\mathbf{Z}}_{ex}$ are the matrix analog of impedances Z_{in} and Z_{ex} , used above). Therefore, we can consider (2-30) as algorithm, which can be classified as {NS/NT/CA/F} (see **Fig. 1-3**).

For acoustical linear microstructure, described above, we define elements $[\hat{\mathbf{Z}}_{ex}]_{s,\ell}$ of matrix $\hat{\mathbf{Z}}_{ex}$:

$$[\hat{\mathbf{Z}}_{ex}]_{s,s} = 4a^2\pi\rho c(k^2a^2 + ika),\ [\hat{\mathbf{Z}}_{ex}]_{s,\ell}(s \neq \ell) = 2a^4\pi\rho ckr_{s,\ell}^{-1}[\sin(kr_{s,\ell}) + i\cos(kr_{s,\ell})], \quad \textbf{(2-31)}$$

where $r_{s,\ell} = 2h|s-\ell|$, ($s,\ell = 1,2,3,..,n$). Vector $\mathbf{E} = \{E_n\}$ (the action of incident wave on the "frozen" elements of absorbing microstructure) consists of elements $E_s = 4\pi a^2 \exp[-i(\mathbf{k}_e\mathbf{r}_s)]$, where $\mathbf{k}_e$ is the wave vector of incident wave, and $\mathbf{r}_s$ is the coordinate of s-th monopole.

Table 2-5: Results of numerical modeling of absorption of incident wave, which propagates along the axis (i.e. $\vartheta = 0$) of microstructure, consisting of $n = 1 \div 8$ monopoles

n	1	2	3	4	5	6	7	8
$\sigma_\otimes/\sigma_0$	1	3.957	8.888	15.793	24.671	35.521	48.345	63.163

Table 2-5 [67] and **Fig. 2-8-b** presents the numerical results (on the base (2-26), (2-27), (2-29), (2-31)) for total relative ACS $\sigma_\otimes/\sigma_0$ ($\sigma_\otimes$–ACS of microstructure in regime MTAP, σ_0–maximum ACS of a lonely monopole) of incident wave running along the axis of microstructure with $n = 1 \div 8$ elements. We obtained (numerically) dependence very close to $\sigma_\otimes/\sigma_0 = n^2$, and see, that all connections between monopoles and sound field (n^2 elements of $\hat{\mathbf{Z}}_{ex}$) gave the identical contribution to total ACS $\sigma_\otimes$ in the regime MTAP, when the wave dimension of microstructure is very small (when $kh = 0.2$). The difference between $\sigma_\otimes/\sigma_0$ and n^2 is decreasing, when $kh \to 0$, but was limited by finite accuracy of calculations in concrete computer.

Let us assume that we may scan phase multiplier $\exp(i\phi)$ (in the range $-\pi < \phi \le +\pi$) of all ($n = 4$) elements of vector $\mathbf{V}_{\text{opt}} \exp(i\phi)$, tuned (when $\phi = 0$) for axial incident wave ($\vartheta = 0$) in the regime MTAP. **Fig. 2-8-c** presents the ACS $\sigma_\otimes/\sigma_0$ as a function of ϕ. Let us consider the scanning the module multiplier A ($\operatorname{Im} A = 0$) of vector $\mathbf{V}_{\text{opt}}A$ ($\mathbf{V}_{\text{opt}}$ is tuned as above) in the range $0 < A \le 2$. Therefore, we obtain the ACS $\sigma_\otimes/\sigma_0$ as a function of A, presented in **Fig. 2-8-d** ($n = 4$). Finally we scan the direction ϑ of incident wave in the range $-\pi < \vartheta \le +\pi$ (when $\mathbf{V}_{\text{opt}}$ is constant and tuned, as above) and then we obtain ACS $\sigma_\otimes/\sigma_0$ as a function of ϑ or the directivity pattern of linear structure in regime MTAP, when $n = 4$ (**Fig. 2-8-e**). We must note, that the one-dimensional case ($n = 1$) gives the same plots **Fig. 2-8-c**, **Fig. 2-8-d**.

Fig. 2-8-a presents the distribution of absorbed (positive sign) and radiated (negative sign) power by each element of structure under different numbers $n = 2 \div 8$ of elements. Note, that for $n = 2$ we have absorption (positive sign) on the element nearest to the source of incident wave and radiation (negative sign) on the farther element. In the rest of the cases ($n = 3 \div 8$), we see the opposite situation with stable qualitative shape (one bipolar oscillation) of power distribution. With the increase of n the power distribution looks antisymmetric (i. e. the sum looks zero). This is due to the smallness of total absorbed power in comparison with the power of interaction between elements. This is due to the weakness of power, absorbed by structure, in comparison with the power of interaction between elements.

The formulas (2-23)-(2-30) can be applied to any oscillatory systems, including electric circuits, for instance. However, the connection $\sigma_\otimes/\sigma_0 = n^2$ presents the specific characteristic of *wave acoustical linear structure.*

We must note, that the directivity pattern of ACS in the regime MTAP does not have zeroes (unlike the directivity patterns of scattering field, see **Fig. 2-11-b**), because it cannot be less than this one of the lonely monopole. In the regime MTAP, we always have $\sigma_\otimes/\sigma_0 \ge 1$ ($0 \le \vartheta < 2\pi$). **Fig. 2-9-a** presents the directivity pattern of ACS $\sigma_\otimes/\sigma_0$ in the regime

MTAP, when $kh = 0.2$, $n = 1 \div 8$. From the theory of antennas, we know that the growth of ACS of arbitrary antenna (MS, MSUR, and any others) is accompanied by more narrow directivity pattern of scattering and absorption [6]:

$$\sigma_A \approx \lambda^2 / \Omega \,, \qquad \textbf{(2-32)}$$

where σ_A is the effective aperture of antenna (or ACS), Ω is the volume corner of the main petal of directivity pattern (we suppose that practically all power streams show the directivity patterns of ACS $\sigma_{\otimes} / \sigma_0$ ($n = 1 \div 8$) in regime MTAP, inserted with deformation of antenna's radiation concentrated inside this corner Ω. **Fig. 2-9-b**) in square to select $2(n-2)$ lateral petals, (besides 2 main petals: forward and backward).

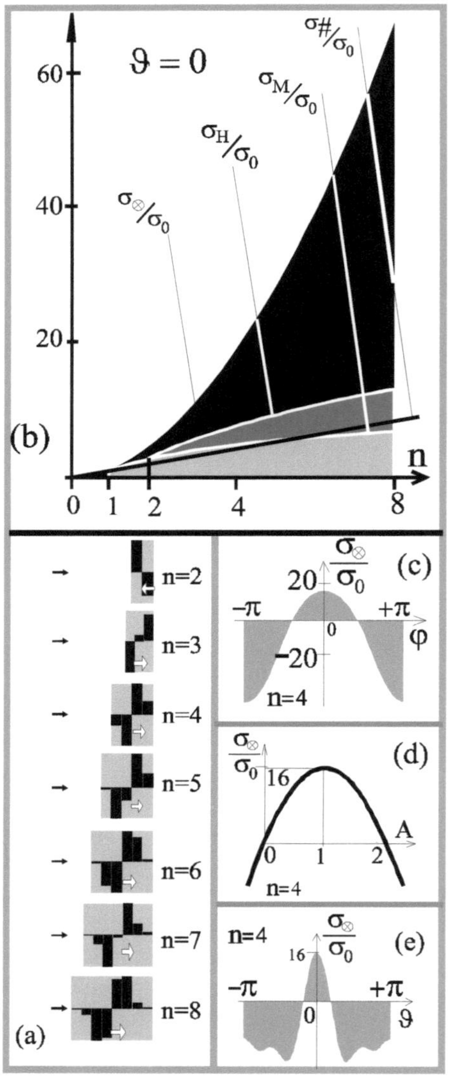

Figure 2-8: (a) The distribution of acoustic powers absorbed (positive), and radiated (negative), being normalized by maximum in regime MTAP ($kh = 0.2$). White arrows denote the direction of energy transfer inside the structure, black arrows denote the direction of incident wave. (b) The normalized ACS of microstructures: $\sigma_{\#} / \sigma_0$ –ACS of linear structure of n uninteracting monopoles (or qualitative plot for lattice of MS or MSUR, section 2.4.); $\sigma_{M,N} / \sigma_0$ –ACS of MS with $n = 2^N$ monopoles; $\sigma_{H,N} / \sigma_0$ –ACS of MSUR with $n = 2^{N+1}$ monopoles; $\sigma_{\otimes} / \sigma_0$ –ACS of linear structure with n monopoles in the regime

MTAP. (c) The action of untuning in phase ϕ of $\mathbf{V}_{\text{opt}}$ (included to all $n = 4$ elements) on the ACS $\sigma_{\otimes} / \sigma_0$ in the regime MTAP. (d) The action of untuning in module A of $\mathbf{V}_{\text{opt}}$ (included to all $n = 4$ elements) on the ACS $\sigma_{\otimes} / \sigma_0$ in the regime MTAP. (e) The action of scanning of incident wave direction ϑ on the ACS $\sigma_{\otimes} / \sigma_0$ in the regime MTAP ($n = 4$).

Fig. 2-10 shows the absorption characteristics of the incident wave propagating normally for the axis of microstructure ($\vartheta = \pi / 2$, $kh = 0.2$), formed by n elements in the regime MTAP. The qualitative oscillatory shape of $\sigma_{\otimes} / \sigma_0$ as the function of kh (for any fixed n) is presented in **Fig. 2-10-a**.

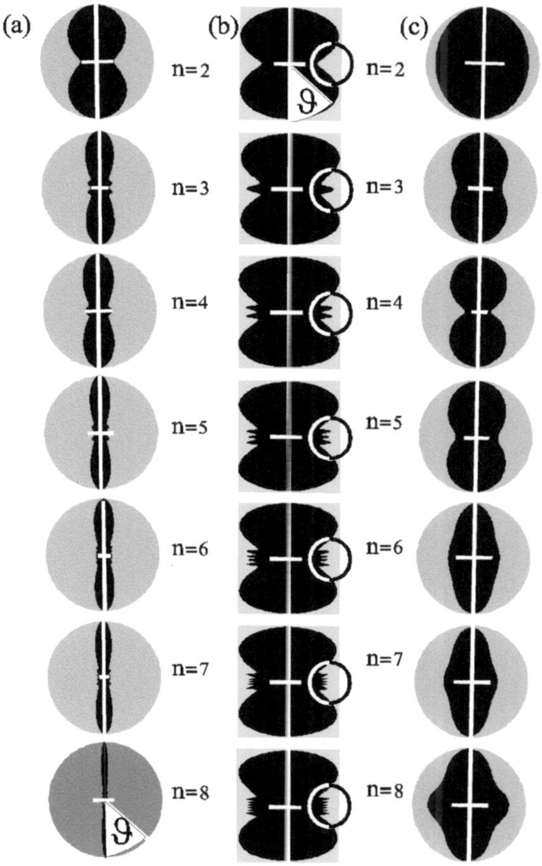

Figure 2-9: (a) Normalized directivity patterns of ACS in regime MTAP ($kh = 0.2$). (b) Directivity patterns of ACS in regime MTAP, inserted (deformed) in square to select lateral petals. (c) Directivity patterns of ACS in regime of individual strategy.

Upper and floor limits of oscillations of the above plot are presented in **Fig. 2-10-b** as a function of number n of elements. The distribution of values V_m / V_0 (when $m = 1 \div n$ and $n = 1 \div 6$) of vector $\mathbf{V}_{\text{opt}}$, when monopoles are spaced closely ($kh = 0.2$) to each other is shown in **Fig. 2-10-c**. In the case of $n = 2$ (due to symmetry of the boundary value problem) the values V_m remains identical for any kh .

If $kh \to 0$ two monopoles meet in one. If $n > 2$ the distribution $\text{Re}(V_m)$ becomes bipolar and symmetrical for mediums. The evolution of elements (exactly speaking, the values $\ln\left(|V_m| / |V_0|\right)$) of vector $\mathbf{V}_{\text{opt}}$ in the range $0.2 < kh < 6$ is shown in **Fig. 2-10-d**. The distribution of $\text{Re}(V_m)$ (under $m = 1 \div n$ and $n = 1 \div 6$) elements V_m of vector $\mathbf{V}_{\text{opt}}$, when monopoles are spaced far ($kh >> 1$) from each other, is shown in **Fig. 2-10-e**, where V_0 is the optimum velocity (2-23) for a lonely monopole.

Fig. 2-11 presents the characteristics of scattering for the case $n=4$ in the regime MTAP (2-26), (2-30). We mean by the scattering field, the field radiated by microstructure with $\mathbf{V}=\mathbf{V}_{opt}$ in the absence of incident waves. The absorption directivity pattern is shown in **Fig. 2-11-a**. Each direction ϑ (arrow denotes the direction of incident wave $0<\vartheta\leq\pi/2$) of incident wave produces the special directivity pattern of scattering (in power) **Fig. 5-b1÷5-b11**. All angular dependences are given in logarithmic scale.

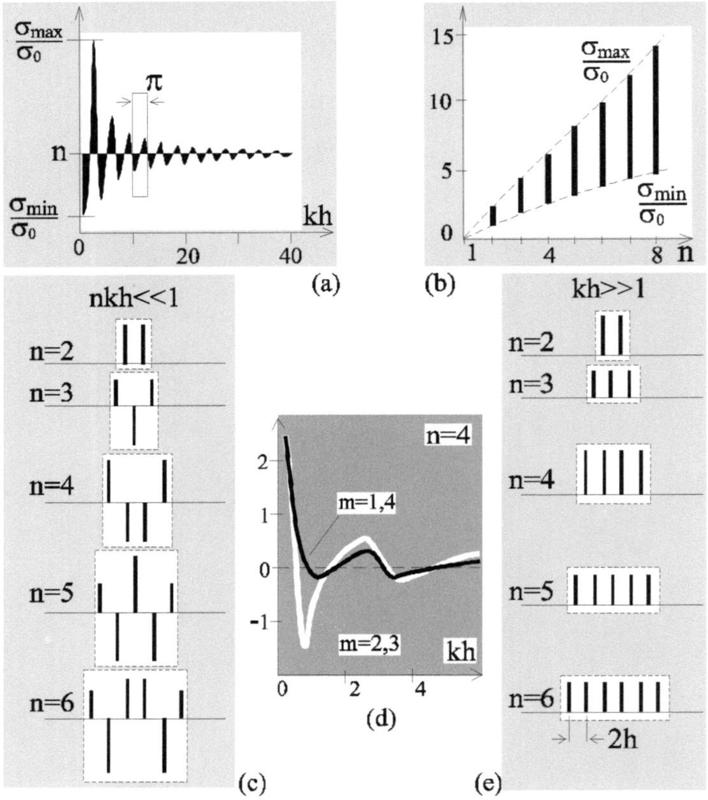

Figure 2-10: Absorption characteristics of the incident wave propagating normally for the axis of microstructure ($\vartheta=\pi/2$, $kh=0.2$), formed by n elements in the regime MTAP: (a) qualitative shape of $\sigma_{\otimes}/\sigma_0$ as the function of kh (for any fixed n); (b) bounds of oscillations of plot (a) as the function of number n of elements; (c) the distribution of V_m/V_0 (under $m=1\div n$ and $n=1\div 6$) elements V_m of vector $\mathbf{V}_{\text{opt}}$, when monopoles are spaced closely ($kh=0.2$) to each other; (d) an example of monopoles amplitudes (exactly speaking, the values $\ln\left(|V_m|/|V_0|\right)$ under $m=1,2,3,4$, where (V_0 –the optimum amplitude of a lonely monopole) evolution with the growth of kh, when $n=4$; (e) the distribution of V_m (under $m=1\div n$ and $n=1\div 6$) elements V_m of vector $\mathbf{V}_{\text{opt}}$, when monopoles are spaced far ($kh>>1$) from each other

Numerically, we found some important characteristics of the acoustical boundary value problem (2-31) for the linear structure of monopoles:

(a) Spatial distribution of modules $|(V_m)_{\text{opt}}|$ of complex components $(V_m)_{\text{opt}}$ of the vector $\mathbf{V}_{\text{opt}}$ is symmetrical for the medium of absorbing microstructure, i. e.

$$\left|(V_{n-m+1})_{\text{opt}}\right|=\left|(V_m)_{\text{opt}}\right|,$$

where $m = 1,2,...,n$. This characteristic is saved under all directions $0 \le \vartheta < 2\pi$ of the incident wave. For phases $\arg[(V_m)_{\text{opt}}]$, symmetry does not exist. So, for instance, in the case $n = 2$, the modules of velocities are identical, but phases are different, if $\vartheta \ne \pm\pi/2$. If $\vartheta = \pm\pi/2$, the spatial distribution of phases $\arg[(V_m)_{\text{opt}}]$ becomes symmetrical for the medium of absorbing microstructure too (due to geometrical symmetry of boundary problem). Moreover, we obtained $(V_m)_{\text{opt}} / V_0 = \text{Re}[(V_m)_{\text{opt}} / V_0]$ (**see Fig. 2-1-c**), where V_0 is the optimum complex amplitude of a lonely absorber.

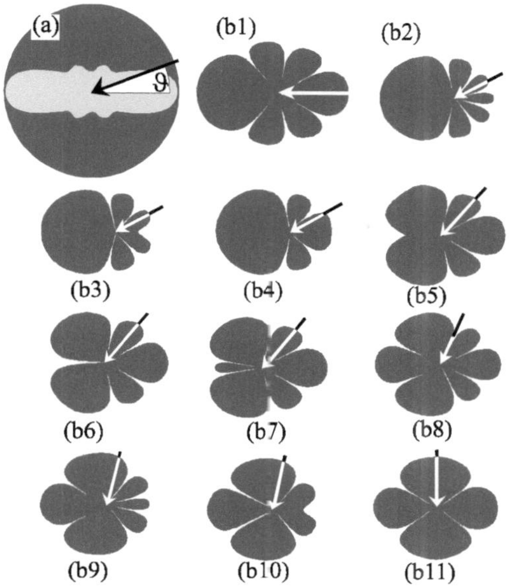

Figure 2-11: (a) Directivity pattern of ACS in the regime MTAF (2-26), (2-30), when ($n = 4$). (b) Scattering power directivity pattern of linear structure in the regime MTAP (11-b1÷11-b11, $n = 4$): arrow denotes the direction of incident wave $0 < \vartheta \le \pi/2$. Each direction ϑ of incident wave defines the special vector $\mathbf{V}_{\text{opt}}(\vartheta)$ and special directivity pattern of scattering power (all angular dependences are given in logarithmic scale).

(b) Vectors $\hat{\mathbf{Z}}_{ex}\mathbf{V}_{\text{opt}}$ and $\mathbf{E}$ are connected by simple ratio $\hat{\mathbf{Z}}_{ex}\mathbf{V}_{\text{opt}} = \alpha\mathbf{E}$, where α is a certain nonzero real number ($\text{Im}\,\alpha = 0$). In other words, $\mathbf{E}$ is eigenvector of the matrix $\hat{\mathbf{Z}}_{ex}\left[\text{Re}\,\hat{\mathbf{Z}}_{ex}\right]^{-1}$.

2.7.1. Individual and Cooperative Strategies of Absorption

The above resonant absorption (2-30) in the system with arbitrary number n of elements referred to us collective resonance and corresponding strategy $\hat{C}$ can be called "*cooperative strategy*" [60], when each element is trying to "help" in the regime MTAP (does not try to grow power absorbed by itself, **Fig. 2-12-a**). This strategy can be realized in both active form (**Fig. 2-12-b**) and passive form (**Fig. 2-12-c**). When we follow the *individual strategy* $\hat{U}$ [60], each element (monopole) tunes its velocity in accordance with the formula (2-26), supposing, that all external acoustical field (besides own acoustical field) is the incident wave field $\mathbf{E}$ (**Fig. 2-12-d**). In this approach we do not take into account acoustical fields, induced by nearby elements, and vector $\mathbf{V}_{\oplus}$ of the optimum pulsation velocities of monopoles is defined by the formula

$$\mathbf{V}_{\oplus} = \hat{\mathbf{Y}}^{-1}\mathbf{E}, \tag{2-33}$$

where $\hat{\mathbf{Y}} = [Y_{s,\ell}]$–matrix with elements $Y_{s,\ell} = Z_{s,\ell}$, when $s \ne \ell$ and $Y_{s,s} = 2[\text{Re}\,\hat{\mathbf{Z}}_{ex}]_{s,s}$, when $s = \ell$ (s, ℓ =1, 2, 3, …,n). Hence, the algorithm (2-33) can be classified as {LS/LT/CA/F} (see **Fig. 1-3**). **Fig. 2-9-c** presents the normalized directivity pattern of ACS σ_{ind} / σ_0 of linear microstructure under the individual strategy, when $n = 2 \div 8$ (where σ_{ind} is the absorption cross section of structure under condition (2-33)). **Fig. 2-9-c** shows that the growth of n (number of elements in the linear structure) does not lead to the more narrow directivity pattern σ_{ind} / σ_0 of ACS under individual strategy. And, clearly, it does not lead to the growth of maximum ACS. **Fig. 2-13** illustrates this fact, when $n = 3$: due to reapproachment ($kh \to 0$) of elements, following individual strategy, the total power absorbed approaches to zero ($(\sigma_{ind})_{\max} \to 0$, see **Fig. 2-13-b**). Unlike the cooperative strategy (MTAP (2-26)), when $(\sigma_{coop})_{\max} \to n^2 \sigma_0$ (see **Fig. 2-13-a**). So, under individual strategy we can get the total maximum absorbed power $(\sigma_{ind})_{\max} = n\sigma_0$ only for uninteracting ($kh \to \infty$). The same limit has, of course, total absorbed power in the regime MTAP (cooperative strategy), if $kh \to \infty$.

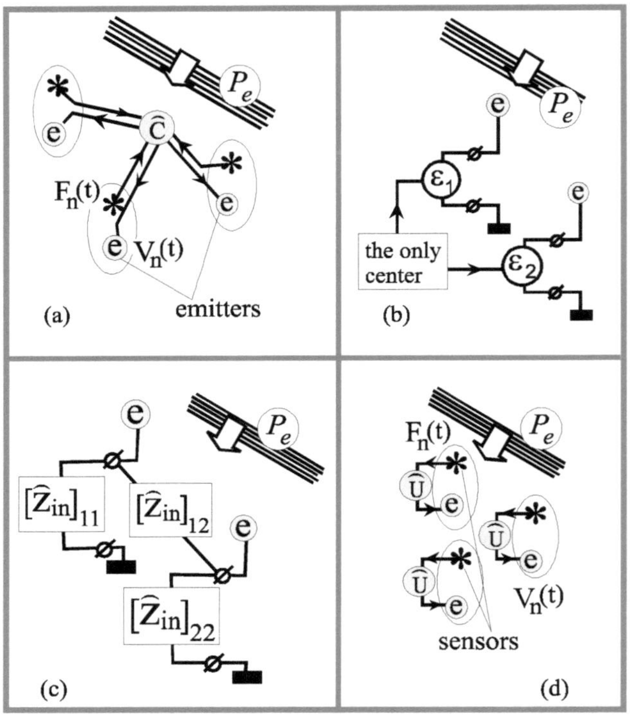

Figure 2-12: Cooperative (MTAP) control strategy $\widehat{C}$ (a, b) for emitters "e" with active ((2-26), (b)) and passive ((2–30), (c)) versions, individual control strategy $\widehat{U}$ ((2-30), (d)).

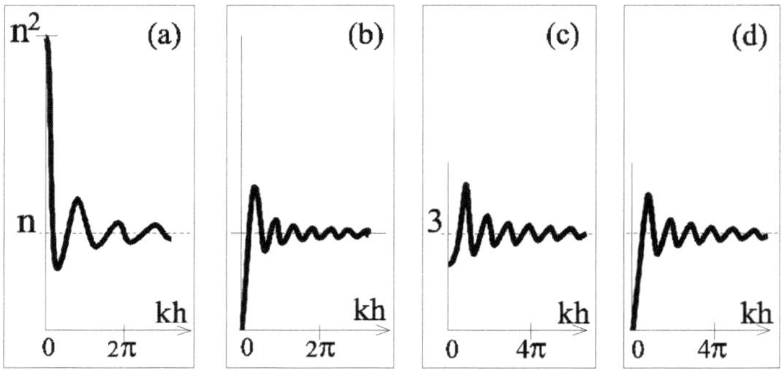

Figure 2-13: The example of total ACS as a function of wave dimension kh of spatial interval between elements: when axial ($\vartheta = 0$) propagation of incident wave (a, b) and normal ($\vartheta = \pi / 2$) propagation of incident wave in the case of linear system of $n = 3$ monopoles, when we have cooperative (MTAP) control strategy (a, c) or individual control strategy (numerical results).

2.7.2. Sound Absorption in the Infinite Lattices of Emitters

Now we consider cooperative and individual strategy on some simple examples of acoustical systems with infinite numbers of the freedom degrees (or infinite numbers of elements $n \to \infty$) in the regime MTAP. Let us consider two simplest variants [60] of the boundary value problem:

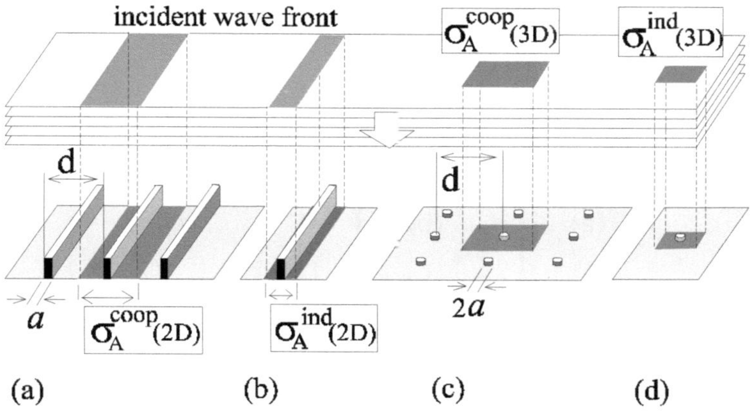

Figure 2-14: The boundary value problems for infinite two-dimensional (a, b) and three-dimensional (c, d) cases. Lonely piston (b, d; individual strategy) and piston lattices (a, c; cooperative strategy). Pistons are spaced in rigid screens, and incident wave propagation is normal to the screens.

1) Plane incident wave is characterized by wavelength λ and by the amplitude of longitudinal particle velocities υ, propagating normally to the rigid plane screen. On the screen, (are spaced periodically with spatial period d) two-dimensional pistons (with width a, **Fig. 2–14,a,b**) and three-dimensional circular pistons (with radius a, **Fig. 2-14,c,d**). It is obvious, that when spatial periods of pistons satisfy the relation $d \le \lambda / 2$ and amplitudes of their normal velocities $V = (d / a)\upsilon$ and $V = \pi^{-1} d^2 a^{-2}$ (for two-dimensional and three-dimensional cases respectively), the total absorbed power is maximum (this is regime MTAP or cooperative strategy, inevitable), and the infinite lattice fully absorbs the incident wave without any reflection. When $d = \lambda / 2$, we obtain ACS $\sigma_A^{coop}(2D) = \lambda / 2$ and $\sigma_A^{coop}(3D) = \lambda^2 / 4$ (ACS per one element), corresponding to one two-dimensional and one three-dimensional piston of lattice with spatial period $d = \lambda / 2$.

2) in the same screen we place the only (two-dimensional or three-dimensional) piston, which can follow only the individual strategy, inevitably. Optimum amplitude of piston velocity is defined by the formula (2-26). Therefore, we obtain ACS $\sigma_A^{ind}(2D)$ и $\sigma_A^{ind}(3D)$ for two-dimensional and three-dimensional cases, respectively.

It is easy to check, that, when $d = \lambda / 2$, ACS $\sigma_A^{coop}(3D)$, $\sigma_A^{ind}(3D)$, $\sigma_A^{coop}(2D)$, $\sigma_A^{ind}(2D)$ are connected by the following ratio:

$$\sigma_A^{coop}(3D) / \sigma_A^{ind}(3D) = \sigma_A^{coop}(2D) / \sigma_A^{ind}(2D) = \pi / 2. \quad \textbf{(2-34)}$$

2.7.3. Narrowband Characteristics of Linear Absorbing System in the Area of Spatial Frequencies

Classical works on the antennas technique (for instance [72]) give the exact ratio between the effective aperture of antenna (absorption cross-section σ) and its coefficient of directional action (CDA) Λ in three-dimensional and two-dimensional cases $\sigma(3D)$, $\Lambda(3D)$ и $\sigma(2D)$, $\Lambda(2D)$. For three-dimensional case [72] we have

$$\sigma(3D) = (\lambda^2 / 4\pi)\Lambda(3D), \quad \textbf{(2-35)}$$

$$\Lambda(3D) = 4\pi\left[\int_0^{2\pi} d\phi \int_0^{\pi} \sin(\vartheta)\bar{\Psi}(\phi,\vartheta)d\vartheta\right]^{-1}, \qquad \textbf{(2-36)}$$

and for the two-dimensional case respectively

$$\sigma(2D) = (\lambda / 2\pi)\Lambda(2D), \qquad \textbf{(2-37)}$$

$$\Lambda(2D) = 2\pi\left[\int_0^{2\pi} \bar{\Psi}(\phi)d\phi\right]^{-1}, \qquad \textbf{(2-38)}$$

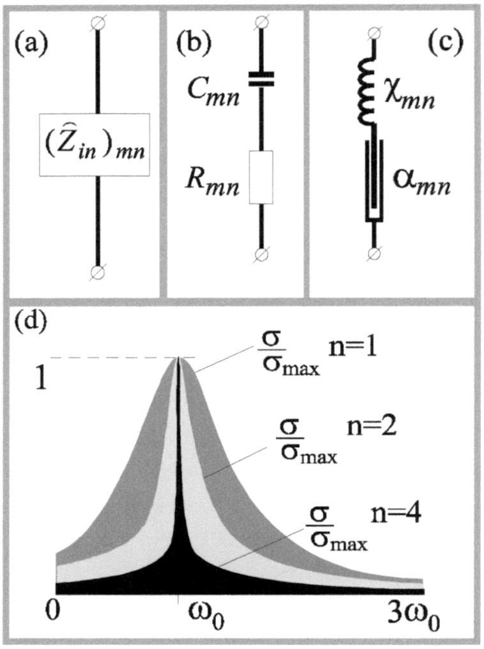

Figure 2-15: The element $(\hat{\mathbf{Z}}_{in})_{nm}$ of matrix $\hat{\mathbf{Z}}_{in}$ of passive correcting load of linear system one, two and four monopoles (a); electric analog "active resistance R_{nm} plus capacitance C_{nm}" (b) of the acoustic element $(\hat{\mathbf{Z}}_{in})_{nm}$ "viscous dissipation δ_{nm} plus elasticity χ_{nm}" (c); the normalized ACS $\sigma / \sigma_{\max}$ of one ($n = 1$), two ($n = 2$) and four ($n = 4$) monopoles, loaded by matrix (2–40), as the function of temporal frequency ω under the resonant condition (2–30) on the frequency $\omega_0 = 134.4$ (radian/s) of incident wave, if $kh = 0.2$.

where $\bar{\Psi}(\phi,\vartheta) = \left\{\Pi(r,\phi,\vartheta)\left(\max_{(\phi,\vartheta)}\Pi(r,\phi,\vartheta)\right)^{-1}\right\}_{r\to\infty}$ is the directivity pattern in power, normalized by maximum for three-dimensional antenna, $\bar{\Psi}(\phi) = \left\{\Pi(r,\phi)\left(\max_{\phi}\Pi(r,\phi)\right)^{-1}\right\}_{r\to\infty}$ is the directivity pattern in power, normalized by maximum for two-dimensional antenna, $\Pi(r,\phi,\vartheta)$ and $\Pi(r,\phi)$ −radiation power stream density of antenna in three- and two-dimensional cases respectively $0 \le \phi < 2\pi$, $0 \le \vartheta < \pi$. For instance, for acoustical linear multipoles (section 2.2.) we obtain $\sigma(3D) = (\lambda^2 / 4\pi)(2N+1)$, $\sigma(2D) = (\lambda / 2\pi)2^N$ (see **Tables 2-1, 2-3**). In addition to (2-32) we obtain the ratio between the width $\Omega(2D)$ main petal of directivity pattern of two-dimensional antenna and its absorption cross-section $\sigma_A(2D)$:

$$\sigma_A(2D) \approx \lambda / \Omega(2D) \;\; (0 < \Omega \le 2\pi). \tag{2-39}$$

We observe from (2-32), (2-39), that the maximum absorption cross-section (ACS) of antenna (multipole) is strongly defined only by the shape of its directivity pattern in power, normalized by maximum, not depending on the concrete construction of antenna. This is why the absorption cross-sections of MS (2-8) and MSUR (2-16) can be obtained from (2-35) and (2-37). In the same way, we could obtain absorption cross-section of linear structure (**Table 2-5**) of n degrees of the freedom, if we know the corresponding radiation directivity pattern, which ensures maximum absorption.

On the other hand, the vector of optimum complex amplitude $\mathbf{V}_{\text{opt}}(\mathbf{E})$ of antenna excitation or its optimum $\hat{\mathbf{Z}}_{in}$ matrix of load is defined by concrete construction. This requires to solve total boundary problem, as, for instance, in the case of maximization of the coefficient of the directivity action of antenna.

2.7.4. Narrowband Characteristics of Linear Absorbing System in the Area of Temporal Frequencies

Now, we consider ACS of linear structure of acoustical monopoles, as a function of temporal frequency of the incident wave. Numerical analysis of matrix $\hat{\mathbf{Z}}_{ex}$ on the incident wave frequency $\omega_0 = 134.4\,\text{(rad/s)}$ shows, that the reactance of all elements has the same sign and can be interpreted as inertial mass. This constancy of sign is caused by small wave-dimensions of absorbing structure. In the simplest correcting circuits (**Fig. 2-15-b, 2-15-c**) dissipative coefficient δ_{nm} and elastic coefficient χ_{nm} do not depend on frequency ω. But $(\text{Re}\,\hat{\mathbf{Z}}_{ex})_{nm}$ and $(\text{Im}\,\hat{\mathbf{Z}}_{ex})_{nm}$ depend on ω. Therefore, the satisfaction of resonant conditions (2-30) is possible on the only incident wave frequency: for instance, on the incident wave frequency ω_0 (if this wave is monochromatic)

$$(\text{Re}\,\hat{\mathbf{Z}}_{in})_{nm} = \delta_{nm}, \;\; (\text{Im}\,\hat{\mathbf{Z}}_{in})_{nm} = \chi_{nm} / (i\omega) \;. \tag{2-40}$$

We see in **Fig. 2-15**, that the above simplest construction of matrix $\hat{\mathbf{Z}}_{in}$ of correction loads leads to the following qualitative connection

$$(\Delta\omega_0 / \omega_0) \sim \lambda^2 / \sigma \;, \tag{2–41}$$

between relative width $(\Delta\omega_0 / \omega_0)$ of absorption range, λ (or ω) and $\sigma / \sigma_{\max}$, analogously to (2-32). As follows from **Fig. 2-15**, the growth in amount n of elements of linear absorbing system the simplest correction matrix (2-40) of loads, the absorption becomes more narrowband. The growth of complexity of elements of matrix $\hat{\mathbf{Z}}_{in}$ allows to ensure conditions (2-30) in the group of closely spaced "knots" $\omega = \omega_{0m}$ $(m = 1, 2, ..., N_\omega)$ of frequency and can make the absorption frequency range wider. It is important, that the passive control of monopoles remains causal, as the correction matrix $\hat{\mathbf{Z}}_{in}$ of loads presents the combination of circuits, which can be realized physically.

2.7.5. Sensor Antennas and Wave Antennas in the Detection Problem

Receiving antennas can be divided into two kinds:

a) "Dynamical antennas" (**Fig. 2-16-a**) convert incident wave into an output signal (in acoustics for instance, microphone converts acoustical pressure into electric signal) at the ends of electric wires. Hereinafter, coordinated load absorbs this electric energy and converts it into heat. In this way we obtain the distortion of initial field of power stream lines in the incident wave as in **Fig. 2-1**;

b) "sensor antennas" (as a rule - lattices) consist of a lot of sensors (**Fig. 2-16-b**). Output electric signal of each sensor describes concrete current local values of any component of the incident wave field. The sensors, as a rule, have very small wave dimensions and produce the negligible scattering. Sensor antennas, due to the signal processing, can have very high CDA (see section 2.7.3.) without absorption of energy of incident waves. Thus, the effective receiving surface (2-35), (2-37) cannot be interpreted as ACS, and this antenna does not distort power stream lines of incident wave field. The advantage of sensor antennas is given by the fact, that sensor antenna does not produce reactive field at all (reactive field exists only virtually inside the computer).

It is clear, that in the case of assigned equidistant linear microstructure formula (2-26) provides the most narrow directivity pattern of ACS (normalized by σ_0) for dynamical antenna (**Fig. 2-16-a**) and sensor antenna (**Fig. 2-16-b**) with the output signal $\Phi_n(\vartheta) = 2^{-3}\left(\sigma_0 \pi a^2 \rho^{-1} c^{-1} |\tilde{\mathbf{E}}|^2\right)^{-1} \mathrm{Re}\left[(\tilde{\mathbf{E}}^T)^*, \mathbf{V}_{\mathrm{opt}}(\mathbf{E})\right]$, where:

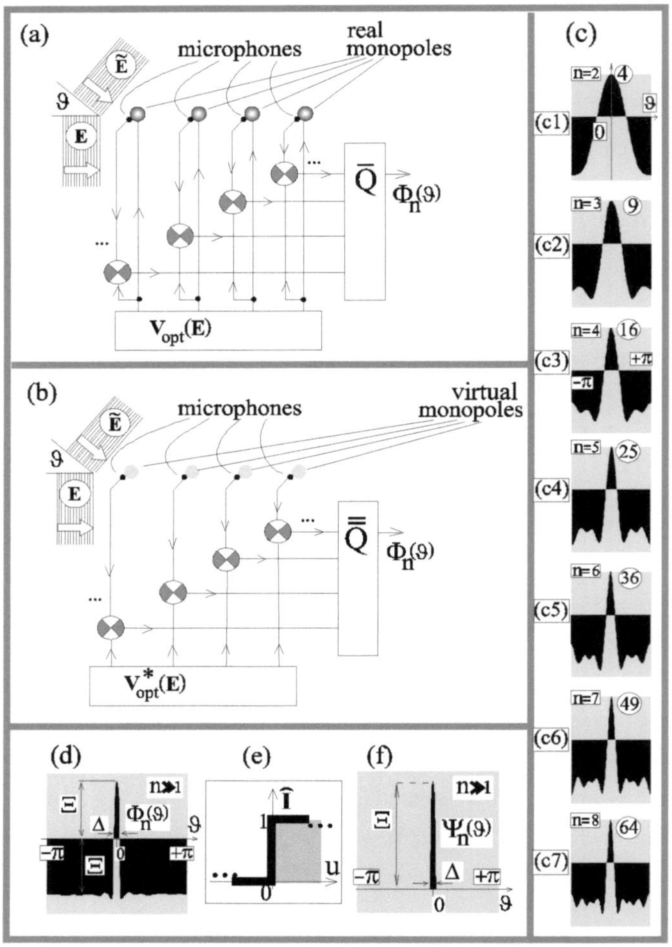

Figure 2-16: Dynamical (for absorption) antenna (a) and sensor antenna (for signals processing) (b), both based on principle MTAP (symbol ⊗ means the product of signals), **E** –tuning wave, $\tilde{\mathbf{E}}$ –incident wave, $\bar{\mathbf{Q}}$ –matrix of dynamical connections, $\bar{\bar{\mathbf{Q}}}$ –matrix of signal processing. Several directivity patterns (c1)–(c7) of ACS (positive values mean absorption with peaks n^2, negative values denote radiation) by structure consisting of $n = 2 \div 8$ monopoles respectively ($-\pi < \vartheta < +\pi$). Qualitative shape (d) of directivity pattern $\Phi_n(\vartheta)$, if $n >> 1$. Step-function $\hat{I}(u)$ (e). The shape (f) of the final directivity pattern $\hat{I}[\Phi_n(\vartheta)]$.

E is the vector of complex amplitudes of sound pressure forces, induced on n elements of microstructure by the "tuning wave" with wave vector $\mathbf{k}(\mathbf{E})$ parallel with the axis "x" of microstructure; $\mathbf{V}_{\mathrm{opt}}$ is the vector (2-26) of pulsating velocities of n elements of microstructure, providing maximum of absorbed power of the wave $\mathbf{k}(\mathbf{E})$; $\tilde{\mathbf{E}}$

is the vector of complex amplitudes of sound pressure forces, induced on n elements of microstructure by the *real incident wave*; ϑ is the angle between wave vectors $\mathbf{k}(\tilde{\mathbf{E}})$ and $\mathbf{k}(\mathbf{E})$.

Fig. 2-16-c presents the directivity patterns $\Phi_n(\vartheta)$ (normalized by maximum, when $n = 2 \div 8$) of absorbed power, where negative values mean radiation, and positive values mean absorption. We assumed in **Fig. 2-16-c** that $|\mathbf{E}| = |\tilde{\mathbf{E}}|$. As shown in **Fig. 2-16-c** and **Fig. 2-16-d**, the directivity pattern of ACS of microstructure consists of two petals (we call "petal" the angular interval, within which the sign of ACS remains constant): $|\vartheta| \le \Delta_n / 2$ ($\Phi_n(\vartheta) \ge 0$), $(\Delta_n / 2) < |\vartheta| < \pi$ ($\Phi_n(\vartheta) < 0$), where $\pm\Delta_n / 2$ are the roots of the function $\Phi_n(\vartheta)$, and Δ_n is the width of petal. It was observed numerically, that $\Delta_n / \Delta_2 = 2 / n$, where $\Delta_2 = 2.388$, $n = 2, 3, 4, \ldots$ **Fig. 2-16-d** presents the qualitative shape of $\Phi_n(\vartheta)$, when $n >> 1$, where $\Xi = n^2$. After processing by step operator $\hat{I}(u)$ (**Fig. 2-16-e**) we obtain the final directivity pattern $\hat{I}[\Phi_n(\vartheta)]$ with the only narrow peak (**Fig. 2-16-f**). The above results can be used in the problems of signal detection.

2.8. VARIATION APPROACH TO THE MAXIMUM ABSORPTION BY A SOLITARY EMITTER AND ITS TEMPORAL REPRESENTATION

The impedances of many types of emitters of small wave dimensions may be presented as power series kD or $i\omega$ (low-frequency approximation, $k = \omega / c$, D –characteristic geometrical dimension of emitter)

$$Z_{ex}(\omega) = \sum_n b_n (i\omega)^n, \tag{2-42}$$

which in temporal representation corresponds to the operator $\tilde{Z}_{ex} = \sum_n b_n (d / dt)^n$, and the action force $f(t)$ of media on the emitter has the form [4]:

$$f(t) = e(t) - \sum_n a_n (d / dt)^n \upsilon(t), \tag{2-43}$$

where $\upsilon(t)$ is the instantaneous velocity ($V(\omega)$ and $F(\omega)$, used above, are the Fourier-transformation of $\upsilon(t)$ and $f(t)$ at the incident wave frequency $V(\omega)$ respectively). The instantaneous absorbed power here equals $w(t) = f(t)\upsilon(t)$. The average power absorbed at some time interval $[t_1, t_2]$ has the functional form

$$\Psi = \int_{t_1}^{t_2} w(t)dt / (t_2 - t_1) = \int_{t_1}^{t_2} \left[e(t) - \sum_n a_n (d / dt)^n \upsilon(t) \right] dt / (t_2 - t_1).$$

It is wellknown that the functional expressions of such type (more general form

$$H[y] = \int_a^b \xi\{x, y(x), y'(x), \ldots, y^{(n)}(x)\}dx\ )$$

are characterized by extremum trajectory $y(x)$, satisfying the Euler's equation

$$(\partial\xi / \partial y) + \sum_{n=1} (-1)^n \{d^n / dx^n\}\{\partial\xi / \partial y^{(n)}\} = 0 \tag{2-45}$$

or (in our symbols) the trajectory $\upsilon(t) = \upsilon_A(t)$, which gives us the maximum of Ψ and satisfies the Euler's equation

$$(\partial\Psi / \partial\upsilon_A) + \sum_n b_n (-1)^n (d / dt)^n (\partial\Psi / \partial\upsilon_A^{(n)}) = 0. \tag{2-46}$$

Substituting the concrete form of functional Ψ into Euler's equation (2-46), we get the equation (algorithm) of optimum control

$$e(t) - 2\sum_n b_n \upsilon_A^{(n)}(t) = 0 \;\; (n = 2m\,,\; m = 0,1,2,...). \qquad \textbf{(2-47)}$$

The interval $[t_1, t_2]$ is arbitrary and Ψ may describe both the instantaneous power ($t_2 - t_1 \to 0$) and average power per period $2\pi/\omega$. Thus, for the local emitters admitting the finite presentation (2-42) of the impedance $Z_{ex}(\omega)$, the criteria of the maximum of the instantaneously absorbed power (MIAP) and absorption power, in average during finite temporal interval $[t_1, t_2]$, are identical. We can observe, that the spectral formulation (2-23) follows from (2-47). But the inverse transition from spectral representation into the differential temporal representation may not be always possible. After substitution of the equation (2-43) for the force $f(t)$ into Euler's equation (2-46), the terms, describing the noneven derivatives of velocity, become mutually annihilated. Therefore, if the radiation resistance $\operatorname{Re} Z_{ex}(\omega)$ does not contain even degrees of frequency, the only spectral approach, formulated in (2-23)-(2-25), can "operate" (and criterion MIAP has not sense in the problem with initial conditions). In this case, radiation resistance $\operatorname{Re} Z_{ex}(\omega)$ is expressed *via* module of frequency (for noneven degrees of ω), *i.e.* $\operatorname{Re} Z_{ex} \sim |\omega|^{\nu}$ (when $\nu = \pm 1,\ \pm 3,\ \pm 5,...$), which is not an analytical function of ω. Radiation resistance of some types of emitters is characterized by noneven degree of frequency (see **Tables 2-1, 2-2, 2-3**) and does not permit control with the criterion MIAP: two-dimensional acoustical emitters, and two-dimensional and three-dimensional emitters of surface gravitational waves in liquid, two-dimensional electromagnetic emitter.

Thus, for the causal control emitter radiation resistance must be presented only by the even degree of frequency. Below, due to the emitter locality (small wave dimensions) we retain only the lowest even order j, different from zero in sum (2-47):

$$\upsilon_A^{(j)}(t) = e(t)/2b_j. \qquad \textbf{(2-48)}$$

We supplement (2-48) by the expression $e(t)$ from (2-43) with non-zero terms of all orders up to a certain $n > j$ (j – even number), giving the expression of the optimum velocity $\upsilon_A(t)$ *via* the measured instantaneous values of the force $f(t)$ affecting the emitter, and its instantaneous velocity:

$$\upsilon_A^{(j)}(t) = (2a_j)^{-1}\left[f(t) + \sum_{n<j} b_n (d/dt)^n \upsilon(t)\right] \qquad \textbf{(2-49)}$$

Therefore, the algorithm (2-49) can be classified as {LS/LT/CA/F} (see **Fig. 1-3**). Further, we assume, that some servo-drive provides the forming of arbitrary assigned trajectory $\upsilon_A(t)$, *i.e.* minimum temporal scale τ_c of control satisfies the condition:

$$\tau_c \ll \tau_w, \qquad \textbf{(2-50)}$$

where τ_w is the minimum temporal scale of wave to be absorbed. The relation (2-50) denotes that the damping system perceives the incident wave as some temporally constant parameter of field. We note that the amount of coefficients $\{b_n\}$ in (2-47) must be sufficiently large to compensate force, caused by fast technological motion of servo-drive with temporal scale $\sim \tau_c$ (see **Fig. 2-17**). We note that the number of coefficients $\{b_n\}$, which are included in (2-47), must be sufficiently large to compensate force, caused by fast "technological" motion of servo-drive with temporal scale $\sim \tau_c$ (see **Fig. 2-17**).

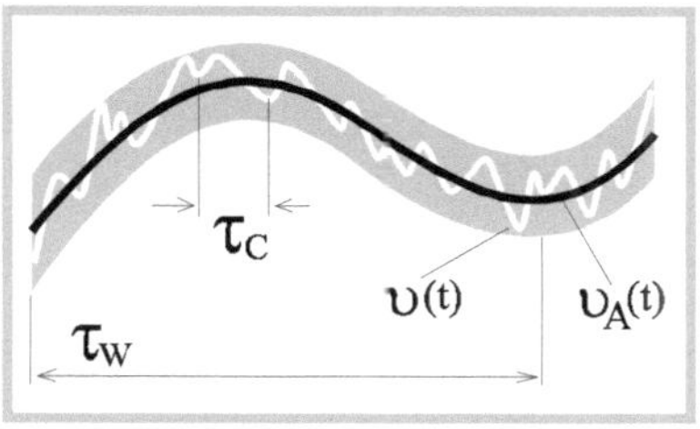

Figure 2-17: Optimum trajectory $\upsilon_A(t)$ of emitter velocity (with temporal scale τ_w), real trajectory $\upsilon(t)$ supported by servo-drive (with temporal scale τ_c).

If we neglect the limitations of the temporal scales of sensors and servo-drives, we can interpret (2-49) as the control algorithm, where $\upsilon_A^{(j)}(t)$ is the optimum (assigned to servo-drive) value of j-th derivative of emitter velocity, $\upsilon(t)$ and $f(t)$ –real (measured) velocity of emitter and force, acting on it from the waveguiding media respectively. Here we assume that coefficients $\{b_n\}$ are known or were measured earlier. "Autoresonant" algorithm (2-49) provides the real trajectory of the emitter closing to the assigned trajectory $\upsilon_A(t)$ of maximum absorption (2-48) during the time, much smaller than the period of absorbed wave, in other words –"without knowing" it. The representation (2-49) is not the transition $\omega \to \partial/\partial t$ from spectral to temporal form only. This means, that the emitter can begin the motion under the control (2-49) at any moment $t \geq t_0$ (and under arbitrary initial conditions). On the other hand (2-23)-(2-25) ensures maximum of the power absorbed per period.

2.8.1. Solitary Absorbing Monopole in the Field of a Wideband Incident Wave

Now we consider the following problem of the plane incident wave propagating normally to the rigid plane screen. Circular plane piston of radius a is placed in this screen. Without the screen, the plane incident wave would produce on the piston random acoustical pressure $P_e(t)$, with zero average $< P_e(t) >= 0$ and spectral power density (SPD) $\Pi_e(\omega)$. Then we obtain statistical characteristics of piston controlled by algorithm of MIAP (2-49) [60]:

$$d^2\upsilon_A(t)/dt^2 = a^2\pi P_e(t)/b_2 \qquad \textbf{(2-51)}$$

where $b_2 = \rho\pi a^4 / 2c$ is "radiating" coefficient (see (2–49)) of piston on the screen:

(a) average square of piston's displacement from initial position ($x = 0$)

$$< x_A^2(t) >= (\pi a^2 / b_2)^2 (2\pi)^{-1} \omega^{-6} \int_{-\infty}^{+\infty} \Pi_e(\omega) d\omega, \qquad \textbf{(2-52)}$$

(b) average square of piston's velocity

$$< \upsilon_A^2(t) >= (\pi a^2 / b_2)^2 (2\pi)^{-1} \omega^{-4} \int_{-\infty}^{+\infty} \Pi_e(\omega) d\omega,$$

(c) average power absorbed

$$< w_A(t) >= (\pi^2 a^4 / b_2)(2\pi)^{-1} \omega^{-2} \int_{-\infty}^{+\infty} \Pi_e(\omega) d\omega,$$

when average power stream density in the incident wave is

$$\overline{S}_e =< P_e^2(t) > / \rho c = (2\pi\rho c)^{-1} \int_{-\infty}^{+\infty} \Pi_e(\omega) d\omega.$$

We see from (2-52), that piston's coordinate is localized near its initial position $x = 0$ (i.e. $< x_A(t) >= 0$, and $< x_A^2(t) >$ −finite value), when SPD $\Pi_e(\omega)$, for instance, satisfies the condition

$$\lim_{\omega \to 0} \Pi_e(\omega) = s_0 \omega^\alpha \; (s_0, \alpha = const, \alpha > 5). \qquad \textbf{(2-53)}$$

Further we use the model SPD of the form:

$$\Pi_e(\omega) = (8\sqrt{\pi}/15)\tau_e^7 \omega^6 < P_e^2 > \exp(-\tau_e^2 \omega^2) \qquad \textbf{(2-54)}$$

(τ_e −time of sound pressure $P_e(t)$ correlation). For the chosen model SPD (2-54) we obtain:

$$< x_A^2(t) >= (16/15)c^2 \tau_e^6 \rho^{-2} a^{-4} < P_e^2 >,$$

$$< \upsilon_A^2(t) >= (8/15)c^2 \tau_e^4 \rho^{-2} a^{-4} < P_e^2 >,$$

$$< w_A(t) >= (2/5)\pi c \tau_e^2 \rho^{-2} < P_e^2 >.$$

Average ACS is equal to $< \sigma_A >=< w_A(t) > / \overline{S}_e = \pi(2/5)c^2 \tau_e^2$. Note that the correlation radius $r_e = c\tau_e >> a$ of the random sound field plays the same role in ACS, as the wavelength λ of monochromatic incident wave. Besides this $< \sigma_A >$ does not depend on the piston's radius a as described above. Taking into account a finite minimum temporal scale $\tau_c << \tau_e$ of active control (tuning) of emitter we formulate the condition

$$32a / c\tau_c << 1, \qquad \textbf{(2-55)}$$

which provides the correctness of low frequency representation (2-42) of the emitter's impedance by two first terms only.

2.9. ON THE ACTIVE SOLUTION OF THE PROBLEM OF MAXIMUM INSTANT ABSORPTION POWER BY THE SYSTEM WITH N DEGREES OF FREEDOM

Active control (unlike the conditions (2-30) of passive loading) assumes formation of the optimum vector

$$\mathbf{V}_{\text{opt}} = 2^{-1} \left[\mathrm{Re}\, \hat{\mathbf{Z}}_{ex} \right]^{-1} \left(\mathbf{F} - \hat{\mathbf{Z}}_{ex} \mathbf{V} \right) \qquad \textbf{(2-56)}$$

of complex amplitudes of velocities of absorbing system elements. These amplitudes are produced by some servo-drive or actuator. $\mathbf{V}$ is the vector of complex amplitudes of real measured velocities, and $\mathbf{F}$ is the vector of complex amplitudes of real measured forces, which act on the absorbing system elements. Now, we consider more thoroughly the temporal representation $\hat{\mathbf{H}}$ of matrix $\hat{\mathbf{R}} = \mathrm{Re}\, \hat{\mathbf{Z}}_{ex}$. We do not try to give rigorous analysis and consider the simplest example of two-element structure. This can be described by the following expression:

$$\begin{bmatrix} \upsilon_1(t) \\ \upsilon_2(t) \end{bmatrix} = \hat{\mathbf{H}} \begin{bmatrix} e_1(t) \\ e_2(t) \end{bmatrix}, \qquad \textbf{(2-57)}$$

where $\hat{\mathbf{H}} = \begin{bmatrix} \hat{H}_{11} & \hat{H}_{12} \\ \hat{H}_{21} & \hat{H}_{22} \end{bmatrix}$. Let us write (2–57) more thoroughly as

$$\upsilon_1(t) = \hat{H}_{11}[e_1(t)] + \hat{H}_{12}[e_2(t)], \; \upsilon_2(t) = \hat{H}_{21}[e_1(t)] + \hat{H}_{22}[e_2(t)], \qquad \textbf{(2-58)}$$

where

$$\upsilon_1(t) = \int_{-\infty}^{+\infty} H_{11}(t-t')e_1(t')dt' + \int_{-\infty}^{+\infty} H_{12}(t-t')e_2(t')dt' , \quad \textbf{(2-59)}$$

$$\upsilon_2(t) = \int_{-\infty}^{+\infty} H_{21}(t-t')e_1(t')dt' + \int_{-\infty}^{+\infty} H_{22}(t-t')e_2(t')dt' . \quad \textbf{(2-60)}$$

Note, that (2–59) и (2–60) have the following form in spectral representation

$$V_1(\omega) = \tilde{H}_{11}(\omega)E_1(\omega) + \tilde{H}_{12}(\omega)E_2(\omega) , \quad \textbf{(2-61)}$$

$$V_2(\omega) = \tilde{H}_{21}(\omega)E_1(\omega) + \tilde{H}_{22}(\omega)E_2(\omega) , \quad \textbf{(2-62)}$$

where $\tilde{H}_{11}(\omega)$, $\tilde{H}_{12}(\omega)$, $\tilde{H}_{21}(\omega)$, $\tilde{H}_{22}(\omega)$ are the elements of matrix

$$\begin{bmatrix} \tilde{H}_{11} & \tilde{H}_{12} \\ \tilde{H}_{21} & \tilde{H}_{22} \end{bmatrix} = \hat{\mathbf{R}} = \begin{bmatrix} R_{11} & R_{12} \\ R_{21} & R_{22} \end{bmatrix},$$

and are defined by following expressions

$$\tilde{H}_{11}(\omega) = \int_{-\infty}^{+\infty} H_{11}(t)\exp(-i\omega t)dt , \quad \tilde{H}_{12}(\omega) = \int_{-\infty}^{+\infty} H_{12}(t)\exp(-i\omega t)dt , \quad \textbf{(2-63)}$$

$$\tilde{H}_{21}(\omega) = \int_{-\infty}^{+\infty} H_{21}(t)\exp(-i\omega t)dt , \quad \tilde{H}_{22}(\omega) = \int_{-\infty}^{+\infty} H_{22}(t)\exp(-i\omega t)dt . \quad \textbf{(2-64)}$$

Values $V_1(\omega) = \int_{-\infty}^{+\infty} \upsilon_1(t)\exp(-i\omega t)dt$, $V_2(\omega) = \int_{-\infty}^{+\infty} \upsilon_2(t)\exp(-i\omega t)dt$ $E_1(\omega) = \int_{-\infty}^{+\infty} e_1(t)\exp(-i\omega t)dt$, $E_2(\omega) = \int_{-\infty}^{+\infty} e_2(t)\exp(-i\omega t)dt$ are the elements of vectors $\mathbf{V} = \begin{bmatrix} V_1 \\ V_2 \end{bmatrix}$ and $\mathbf{E} = \begin{bmatrix} E_1 \\ E_2 \end{bmatrix}$ respectively. From $\mathrm{Im}\hat{\mathbf{R}} = 0$ we obtain $\mathrm{Im}[\tilde{H}_{11}(\omega)] = 0$, $\mathrm{Im}[\tilde{H}_{12}(\omega)] = 0$, $\mathrm{Im}[\tilde{H}_{21}(\omega)] = 0$, $\mathrm{Im}[\tilde{H}_{22}(\omega)] = 0$ or

$$H_{11}(+t) = H_{11}(-t) , \; H_{12}(+t) = H_{12}(-t) , \quad \textbf{(2-65)}$$

$$H_{21}(+t) = H_{21}(-t) , \; H_{22}(+t) = H_{22}(-t) ,$$

if $t \neq 0$, but this contradicts the sense of values H_{11}, H_{12}, H_{21}, H_{22} какas a pulse reply of linear system to the input delta pulse. Therefore, it is impossible to ensure MIAP by causal operations (i. e. tuning in real time in the problem with initial conditions) in the system with more than one degree of freedom. The algorithm (2-56) can be classified as {NS/NT/NC/F} (see **Fig. 1-3**).

CHAPTER 3

Black Body Approach: Parametric Version

Abstract: In this chapter, we use simple models for acoustical waves (Section 3.1) [59], [61], [63], [64], [59], for water surface waves (Section 3.2) [62], [59], [61], [50], and for electromagnetic waves (Section 3.3) [65]. On the basis of these models, we consider the new concept of parametric "black body" with conceptual possibility of designing an active absorbing (nonreflecting) coatings in the form of a thin layer with small-scale stratification and fast temporal modulation of parameters. Algorithms for spatial-temporal modulation of the controlled-layer structure are studied in detail for a one-dimensional boundary-value problem. These algorithms do not require wave-field measurements, which eliminate self-excitation problem, that is the characteristic of traditional active systems. The majority of the considered algorithms of parametric control transforms the low-frequency incident wave to high-frequency waves of the technological band for which the waveguiding medium inside the layer is assumed to be opaque (absorbing). The efficient conditions of use are found for all the algorithms. It is shown that the absorbing layer can be as thin as desired with respect to the minimum spatial scale of the incident wave ensuring efficient absorption in a wide frequency interval (starting from zero frequency) that is bounded from above only by a finite space-time resolution of the parameter-control operations. The structure of a three-dimensional parametric "black" coating, whose efficiency is independent of the angle of incidence of an incoming wave is developed on the basis of the studied one-dimensional problems. The general solutions of the problem of diffraction of incident waves from such coatings are obtained. These solutions are analyzed in detail for the case of a disk-shaped element.

Let us consider the following problem: flat wave (acoustic wave, electromagnetic wave,...etc) with spectral (on the frequency ω) power stream density $\mathbf{W}_i$ falls on a certain body occupying a space area $\hat{\Theta}$ and bounded by surface S_Θ (**Fig. 3-1-a**). The body $\hat{\Theta}$ has geometric projection S_g to the planar front of incident wave (**Fig. 3-1-b**). The body $\hat{\Theta}$ is equipped with some absorbing coating with external surface $\overline{S}_A$ and internal surface $\overline{\overline{S}}_A = S_\Theta$ (**Fig. 3-1-c**). Body $\hat{\Theta}$ produces the spectral power stream density $\mathbf{W}_S$ of the scattered field both to the face half-space (half-sphere S_F of radius R) and to the rear half-space (half-sphere S_B of radius R). For instance in acoustic case we have $\mathbf{W}_S = (1/2)\,\mathrm{Re}[\overset{*}{p}_S \boldsymbol{\upsilon}_S]$, where S_g is the pressure (scalar) and $\boldsymbol{\upsilon}_S$ is the particle velocity (vector) of the scattered field. The sphere $(S_F + S_B)$ overlaps the body $\hat{\Theta}$. Further, we define the values Γ_{forw}, Γ_{back}, Π_w: $\Pi_{back} = \iint\limits_{S_B} \mathbf{W}_S \mathbf{n}\, dS_B$ is the power stream of scattered field *via* half sphere S_B with radius $R \to \infty$; $\Pi_{forw} = \iint\limits_{S_F} \mathbf{W}_S \mathbf{n}\, dS_F$ is the power stream of scattered field *via* half sphere S_F; $\Pi_w = \iint\limits_{S_g} \mathbf{W}_i \mathbf{n}\, dS_g$ power stream of the incident wave *via* area S_g, $\mathbf{n}$ is the normal to S_B, S_F ($|\mathbf{n}| = 1$). We will characterize scattering by coefficient of back scattering $\Gamma_{back} = \Pi_{back}/\Pi_w$ and coefficient of forward scattering $\Gamma_{forw} = \Pi_{forw}/\Pi_w$. A variety of definitions of black body already exists [77-81]. In this chapter, we will call black body some body, characterized by zero value of back scattering power stream or $\Gamma_{back} = 0$. So in this chapter, we pursue the goal: the reduction of the backscattering coefficient

$$\Gamma_{back} \Rightarrow 0. \qquad \textbf{(3-1)}$$

We will try to minimize Γ_{back} by means of some absorbing coating of thickness L (see **Figs. 3-1-b, 3-1-c**). Concept of coating assumes the absence of tangential functional connections between its parts. In other words, one could cut (**Fig. 3-1-d**) this coating into a number of pieces and paste together without any damage to the absorption. Parameters of waveguiding medium are variable in space (along normal $\mathbf{n}$ to the surface $\overline{S}_A$ of coating or to the surface S_g) with scale r_c and in time with scale τ_c.

Vladimir V. Arabadzhi

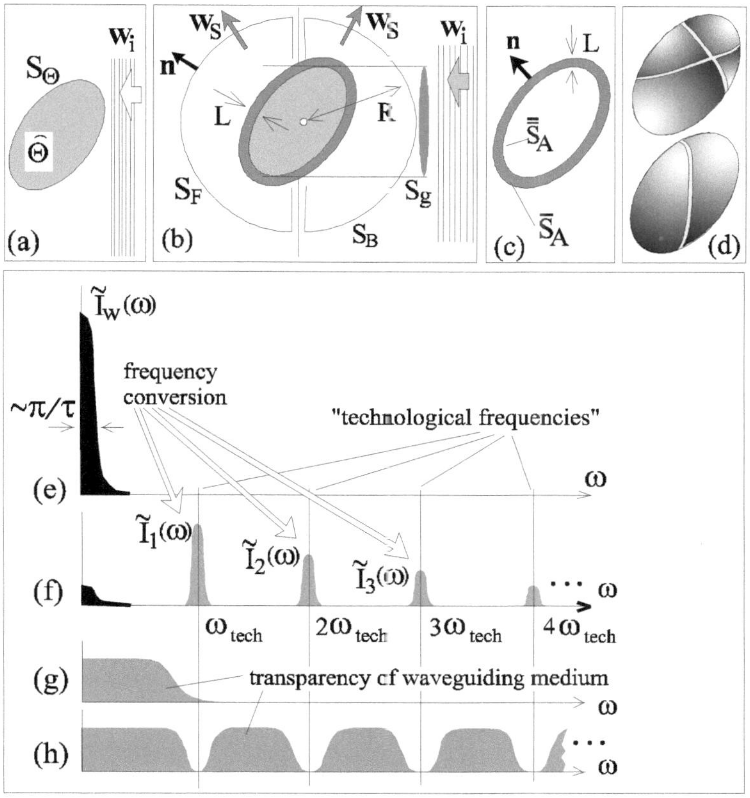

Figure 3-1: Geometry of backscattering suppression problem: (a) protected body $\hat{\Theta}$ with surface S_{Θ} and incident wave; (b) face surface S_F and rear surface S_B, embracing the body $\hat{\Theta}$ equipped with absorbing coating ($\overline{\overline{S}}_A = S_{\Theta}$), S_g geometric projection of body $\hat{\Theta}$ to the planar front of incident wave, $\mathbf{W}_S$ and $\mathbf{W}_i$–power stream densities of the scattered and incident waves respectively; (c) absorbing coating with internal surface $\overline{\overline{S}}_A$ and external surface $\overline{S}_A$ and thickness L; (d) cutting of the absorbing coating (without tangential connections). On the transformation of spectral power density of the incident wave by the parametric control: (e) initial power spectral density $\tilde{I}_w$ of incident wave; (f) conversion of the PSD $\tilde{I}_w$ of incident wave into PSD $\tilde{I}_1, \tilde{I}_2, \tilde{I}_3$ of high frequency technological waves, concentrated near the technological frequencies $n\omega_{tech} >> 2\pi/\tau$, where $n = 1, 2, 3, \ldots$, τ –characteristic temporal scale of incident wave; (g), (h) types of desired transparency of waveguiding media.

3.1. BOUNDARY-VALUE PROBLEM FOR SOUND

Let us consider a one-dimensional boundary-value problem (analog of absorbing coating in **Figs. 3-1-b,c,d**) for scalar dispersionless waves, namely, for a field $\varphi(x,t)$ of longitudinal displacements of particles of an infinite ($-\infty < x < +\infty$) elastic rod (the boundary-value problem is formulated in the same manner for sound waves in a compressible media). For $t \geq 0$, the wave field is described by the equation,

$$c^2 \varphi''_{xx} - \varphi''_{tt} = (\rho c)^{-1} f_{visc}, \qquad \textbf{(3-2)}$$

where $c = \sqrt{\chi / \rho}$ is the wave velocity, ρ and σ are the mass density and the cross-sectional area of the rod, respectively, χ is Young's modulus for the rod material, and $f_{visc} = 2\overline{\gamma}[\varphi'_t]$ is the linear density of the viscous-resistance force as the linear functional $2\overline{\gamma}$ of the velocity $\varphi(x,t)$. In the majority of cases, the Fourier spectra $\overline{f}_{visc}(x,\omega)$ and $\overline{\varphi}(x,\omega)$ of the force $f_{visc}(x,t)$ and the wave field $\varphi(x,t)$, respectively, are related as $\overline{f}_{visc}(x,\omega) \approx i\omega\gamma\sigma\rho\overline{\varphi}(x,\omega)$, where

$$\gamma = \alpha|\omega|^{\beta}, \qquad \textbf{(3-3)}$$

$\alpha = const \geq 0$, $\beta = const > 0$. Let the traveling wave $\phi(x+ct)$ with the longitudinal particle-displacement amplitude Ξ and the minimum time scale τ and spatial scale $\lambda = c\tau$ scales be incident on the layer $x \in [0, L]$ from the right (this is an incident wave with power spectral density $\tilde{I}_w$ as in **Fig. 3-1-e**). Assume for definiteness that $\varphi(x,t)$ in the layer $x \in [0, L]$ for $t \leq 0$. Assuming that $\alpha(2\pi/\tau)^{\beta} L/c << 1$, *i.e.*, the viscous forces for such a wave are much smaller than inertia and elastic forces, we ignore the incident wave decay within the layer (the layer is assumed to be transparent). The parameters are controlled in the range of high (technological) frequencies: $\omega \sim m\omega_{tech} >> 2\pi/\tau$, where $m = 1, 2, ...$ (**Fig. 3-1-f**). The scales of time and space resolution of control tools are of the order $\tau_c << 2\pi/(m\omega_{tech})$ and $r_c = c\tau_c$, respectively. The waveguiding medium can be assumed opaque (absorbing) in a path of length L for the above frequency interval under the condition $\alpha|\omega_{tech}|^{\beta} L/c >> 1$, *i.e.*, for $\exp[-\alpha|\omega_{tech}|^{\beta} L/c] << 1$, which is considered below (see for example **Fig. 3-1-g, 3-1-h**). The desired space-time parameter distribution in a thin active layer $x \in [0, L]$ ($L << \lambda$) should ensure nonreflecting absorption, *i.e.*, zero total field for $x < 0$ (behind the active layer) and zero reflected field for $x > L$ (in front of the active layer).

The waveguiding medium is weakly absorbing on the path with a length of order $2\pi c/\omega_{tech}$ for reflected waves at the technological frequencies, which means the weak influence of viscosity of the medium on the dynamics of the boundary $x = 0$. However, the medium is opaque for the process-frequency waves in the path with a length of order L. However, the layer $x \in [0, L]$ remains transparent and thin with respect to the low-frequency incident wave ($L << \lambda$).

3.1.2. Parametric Solutions of the Wave Suppression Problem

The solutions of the problem formulated in Sec. 3.1, which are discussed below, cannot be reduced to a combination of electric circuits with constant parameters as is done in the theory of active wave quenching. Unlike the conventional control of the slow amplitudes of wave fields at the frequency $\omega \approx 2\pi/\tau$ of the incident wave, the proposed method requires allowance for the wave-system dynamics at the frequencies ω that are both much smaller than the incident-wave frequency ($\omega << 2\pi/\tau$, including zero frequency) and much higher than the incident-wave frequency ($\omega \sim 2\pi/\tau_c >> 2\pi/\tau$). The solutions described in Sec. 3.1.2. - 3.1.6. are algorithms for controlling the space-time parameter distributions of the waveguiding medium in a thin layer $x \in [0, L]$ using a specified procedure, *i.e.*, without any measurements of the wave field. This fact along with the opacity of the waveguiding medium at the technological frequencies and the absence of resonance relationships between the incident-wave frequency and the parametric-control frequencies remove the stability problem (self-excitation) that is typical of active systems. The fine-scale nature of the parametric structure of the absorbing layer means (a) a split of the boundary-value problem in time (as opposed to the conventional control of the slow amplitudes of sources), *i.e.*, control with resolution $\tau_c << \tau$ and (b) a splitting of the boundary-value problem in space, *i.e.*, control with a high space resolution in the direction of the normal to the layer: $r_c \sim c\tau_c << L << \lambda = c\tau$. In this case, the finite wave-propagation velocity c which is usually used only for estimating the wave size of objects in active wave suppression is of prime importance. We also note that the use of a high space-time resolution in the proposed solutions is not related to the well-known problems of data extrapolation or antenna superdirectivity. The parametric solutions considered below use a lumped control in time and both lumped and distributed controls in space. In the solutions presented, the process of incident-wave absorption has both the time-distributed character (oscillatory damping) and time-lumped character (inelastic impact).

Let us note that all of the considered wave suppressing systems treat the wave field as a quasistationary field of parameters and do not require any information on the quenched wave or the boundary-value problem outside the absorbing layer. Moreover, the algorithms under study have the same efficiency no matter whether they start to work before the incident-wave arrival or after the reflected wave becomes steady-state. Control is assumed to be binary, *i.e.*, it is reduced to a fast (within a time scale of order $\tau_c << L/c$) switching (commutation) of the parameters of the boundary-value problem between two possible values at the specified time at the given point of the layer.

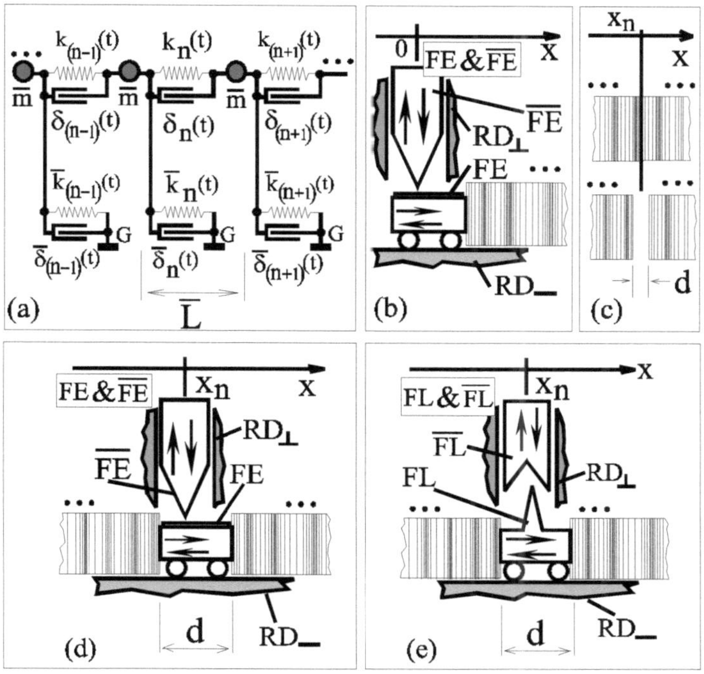

Figure 3-2: Tools for control of parameters of the boundary-value problem: (a) microstructure of discrete model of continuous compressible rod with controlled parameters ($k_n(t)$, $\bar{k}_n(t)$ –controlled elasticity; $\bar{m}$ –elementary mass; $\delta_n(t)$, $\bar{\delta}_n(t)$ –controlled viscosity or dissipation; $\bar{L}$ –spatial period; G –mechanical support (ground or "vibrostat"); (b) Euler-wall-fixing (without reset) device for semi-infinite rod ($0 \le x < \infty$, friction brake $\overline{FE}$ with friction plate FE and vertical rigid directions $RD_{\perp}$ and horizontal rigid directions RD_{-}); (c) breakup in the continuous infinite rod ($-\infty \le x < \infty$) for inserting of fixing-reset devices of length d in the point $x = 0$; (d) Euler-wall-fixing device in the point x_n ; (e) Lagrange-wall-fixing-reset device in the point x_n (reset-fixing-brake $\overline{FL}$ with reset-fixing-teeth FL).

In **Fig. 3-2-a** we show a discrete model of the waveguiding medium in the layer $x \in [0, L]$, where $\bar{m}$ is the mass element, $k_n(t)$ and $\bar{k}_n(t)$ are the elements of controlled elasticity, $\delta_n(t)$ and $\bar{\delta}_n(t)$ are the elements of controlled viscosity, G is the immobile support (vibrostat), and $\bar{L}$ is the spatial period of the discrete model. The parameters $\delta_n(t) = 0$, $\bar{\delta}_n(t) = 0$, $\bar{k}_n(t) = 0$, $k_n(t) = (c / \bar{L})^2 \bar{m} / 4$, and $\bar{L} << c\tau$ correspond to matching with the external medium ($x \notin [0, L]$) for frequencies of order $2\pi / \tau$. Since the discrete model ensures a more accurate correspondence to the boundary-value problem in some cases, the description of the parametric-control algorithms will be accompanied by interpretation in terms of this model. We consider the processes of wave emission and energy accumulation accompanying the switching of parameters of the boundary-value problem for each algorithm of binary control of the parameters $\delta_n(t)$ and estimate the efficiency of the solutions together with the conditions of their technical implementability.

3.1.3. Algorithm of the Reflection Coefficient Modulation

Let us start with the simplest use of high-resolution (high resolution in time) control instruments: binary balanced switching with period T_B between the values $+1$ and -1 of the reflection coefficient Γ of the incident wave at the boundary $x = 0$ of a semi-infinite rod ($0 \le x < \infty$, which corresponds to the boundary condition $a(t)\varphi_t'(0,t) + b(t)\varphi_x'(0,t) = 0$, where $a = 1 + \mathrm{sgn}[\cos(\omega_{tech} t)]$, $b = 1 - \mathrm{sgn}[\cos(\omega_{tech} t)]$, and $\omega_{tech} = 2\pi / T_B$. This is equivalent to the transformation of the incident wave $\phi = \phi(x + ct)$ with spectral power density (SPD) $S_\phi(\omega)$ to the reflected wave $\psi(x - ct) = \phi(ct)\,\mathrm{sgn}\cos[\omega_{tech} t - \omega_{tech}(x / c)]$ with an SPD of the form $S_\psi(\omega) = \sum_{n \ne 0} (n\pi)^{-2} S_\phi(\omega - 2n\omega_{tech})$. If we assume that the spectral power density $S_\phi(\omega)$ of the incident wave (with width of order $2\pi/\tau$) is concentrated near zero frequency (**Fig. 3-1-e**), we obtain $S_{\psi}(\omega)\approx 0$ for the frequencies $\omega\sim 2\pi/\tau$ because of the almost zero average value of $<\psi(0,t)>_{\tau}$. For a sufficiently high quenching efficiency, the reflection coefficient $|\Gamma|<<1$ of the incident wave at low frequency is written as sum of two qualitatively different factors: $|\Gamma|\le|\Gamma_1|+|\Gamma_2|$. Here $|\Gamma_1|\approx 2\pi\alpha\omega_{tech}^{\beta-1}$ corresponds to the influence of the frequency-dependent viscosity (3–3) on the dynamics of the boundary (the ratio between the average (over the period T_B) absolute values of viscous stresses to the dynamic stresses at the technological frequency) and $|\Gamma_2|\approx\sum_{n\neq 0}(\pi n)^{-2}S_{\phi}(2n\omega_{tech})/(S_{\phi})_{\max}$ describes the influence of the finite reduction of the incident-wave SPD at high frequencies. Thus, for example, for a finite spectral power density $S_{\phi}(\omega)$ with width $\Delta\omega_{\phi}$, we obtain $|\Gamma_2|=0$ for $\omega_{tech}>\Delta\omega_{\phi}$. The balanced modulation of the reflection coefficient at the controlled boundary $x=0$ with a frequency of order ω_{tech} transforms the long incident wave (with a wavelength about $\lambda=c\tau$) to short reflected waves (with a length of order $2\pi c/\omega_{tech}<<\lambda$) at the technological frequencies (**Fig. 3-1-f**). The above control option is technically implemented if we remove the left-hand side ($-\infty<x\le 0$) of the rod and attach the fixing node $FE\,\&\,\overline{FE}$ shown in **Fig. 3-2-b** to the free end $x=0$ of a semi-infinite rod ($0\le x<\infty$). The node consists of a rigid weightless carriage FE and brake $\overline{FE}$, which can move without friction along the rigid horizontal guides RD_{-} and vertical guides $RD_{\perp}$.

Pressing the brake $\overline{FE}$, we ensure fast fixing of the carriage FE at the current point with dry friction of rest and obtain the reflection coefficient $\Gamma=-1$ (for longitudinal displacement), whereas lifting the brake we ensure fast release of the carriage and $\Gamma=+1$. Both operations take time of the order of τ_c ($\tau_c<<T_B<<\tau$), are not related to performing work in the horizontal direction, and, therefore, do not result in radiation or absorption of waves. Energy is not accumulated in proximity to the boundary $x=0$ after switching. Let us call the point $x=0$ with the above fixing along the Euler coordinate the $T\Leftrightarrow E$ wall (Euler-wall-fixing (without reset)). In terms of the discrete model shown in **Fig. 3-2-a**, the $T\Leftrightarrow E$ wall means switching of the viscosity element $\overline{\delta}_n(t)$ between the values $\overline{\delta}_n(t)=0$ and $\overline{\delta}_n(t)=\infty$. An important disadvantage of the algorithm of reflection-coefficient modulation is its inapplicability to two-dimensional and three-dimensional boundary-value problems for bodies with finite and small wave size. This is due to the fact that, for such bodies, the amplitudes of waves scattered by their surface with zero pressure (pliable surface) and by the same surface but with zero velocity (rigid surface) are not antisymmetric. Therefore, the scattered-field modulation is no longer of balanced type, which means the finite field of scattering at the low frequency of the incident wave. The above algorithm can be classified as {LS/LT/CA/M} (see **Fig. 1-3**). The modulation algorithm assume s the presence of a vibrostat (direction devices $RD_{\perp}\,\&\,\overline{FE}$ in **Fig. 3-2-b** and in **Fig. 3-2-d** of G in **Fig. 3-2-a**), which can ensure a maximum reaction force of the support of order $\Xi\sigma\sqrt{\rho\chi}/\tau$. For $\tau\sim 2\pi/\omega_{tech}$, the modulation algorithm permits a heterodyne transformation of the incident wave to the lower-frequency waves, which neutralize the mechanism of high-frequency absorption. The efficiency of the quenching system on the basis of the algorithm of reflection coefficient modulation was experimentally confirmed for surface gravity waves in a liquid [50].

3.1.4. Cyclical Wave-Bolt-1 (CWB-1)

In Section 3.1.3. we described the algorithm of reflected-wave transformation suppressing reflections at the low frequency but retaining them at the high frequency. However, it is possible to formulate the problem in which the reflected wave does not appear. For example, this occurs if the protected boundary runs away from the incident wave with supersonic velocity $V>c$. Such an approach can be implemented only with the help of a technical intermediary allowing us to match two requirements that seem to be conflicting at first sight:

a) the incident wave propagating with the acoustic velocity c should recognize the protected surface as a surface that is running away with velocity $V>c$, *i.e.*, it should not "feel" this surface and

b) the actual physical surface is motionless. We show that such an intermediary is possible and that it should work at high technological frequencies $\omega\sim m\omega_{tech}>>2\pi/\tau$, where $m=1,2,\ldots$, in a layer with thickness L.

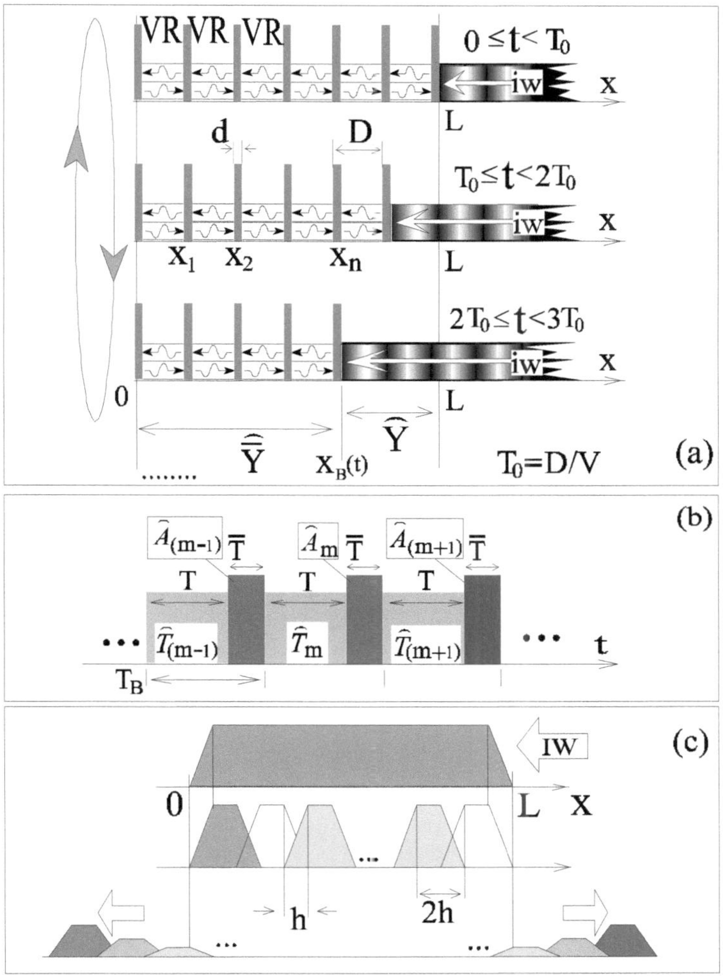

Figure 3-3: Various types of CWB. Spatial-temporal diagrams of CWB-1 (Section 3.1.4.): during the temporal period T_B of CWB, the border between the transparent (passive) zone $\widehat{Y}$ of CWB and opaque (active) zone $\widehat{\overline{Y}}$ of CWB is running away from the incident wave with velocity $V > c$; wave energy is relaxed quickly, being closed between opaque walls (of thickness d) of virtual resonators (VR) (of size D) in active zone $\widehat{\overline{Y}}$; Positions of CWB-1 are presented in three ((a1), (a2), (a3)) consequent time intervals. Temporal diagram of CWB-2 (Section 3.1.5.1.) with commutation of the wave propagation velocity in the active zone of CWB. Spatial diagrams of CWB-3 (Section 3.1.5.1.), which is based on commutation of the viscosity factor $\overline{\delta}_n(t)$.

In this case, as in Section 3.1.3. we assume that the waveguiding medium is opaque (absorbing) at the technological frequencies and has a high-frequency damping factor in the form (3-3).

Let us consider an echelon of walls with controlled transparency (set $\widehat{Y}_B(t)$) located at N points $x_n = nD$ of the layer $x \in [0, L]$ of the waveguiding medium ($D = L / N$, $m = 1, 2, ..., N$). The set $\widehat{Y}(t)$ of transparent (or free) walls in the interval $x_B(t) \leq x \leq L$ and the set $\widehat{\overline{Y}}(t)$ of opaque (or fixed) walls in the interval $0 \leq x \leq x_B(t)$ ($\widehat{Y}_B = \widehat{Y}(t) \cup \widehat{\overline{Y}}(t)$, and $\widehat{Y}(t) \cap \widehat{\overline{Y}}(t) = 0$) are separated by a moving boundary $x_B(t) = L - D \; ent[N_T / T_B] + L \; ent[t / T_B]$, where $ent[\xi]$ denotes the integer part of ξ. At the beginning of each m-th cycle $mT_B \leq t \leq (m+1)T_B$ with duration T_B, all the walls are opaque ($\widehat{Y} = 0$ and $x_B(mT_B) = L$, where $m = 1, 2, ...$). At the same time, the boundary $x_B(t)$ (the front of the "bleaching wave") starts moving in steps from right to left with the average velocity $V > c$ until it reaches the left-hand edge of the CWB. At this moment, all the walls become opaque simultaneously and instantaneously (during a time of the order $\tau_c << T_B / N$). Then all operations repeat with period $T_B = L / V << \tau$. The right-hand edge of the set $\overline{Y}(t)$ travels to the left during each m-th cycle $mT_B < t < (m+1)T_B$ and runs into the incident wave only in the beginning of each cycle within the time interval $\overline{T} = D / V$. The edges of each pair of the opaque walls that are the closest to one another and the interval D between them form a virtual resonator (VR) with eigenfrequencies $\omega_{tech} = \ell\pi c / D$, where $\ell = 1,2,...$ The VRs occupy the following space-time regions: $\{x \in [(n-1)D, nD]\} \cup \{t \in [mT_B, mT_B + \tau_n]\}$, where $\tau_n = (N-n)D/V$ is the lifetime of the n-th VR, *i.e.*, the time interval during which the $(n-1)$-th and the n-th walls are simultaneously opaque, $n = 1,2,...,N$, $m = 1,2,...$ This space-time structure is called the cyclical wave bolt 1 (CWB-1). Now N_E walls that are closest to the incident wave have the type $T \Leftrightarrow E$ control described in Section 3.1.3 (**Fig. 3-2-b**) with the only difference that the control node $FE \& \overline{FE}$ is placed in the break x_n (**Fig. 3-2-c**) of the waveguiding medium. As in Section 3.1.3, the carriage FE is assumed to be perfectly rigid and weightless, which makes its horizontal dimensions insignificant for wave propagation. Therefore, under the fixed condition of the node $FE \& \overline{FE}$, the wall is characterized by the reflection coefficient $\Gamma = -1$ with respect to the longitudinal displacement, while its motion coincides with the particles of the waveguiding medium under the free condition of the node and, therefore, the wall is transparent, *i.e.*, $\Gamma = 0$. Two such fixed neighboring walls form the VR of the type $T \Leftrightarrow E$. The remaining $N_L = N - N_E$ walls are under the $T \Leftrightarrow E$ binary control at points x_n with the fixing with respect to the Lagrangian coordinate. The control includes:

(a) the motion that agrees with the field $\varphi(x,t)$ of displacements of the waveguiding medium ($\Gamma = 0$ as for the $T \Leftrightarrow E$ control) and

(b) the wall fixing at time $t*$ with the reflection coefficient becoming $\Gamma = -1$ simultaneously with the displacement to the distance $\varphi(x_n, t*)$ (reset) to the coordinate x_n corresponding to the state of rest of particles along the entire axis $-\infty < x < +\infty$, *i.e.*, the absence of stresses ($\varphi'_x(x,t) = 0$), velocities ($\varphi'_t(x,t) = 0$), and incident waves ($\varphi(x+ct) = 0$). The switching from the state (b) to the state (a) indicates the wall release and is not related to mechanical work or radiation. The transition (*i.e.*, the fixing) from the state (a) to the state (b) during a time of order τ_c is accompanied by radiation of the switching displacement wave of step type with the amplitude τ_c and energy of order $\varphi^2(x_n, t*)\rho c\sigma / \tau_c$ on both sides of the wall. **Figure 3-2**-e shows the $T \Leftrightarrow L$ control (Lagrangian-fixing-reset) when the fixing node $FL \& \overline{FL}$ consisting of the weightless rigid carriage FL and the rigid fixing element $\overline{FL}$ is inserted in the break x_n of the waveguiding medium (**Fig. 3-2-e**). The fixing element ensures $\Gamma = 0$ in the upper position and returns the carriage to the specified coordinate once it is lowered. In terms of the discrete model of the waveguiding medium, the $T \Leftrightarrow L$ control means switching of the elasticity element $k_n(t)$ between zero and infinity (**Fig. 3-2-a**). Each pair of fixed neighboring walls of $T \Leftrightarrow L$ type form the virtual $T \Leftrightarrow L$ resonator. When the virtual resonator is formed, the current values of the wave field become initial conditions for oscillations inside the VR. Since resonators of both types have no eigenfrequencies below $\omega_{tech} = \pi c / D$, oscillations in the VR due to the initial conditions occur at the technological frequencies with the viscous damping factor $\gamma = \alpha |\omega_{tech}|^{\beta}$ (see (3-3)). In this case, for efficient absorption, the relaxation process should be oscillatory so that the wave can work against viscous stresses in the medium. The viscous damping factor should be sufficiently small so that the lifetime for the majority of VRs amounts to several oscillation periods at the lowest technological frequency, *i.e.*, the VR should have a sufficiently high Q-factor. In the opposite case, the unlimited viscosity would result in "freezing" of oscillations. On the other hand, to ensure efficient absorption, the factor γ should be sufficiently large so that the field becomes negligible at the end of each cycle inside the majority of long-lived VRs located in deep layers of the CWB-1. Both conditions are compatible only for $N >> 1$. As mentioned above, the VR walls of the $T \Leftrightarrow L$ type are fixed at the points with coordinates corresponding to the rest of the entire wave system. Therefore, the longer such a resonator exists, the closer to zero is the energy stored in it and, correspondingly, the energy of waves emitted from the resonator once the walls are released. If the $T \Leftrightarrow E$ virtual resonator is absent in the CWB-1 external layers, the $T \Leftrightarrow L$ resonators would ensure a reflection coefficient $|\Gamma_B| = 1$ at the incident wave frequency for any $N_L > 0$ due to the generation of switching waves.

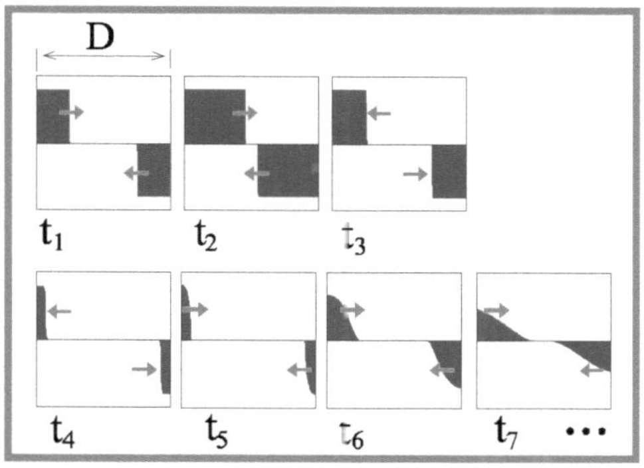

Figure 3-4: Commutative wave of longitudinal shifts, when $L << \tau_w$, inside VR of type $T \Leftrightarrow L$, produced by Lagrange-wall-fixation-reset devices (**Fig. 3-2-e**) at the moments $t_1 < t_2 < t_3 < t_4 < t_5 < t_6 < t_7$ (the wave energy is concentrated in areas of maximum fast change of particles shifts). The initial amplitude of commutative wave is equal to the current value of incident wave in the time interval $t \in \hat{T}_m$.

However, the presence of the $T \Leftrightarrow E$ type VRs prevents the switching radiation (see commutative wave in **Fig. 3-4**) output from the $T \Leftrightarrow L$ virtual resonator during the time $N_E D / V$. This time is sufficient for the attenuation of the field in the $T \Leftrightarrow L$ resonators to a relative level that does not exceed $\exp[-\gamma(\omega_{tech})\Delta t]$ due to the mechanism of high-frequency damping. Therefore, the contribution $|\Gamma_A|$ to the CWB-1 reflection coefficient from the fields which did not decay in the $T \Leftrightarrow L$ resonators can be estimated approximately as the sum of the geometric progression: $|\Gamma_A| \le \sum_{n=1}^{N_L} \exp[-\chi(N_E + n)]$, where $\chi = \gamma(\omega_{tech}) D / c$, $\omega_{tech} \ge \pi c / D$. If a spatially homogeneous stress exists at the time when the wall of the $T \Leftrightarrow E$ virtual resonator is fixed (which is inevitable in the case $L << r_{iw}$), the potential energy corresponding to this stress remains unchanged as long as desired, which is not the case for the $T \Leftrightarrow L$ resonator. When the walls are released, this stored potential energy is emitted in the form of a wave, and such a resonator is ineffective as an absorber. Ignoring energy absorption in the $T \Leftrightarrow E$ resonators, we estimate their contribution to the reflection coefficient as $|\bar{\Gamma}| \le N_E D / L$. The finite quantity $0 < |\bar{\Gamma}| << 1$ is an inevitable compensation for repeatability (periodicity) of CWB-1 operations as an intermediary between the incident wave and protected surface. In the case $|\Gamma_B| << 1$, which is of interest, the absolute value $|\Gamma_B|$ of the reflection coefficient is written as the sum of two qualitatively different quantities $|\bar{\Gamma}| << 1$ and $|\Gamma_A| << 1$:

$$|\Gamma_B| \approx |\bar{\Gamma}| + |\Gamma_A|. \quad \textbf{(3-4)}$$

Let us note that the coefficient $|\bar{\Gamma}|$ increases linearly with N_E, whereas $|\Gamma_A| \le \exp(-\gamma N_E) / [1 - \exp(-\chi)]$ decreases exponentially. This means that the minimum reflection coefficient of the CWB-1

$$|\Gamma_B| = |\Gamma|_{\min} \approx (N\chi)^{-1}\{1 + \ln[(1 - \exp(-\chi))^{-1}(N\chi)]\} \quad \textbf{(3-5)}$$

is reached for $N_E = (N_E)_{opt} = \chi^{-1} \ln\{\chi^N / [1 - \exp(-\chi)]\}$ and can be as small as desired in the case of CWB-1 structure miniaturization (*i.e.*, for $N \to \infty$) without increase in $L << \lambda$. Assuming that $\beta = 2$ for acoustic waves [63] at the boundary of oscillatory and exponential damping, *i.e*, in the case of the minimum permissible Q-factor of the VR, we obtain $|\Gamma_B|_{\min} \approx (N\pi)^{-1}\{1 + \ln[(1 - \exp(-\pi))^{-1}(N\pi)]\}$, while far from this boundary, *i.e.*, for a large number of high Q-factor VRs, we obtain $|\Gamma_B|_{\min} \approx L(\pi^2 c\alpha)^{-1} N^{-2}(1 + \ln N)$. In this expression, α and c are the parameters of the medium and N and L are free parameters for design. CWB-1 requires much more peak reaction force of the vibrostat (of the order of $\Xi\sigma N_L T_B \tau_c^{-2}\sqrt{\rho\chi}$) as compared with the modulation algorithm described in Section 3.1.3. In CWB-1 we can also use the limiting velocity relationship $V = c$ but in this case, coordinate overtaking is required rather than velocity overtaking. If we follow this limiting trajectory, all resonators of the $T \Leftrightarrow E$ type remain closed at the beginning of each m -th CWB-1 cycle $mT_B \le t \le (m+1)T_B$ in the interval $mT_B \le t \le mT_B + (N_E D / c)$. Then all of them become transparent simultaneously and instantaneously at time $t = mT_B + (N_E D / c)$, and the "bleaching" wave travels along N_L resonators of the $T \Leftrightarrow L$ type with the average velocity $V = c$. The $T \Leftrightarrow E$ virtual resonators actually do not participate in the absorption process, only blocking switching radiation of the deep $T \Leftrightarrow E$ resonators and delaying their opening. Therefore, we can use only one $T \Leftrightarrow E$ wall at the point $x = N_E D$ instead of N_E walls and open it in the time interval $N_E D / V$ after the beginning of each cycle. Unlike the modulation algorithm, the CWB-1 cannot transform an incident wave to lower-frequency waves and may be used as a parametric coating (see Section 3.1.8) not only in one-dimensional but also two-dimensional and three-dimensional boundary-value problems. The efficiency of the CWB-1 based quenching system was experimentally confirmed for surface gravity waves in a liquid [62], [59], [61]. The above algorithm can be classified as {LS/LT/CA/M} (see **Fig. 1-3**).

3.1.5. Cyclical Wave-Bolt with Separation of Variables

The above "traveling" option of the CWB-1 with inseparable spatial and time control operations allows us to attain the maximum VR lifetime in deep CWB-1 layers and, hence, maximum field damping in such resonators in one cycle. However, simplified CWB-1 versions consisting of one or several structurally homogeneous regions switched on and off simultaneously (**Fig. 3-3-b**) on the entire segment $x \in [0, L]$ can also be efficient in the case of sufficiently high operation speed and control miniaturization. The CWB structure is transparent and ensures reflectionless transmission of the incident wave in the time intervals $\hat{T}_m = \{mT_B \le t < (m+1)T_B - \overline{T}\}$ of duration $T < T_B$, where $m = 1,2,...$ In the time intervals $\hat{\hat{T}}_m = \{(m+1)T_B - \overline{T} \le t < (m+1)T_B\}$ of duration $\overline{T} << T < T_B$ (where $\overline{T} + T = T_B$), the structure is opaque, *i.e.*, its reflection coefficient is $|\overline{\Gamma}| \approx T / T_B << 1$ at the low frequency of the incident wave. In this case, the wave gate must ensure an efficient absorption of the incident wave, which has penetrated (during intervals $\hat{T}_m$) into it, in the time intervals $\hat{\hat{T}}_m$ on the time scale of the order $\overline{T}$. As a result, the undamped waves ensure a sufficiently small reflection coefficient $|\Gamma_A| << 1$ at the low frequency of an incident wave during the subsequent bleaching of the CWB (in the time interval $\hat{T}_m$). Primarily, this means that the CWB period T_B cannot be greater than L / c. In this case, the total low-frequency reflection coefficient is also determined by (3–4). In particular, using such simplification of the structure, we obtain $|\Gamma_A| \le (N - N_E)\exp(-\gamma N_E)$ instead of $|\Gamma_A| \le \exp(-\gamma N_E) / [1 - \exp(-\chi)]$ for the CWB–1 algorithm. It can readily be shown that this option also results in $|\Gamma_B| \to 0$ in the case of an unlimited increase in the space-time resolution of the CWB structure. Now consider briefly two more versions of the structures of this type, which are qualitatively different from the CWB-1.

3.1.5.1. Cyclical Wave Bolt 2 (CWB-2)

A low-frequency incident wave can be transformed to waves at technological frequency by not only decreasing the dimensions of the virtual resonators ($N >> 1$) but also controlling the wave propagation velocity $\mu(x,t)$ in a single VR with length of order L ($\mu(x,t)$ stands for $\overline{c} >> c$ in (3–2). In the CWB-2 segment $x \in [D_0, L - D_0]$, where $D_0 << L$, a high wave-propagation velocity $\overline{c} >> c$ is ensured in a short time interval $\overline{T} \ge \tau_c$ with period T_B and the $T \Leftrightarrow E$ walls appear at the points $x = 0$ and $x = L$. These walls block the switching radiation generated when the parameter $\mu(x,t)$ is switched from c to $\overline{c}$. The energy of the switching radiation during the switching to $\overline{c} >> c$ is the same as in the case of fixing of the $T \Leftrightarrow L$ wall (see Section 3.1.4.5). The VR boundaries $x = D_0$ and $x = L - D_0$ in the CWB-2 have small wave transmission coefficient (of the order of $c / \overline{c} << 1$), while oscillations inside the VR occur at frequencies much higher than those for the incident wave. Therefore, we ignore the boundary transparency during the intervals $\hat{\hat{T}}_m$. Oscillations inside such a VR are similar to oscillations of a rod, one end of which is fixed, while the other end is free.

The minimum eigenoscillation frequency of such a rod is equal to $(\omega_{tech})_{\min} \approx \pi\bar{c}/[2(L-D_0)]$. In terms of the discrete model of the medium (**Fig. 3-2-a**), the control of the wave propagation velocity corresponds to switching of the elasticity parameter $k_n(t)$ and the CWB-2 structure is fully determined by the following relations inside the m-th cycle $t \in [mT_B, (m+1)T_B]$: a) in the time intervals of transparency $\hat{T}_m$ we have $\bar{\delta}_n = 0$ for $n = n(x=0)$ and $n = n(x=L)$ and $k_n(t) = (c/\bar{L})^2\bar{m}/4$ for $n = n(x = D_0) \le x < n = n(x = L - D_0)$; (b) in the absorption intervals $\hat{A}_m$ we have $\bar{\delta}_n = \infty$ for $n = n(x = D_0)$ and $n = n(x = L - D_0)$ and $k_n(t) = (\bar{c}/\bar{L})^2\bar{m}/4$ for $n = n(x = D_0) \le x < n = n(x = L - D_0)$, where $n(x)$ is the number of the discrete-model element, which is the closest to the point x. Hereafter, we assume that the omitted elements of the discrete model of the waveguiding medium in the segment $x \in [0, L]$ are consistent with the parameters of the external medium ($x \notin [0, L]$) for both $t \in \hat{T}_m$ and $t \in \hat{\hat{T}}_m$ for any m. It is desirable to fulfill the relation $D_0/c << (L - 2D_0)/\bar{c}$ for the efficient blocking of switching radiation.

Fulfilling the condition $\alpha|\omega_{tech}|^{\beta} < (\omega_{tech})_{\min}$ of oscillatory damping of oscillations inside the VR and the condition of efficient damping during the time $\bar{T}$ (*i.e.*, $\exp(\gamma\bar{T}) = \exp\left(\alpha|(\omega_{tech})_{\min}|^{\beta}\bar{T}\right) << 1$), we obtain the estimates and $|\Gamma| \le \bar{T}/T_B$. In accordance with (3-5), the minimum coefficient $|\Gamma_B|_{\min} \approx [1 + \ln(-\gamma T_B)]/(\gamma T_B) << 1$ of reflection from CWB-2 at the incident-wave frequency is attained for $\bar{T} = \gamma^{-1}\ln(\gamma T_B)$, where $\gamma = \gamma[(\omega_{tech})_{\min}]$. As in the CWB-1, this algorithm can be used in two- and three-dimensional problems, but it requires a peak reaction force from the vibrostat that is a factor of N_L smaller (of the order of $\Xi\sigma T_B\tau_c^{-2}\sqrt{\rho\chi}$). The above algorithm can be classified as {LS/LT/CA/M} (see **Fig. 1-3**).

3.1.5.2. Cyclical Wave Bolt 3 (CWB-3)

It was mentioned in Sections 3.1.3 and 3.1.4, that the $T \Leftrightarrow E$ wall does not radiate or absorb energy at both the release and fixing. However, only E-type fixing (**Fig. 3-2-b**) applied simultaneously to all particles of the layer $x \in [0, L]$ of the waveguiding medium results in absorption of a finite energy. In terms of the discrete model of the medium (**Fig. 3-2-a**), this operation corresponds to switching of the viscosity elements $\bar{\delta}_n(t)$ within time of order τ_c from zero to $\bar{\delta}_n(t) = \bar{\delta}_{\max} \to \infty$ for any $n = n(x)$, where $0 \le x \le L$. Such an energy absorption occurs in the case of either a perfectly inelastic fast interaction (impact) between the wave and vibrostat G (**Fig. 3-2-a**) or fast zeroing of the kinetic component $\sigma\rho\int_0^L(\varphi_t')^2\,dx/2$ of the wave energy in the layer $x \in [0, L]$ without preliminary transformation of this field to high-frequency waves of the technological range. Impact interaction between the wave and the vibrostat assumes a peak reaction force (of the order of $2\pi\Xi\rho\sigma L/(\tau\tau_c)$). Simultaneous release of all particles of the layer $x \in [0, L]$ transforms the elastic-stress energy $\sigma\rho c^2\int_0^L(\varphi_x')^2dx/2$ stored in the above layer into waves running away on both sides. Let us consider a CWB-3 whose space-time structure is fully determined by parameter $\bar{\delta}_n$ of the viscous connection between the particles of the medium and the vibrostat G (**Fig. 3-2-a**):

(a) in the transparency intervals $\hat{T}_m$, $\bar{\delta}_n = 0$ for $n(x=0) \le n < n(x=L)$;

(b) in the absorption intervals $\hat{\hat{T}}_m$, $\bar{\delta}_n = \infty$ for $n(x=0) \le n < n(x=L)$.

The actual value of the viscosity parameter $\bar{\delta}_m(t)$ in the intervals $\hat{\hat{T}}_m$ should be sufficiently large so that the brake path Δx of a particle during the "freezing" time τ_c of the medium does not exceed the acoustic wave path, *i.e.*, $\Delta x < c\bar{T}$, or $[\bar{\delta}(t)]_{\max} > \sigma\rho\Xi T_B\tau^{-1}\bar{T}^{-1}$. The duration of the transparency intervals $\hat{T}_m$ of the layer is equal to $T_B - \bar{T}$, while that of absorption (fixing) intervals $\hat{\hat{T}}_m$ is $\bar{T} \ge \tau_c$. To simplify the analysis, the incident wave pulse (**Fig. 3-3-**

c1) is written as a group of $\overline{N}$ equidistant elementary pulses (**Fig. 3-3-c2**) whose location period $2h$ is related to the active-region size and the viscosity-switching period T_B by $L = 2h[\overline{N} + (1/2)]$ and $2h = cT_B$. In the case considered, the time at which such an incident-wave pulse completely enters (see **Fig. 3-3-c1**) the transparent layer $0 \le x < L$ coincides with the initial time of the first interval $\overline{\overline{T}}_1$. After the first impact and zeroing of the kinetic component of the field energy inside the layer, the problem becomes symmetric about the point $x = L/2$, and the initial displacement remaining in each elementary pulse after viscosity switching-off generates two half-amplitude pulses that run apart symmetrically. Each pulse covers the distance $2h_B$ during the time T_B between the viscosity switching-on pulses and is impacted, being located exactly at the place of the adjacent pulse. In this case, the shape of the elementary pulses remains unchanged, which makes the analysis much simpler. We have only pulse distortions at the edges of the active region, but the contribution from these boundary phenomena to the reflection coefficient is proportional to $1/\overline{N}$, where the number $\overline{N}$ can be chosen sufficiently large. This number is limited only by the finite duration $\overline{T} \ge \tau_c$ of viscosity switching-on and the relation $\overline{T} << T_V / N$. In accordance with (3-4), we can obtain the approximate estimates $|\overline{\Gamma}| \approx \overline{T}/T << 1$ and $|\Gamma_A| \approx cT_B / L << 1$, where $|\Gamma_A|$ is determined by the ratio between the total area of leaving pulses (**Fig. 3-3-c3**) whose amplitudes form a decreasing geometric progression with index $1/2$, to the area of the incident wave pulse. For $(T_B)_{opt} = (\overline{T}L/c)^{1/2}$, we obtain the minimum reflection coefficient $|\Gamma_B|_{\min} \approx 2(c\overline{T}/L)^{1/2} << 1$. As the previous two versions of the wave gates, the CWB-3 can be used in three-dimensional boundary-value problems. The above algorithm can be classified as {LS/LT/CA/M} (see **Fig. 1-3**).

3.1.6. Parametric Coating

In the one-dimensional case, the CWB was an echelon of thin, plane and infinitely extended controlled walls parallel to the front of an incident wave. Virtual resonators (VR) were formed by the spatial regions between adjacent walls (see, for example, CWB-1). However, in the case of oblique incidence of a plane quenched wave, such a system is inefficient. In this case, for waves traveling along the walls, the problem has no boundaries, and any eigen-frequencies of the VR, including zero frequency, are possible. Then, using two-dimensional or three-dimensional VRs, we modify the CWB-1 algorithm for parametric coverage of the two-dimensional and three-dimensional protected object, respectively, which occupies the spatial region $\mathbf{r} \in \hat{\Theta}$ bounded by the surface $\overline{S}_B$. Such a coating occupies the region $\hat{L}$ between the external surface $\overline{S}_B$ and internal surface $\overline{\overline{S}}_B$ with the minimum thickness $L = \min[\hat{D}(\overline{S}_B, \overline{\overline{S}}_B)]$, where $\hat{D}(\overline{S}_B, \overline{\overline{S}}_B)$ is the distance between the surfaces $\overline{S}_B$ and $\overline{\overline{S}}_B$ such that the internal surface $\overline{\overline{S}}_B$ coincides with the surface of protected object: $\overline{\overline{S}}_B = S_{\Theta}$. For example, the CWB-1 based coating is a three-dimensional foam-like structure shown in **Fig. 3–5** at subsequent times $0 < t_1 < t_2 < t_3 < t_4 < T_B$. The cavities (or virtual resonators) of this structure have characteristic spatial scale $H << L$ and are separated by walls with controlled transparency. So the total number of VR in a coating is of the order $N_c \sim LS_c H^{-2} >> N >> 1$, where S_c - area of the surface of coating. In this case, the surface $S_B(t) \in \hat{L}$, which separates the region $\hat{Y}$ of transparent walls ($\hat{Y} \subset \hat{L}$ for external side) and the region $\hat{\overline{Y}}$ of opaque walls ($\hat{\overline{Y}} \subset \hat{L}$ for internal side) walls, travels with velocity $V > c$ from the external edge of the coating to the internal edge ($\hat{Y} \cup \hat{\overline{Y}} = \hat{L}$ and $\hat{Y} \cap \hat{\overline{Y}} = \hat{0}$). Once the surface $S_B(t)$ reaches the internal edge of the coating, the region of opaque walls is immediately recovered in the entire coating, and all operations are repeated again.

Continuous synchronization of individual units (parts of the coating that can be "glued on the protected surface) along the normal to the surface is important for a parametric coating, but the units themselves can perform CWB operations in the self-sufficient mode. Let us note that the external layer of virtual resonators should have the $T \Leftrightarrow E$ walls blocking the switching radiation, while deeper layers should have the $T \Leftrightarrow L$ walls. As is obvious from the description, this coating does not contain either elements or couplings sensitive to the structure of the incident-wave front or determining any spatial absorption anisotropy (unlike the cases, described in Chapter 2). In this sense, the angle under which the incident wave enters the CWB active region that is currently transparent is unimportant and, hence, the efficiency of absorption by such a parametric coating is independent of the incidence angle of quenched waves. However, this does not eliminate the problem of diffraction of incident waves from the parametric coating.

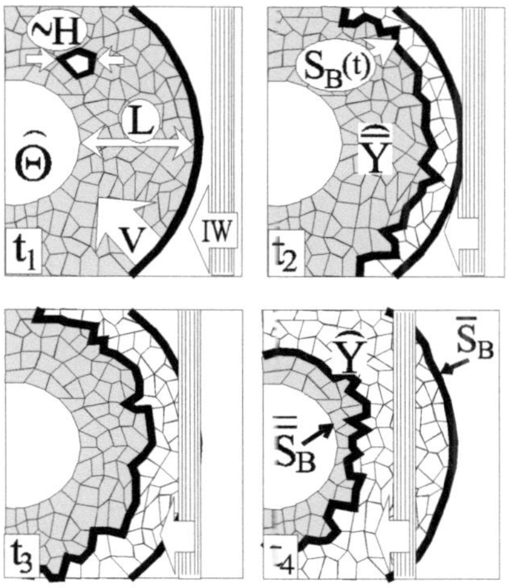

Figure 3-5: Foam like structure of absorbing coating based on CWB-1 ($t_1 < t_2 < t_3 < t_4$), $\widehat{\Theta}$ -the body protected or a empty cavity, $H \sim L/N$ -a characteristic linear dimension of VR, $S_B(t)$ -the boundary surface between the set of $\widehat{Y}(t)$ transparent walls and the set $\widehat{\widehat{Y}}(t)$ of opaque walls (VRs).

3.1.7. General Solution of Diffraction Problem

We now assume that the initial conditions in the CWB-1 active region are zero at the time of the CWB switching-on and that the incident wave has not yet reached region $\widehat{L}$. In accordance with the Poisson formula for free space, the wave field $\varphi(\mathbf{r},t)$ satisfying the equation $\varphi''_{xx} + \varphi''_{yy} + \varphi''_{zz} = c^{-2}\varphi''_{tt}$ (*e.g.*, the acoustic-pressure field in a gas or liquid) is determined by the values $\varphi(\mathbf{r}_0,t_0)$ and $\varphi'_t(\mathbf{r}_0,t_0)$ at time $t_0 < t_N$ on the cone $|\mathbf{r}-\mathbf{r}_0| = c(t_N - t_0)$: $\varphi(\mathbf{r},t) = \widehat{P}[t_0,t_N;\Phi_0,\Psi_0]$, where $\widehat{P}[t_0,t_N;\Phi_0,\Psi_0] = \widehat{U}[\Psi_0] + \partial U[\Phi_0]/\partial t$, $\Phi_0 = \varphi(r,t_0)$, $\Psi_0 = \varphi'_t(r,t_0)$, and [82], [83]

$$\widehat{U}[u] = [4\pi c^2(t_N - t_0)]^{-1} \iint_{|\mathbf{r}-\mathbf{r}_0|=c(t_N-t_0)} u(\mathbf{r}_0)dS(\mathbf{r}_0). \tag{3-6}$$

On the other hand, we can obtain an identical result of recalculation of the field from t_0 to t_N by using the sequence of N calculation steps similar to that described above, but with node moments $t_0 < t_1 < ... < t_n < t_{n+1} < ... < t_{N-1} < t_N$. Here, the first step, $\varphi(\mathbf{r},t) = \widehat{P}[t_0,t_1;\Phi_0,\Psi_0]$, yields new initial conditions $\Phi_1 = \varphi(\mathbf{r},t_1)$, $\Psi_1 = \varphi'_t(\mathbf{r},t_1)$;...; the n-th step, $\varphi(\mathbf{r},t_n) = \widehat{P}[t_{n-1},t_n;\Phi_{n-1},\Psi_{n-1}]$, yields new initial conditions $\Phi_n = \varphi(\mathbf{r},t_n)$, $\Psi_n = \varphi'_t(\mathbf{r},t_n)$; the $(n+1)$-th step, $\varphi(\mathbf{r},t_{n+1}) = \widehat{P}[t_{n-1},t_n;\Phi_n,\Psi_n]$, yields new initial conditions $\Phi_{n+1} = \varphi(\mathbf{r},t_{n+1})$, $\Psi_{n+1} = \varphi'_t(\mathbf{r},t_{n+1})$;...; and the N-th step gives $\varphi(\mathbf{r},t_N) = \widehat{P}[t_{N-1},t_N;\Phi_{N-1},\Psi_{N-1}]$.

In the case of unlimited miniaturization of the coating structure, the action of a perfect CWB can be represented as the covering of a three-dimensional mask described by the linear operator $\widehat{Z} = \widehat{Z}(\widehat{L},t*)$, where $t* = nT_B$, $n = 1,2,...$, at the end of each n-th cycle. This operator makes instantaneously the field $\Phi(\mathbf{r}) = \widehat{Z}[\varphi(\mathbf{r},t*)]$ and its derivative $\Psi(\mathbf{r}) = \widehat{Z}[\varphi'_t(\mathbf{r},t*)]$ zeros (for example, $T_B << \tau_{iw}$ for CWB-1) with accuracy of the order of $|\Gamma_B| << 1$ (see (3-5)) at some time $t*$ in the region $\widehat{L}$ and keeps $\varphi(\mathbf{r},t*)$ and $\varphi'_t(\mathbf{r},t*)$ unchanged outside this region, *i.e.*, $\widehat{Z} = \{1, \mathbf{r} \notin \widehat{L}; 0, \mathbf{r} \in \widehat{L}\}$.

Let us modify the above calculation procedure. We use the operator $\hat{Z}(\hat{L},t_*)$ to act on the initial conditions at the node times $t_0 < t_1 < ... < t_n < t_{n+1} < ... < t_{N-1} < t_N$ coinciding with the times $t_n = nT_B$ of CWB cycle termination. Therefore, we obtain an adequate description of diffraction of an arbitrary incident wave from a parametric coating of arbitrary size and shape (**Fig. 3-6-a**):

$$\varphi(\mathbf{r},t_1) = \hat{P}[t_0,t_1;\Phi_0,\Psi_0],\ \Phi_1 = \hat{Z}\varphi(\mathbf{r},t_1),\ \Psi_1 = \hat{Z}\varphi'_t(\mathbf{r},t_1);...; \quad (3\text{-}7)$$

$$\varphi(\mathbf{r},t_n) = \hat{P}[t_{n-1},t_n;\Phi_{n-1},\Psi_{n-1}],\ \Phi_n = \hat{Z}\varphi(\mathbf{r},t_n),\ \Psi_n = \hat{Z}\varphi'_t(\mathbf{r},t_n);$$

$$\varphi(\mathbf{r},t_{n+1}) = \hat{P}[t_n,t_{n+1};\Phi_n,\Psi_n],\ \Phi_{n+1} = \hat{Z}\varphi(\mathbf{r},t_{n+1}),\ \Psi_{n+1} = \hat{Z}\varphi'_t(\mathbf{r},t_{n+1});$$

$$\varphi(\mathbf{r},t_N) = \hat{P}[t_{N-1},t_N;\Phi_{N-1},\Psi_{N-1}].$$

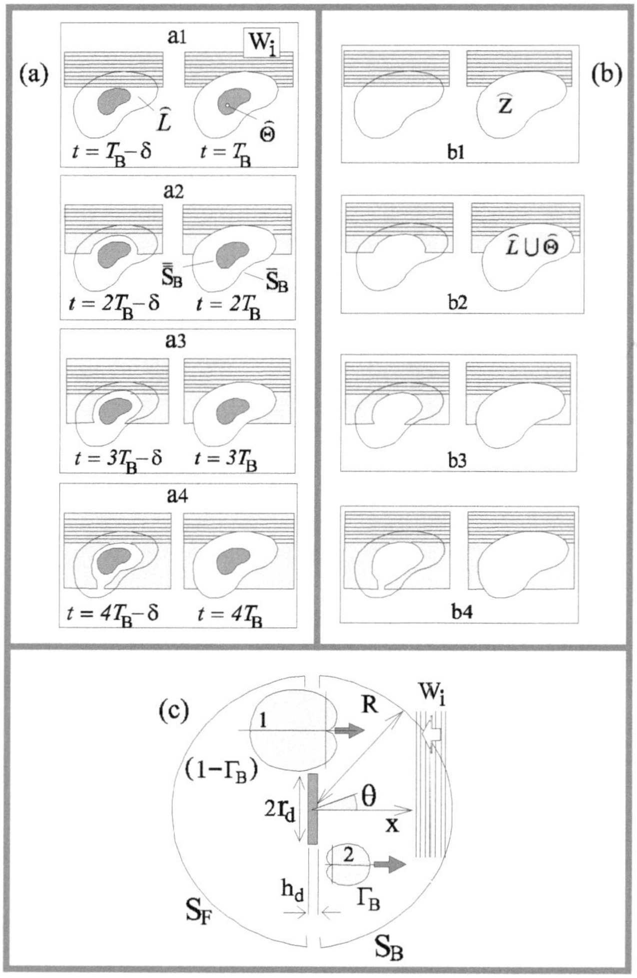

Figure 3-6: Geometry of scattering problem for body $\hat{\Theta}$ with surface S_Θ, equipped with parametric coating $\hat{L}$, based on CWB-1 (Section 3.1.4.), with mask operator $\hat{Z}$ (body $\hat{\Theta}$ with surface S_Θ, equipped with parametric coating $\hat{L}$ of thickness $\geq L$ ((a), temporal sequence a1–a4). Geometry of the diffraction of the same incident wave on the "black body" with surface $\bar{S}_B$ ((b), temporal sequence b1–b4). Scattering by the disk (c): 1–directivity pattern of the forward scattering, 2–directivity pattern of the backscattering, W_i –planar incident wave.

Let us note that in accordance with the concept of a CWB with "traveling" boundary (for example, CWB-1), an incident wave never reaches the internal surface $\overline{\overline{S}}_B$ of the parametric coating $\hat{L}$. Therefore, the use of the Poisson formula for free space is correct and an arbitrary distribution of parameters of the medium in the region $\mathbf{r} \in \hat{\Theta}$ does not affect the scattering field. The latter means that the mask $\hat{Z}$ can be modified: $\hat{Z} = \{1, \mathbf{r} \notin \hat{L} \cup \hat{\Theta}; 0, \mathbf{r} \in \hat{L} \cup \hat{\Theta}\}$. Such a mask $\hat{Z}$ is considered as a "black body" option (**Fig. 3-6-b** that is an alternative to those discussed in [77-81].

3.1.8. Scattering by the "Black" Disk

The value Γ_B characterizes reflection or one-dimensional backscattering. To describe 2-3 dimensional wave scattering problem we will use back scattering coefficient Γ_{back} (3-1), which of course is connected functionally with Γ_B. As the most simple example of three dimensional scattering problem let us consider the following case: plane incident wave $\varphi_w = \Xi \cos(\omega t - kx)$, where $k = \omega / c$, $\omega = (2\pi / \tau_w) << (2\pi / T_B)$) is scattered by thin disk parallel to the incident wave front and equipped with parametric coating a.m. Disk has thickness $h_d = cT_B$, radius $r_d >> h_d$ and occupies space area $\hat{\Theta} = \hat{L}$. Let's assume at first that $\Gamma_B = 0$. CWB is adding (with period T_B) distributions $\overline{\varphi} = -(1-\hat{Z})\varphi_{iw}$, $(\overline{\varphi})'_t = -(1-\hat{Z})(\varphi_{iw})'_t$ to incident wave field φ_w, $(\varphi_w)'_t$ inside $\hat{L}$. Averaging on the time scale $2\pi / \omega$ one gets outside ($\mathbf{r} \notin \hat{L}$) wave field with space diagram in amplitude $f(\vartheta) \approx (2\pi)^{-1/2}[J_1(kr_d \sin\vartheta) / (kr_d \sin\vartheta)]\cos^2(\vartheta / 2)$ equivalent to the field of disk extracted from the plane Huygens surface. Here J_1–Bessel's function of 1-st order, ϑ–angle between disk's axis and direction of observation in far zone, *i.e.* at $R > r_d^2 k / 2\pi$ (**Fig. 3-1-c**). Analogously one will get backscattered field $\overline{\overline{\varphi}}$, caused by finite value $|\Gamma_B| > 0$ with space diagram $f(\vartheta - \pi)$. Scattered field $\varphi_S(\mathbf{r}, t)$ may be represented as a sum $\varphi_S = \overline{\varphi} + \overline{\overline{\varphi}}$ of mutually coherent fields $\overline{\varphi} \sim \Xi(1-\Gamma_B)f(\vartheta)$ (diagram 1 in **Fig. 3-6-c**) and $\overline{\overline{\varphi}} \sim \Xi\Gamma_B f(\vartheta - \pi)$ (diagram 2 in **Fig. 3-6-c**), produced by two disks extracted from plane Huygens surfaces radiating in mutually opposite directions. So one gets the following expression of back scattering coefficient for the disk equipped with parametric coating

$$\Gamma_{back} \approx \int_0^{\pi/2} \left|\Gamma_B f(\vartheta) - (1-\Gamma_B) f(\vartheta - \pi)\right|^2 \sin\vartheta d\vartheta \,. \quad \textbf{(3-8)}$$

Expression (3-8) of disk's Γ_{back} assumes that initial conditions inside the area $\mathbf{r} \in (\hat{\Theta} \cup \hat{L})$ are determined only by current values of incident wave field. This is correct perfectly if "black" disk radius r_d is infinite, *i.e.* if area $\mathbf{r} \in \hat{L}$ presents the space between two planes parallel to the incident wave front. The last means that during time frame h_d / c all perturbations have enough time to leave this space area. Next cycle of CWB "processes" only next undistorted portion of incident wave. For the case of disk of finite radius one need estimate boundary effects. During time h_d / c after instant disk's extracting from infinite Huygens plane the small amount of points inside disk have time to know about this event: points on distance $\le h_d$ from disk's border. The area of initial conditions distorted takes the square in $\sim (r_d / 2h_d) >> 1$ times less than disk's square. Maximum of boundary distortions can't be more than maximum value of undistorted incident wave field which is uniform inside $\hat{L}$. So one can state that relative difference between real wave field and mode assumed in disk's Γ_{back} has the order not more $\sim (2h_d / r_d) << 1$. Note that this simple formulation is possible due to the coincidence of disk's plane with wave front surface of incident wave. Taking $\Gamma_B = 0$ one can get for disk in concrete cases: $\Gamma_{back} = 1/8$ at $kr_d << 1$ (nonzero integral value Γ_{back} in half-space, but zero scattering in back direction immediately) and $\Gamma_{back} \approx (kr_d)^{-4} / 2 << 1$ at $kr_d << 1$. Above we considered scattering by parametric coating based on CWB-1. However, this computational procedure (Section 3.1.9.) also can be applied to the scattering by coatings based on versions CWB-2,3 taking into account their concrete peculiarities. In particular, for parametric coating based on CWB-3 the sequence of time points of the wave field calculation (Section 3.1.7) have the form $t_n = nT$, where $T \le (L / c)$ and $T >> T_B$.

3.2. WATER SURFACE WAVES (EXPERIMENTS)

Now we consider the application of the space-time local algorithms more thoroughly for the case of gravitational surface water waves in the laboratory tank. Below two active damping systems for surface gravitational waves are designed and experimentally tested. These systems are based on: modulation of reflection coefficient of controlled wall (see section 3.1.3 above, [50]) and concept of running cyclical wave-bolt (RCWB) (see section 3.1.4, [63], [64], [62], [59], [61]). In the second case the thickness of active damping system was about one-tenth of the length of the seiche wave but can also be much smaller, than that in accordance with the CWB principle.

Measurement and control of water waves is easier than measurement and control of acoustical waves, due to their slowness. In addition, it is well known, that the viscous damping factor γ (see (3-2) in section 3.1) of surface gravity waves is proportional to the fourth power of their frequency f ($\gamma = 32\pi^4 f^4 \nu / g^2$, where ν is the cinematic viscosity of the water and g is the acceleration due to gravity), which increases the advantages of the modulation methods based on frequency conversion.

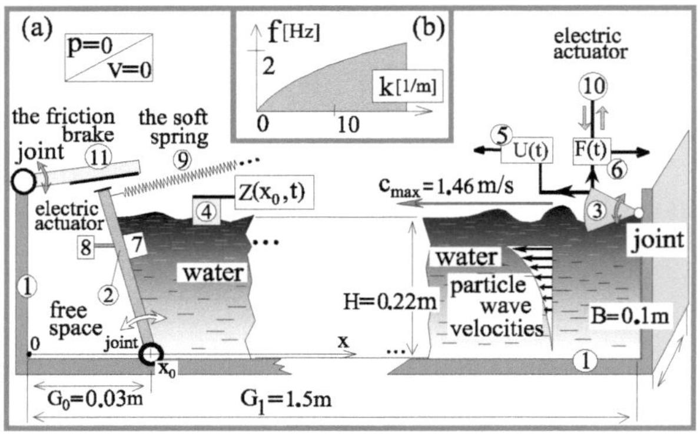

Figure 3-7:. The experimental setup for the investigation of the effect of balance modulation of the controlled wall reflection coefficient (a); dispersion characteristics $f(k)$ of water waves in tank (b).

(a) modulus $|\tilde{Z}(f_w)|$ of spectrum amplitude of the liquid surface oscillations $Z(t)$ (at the boundary x_0) on the incident wave frequency (given by wavemaker 3, $f_W = 2Hz$) at various modulation frequencies f_M;

(b) relaxation of the liquid surface oscillations $Z(t)$ (at the boundary x_0) after sinusoidal source (with wavemaker frequency $f_W = 1.0Hz$) switching off: (b1) without modulation with rigid wall; (b2) with modulated wall, when $f_M = 8.0Hz$;

(c) relaxation of the liquid surface oscillations $Z(t)$ (at the boundary x_0) after pulse source switching off: (c1) without modulation with rigid wall 2 (brake 11 in lower position); (c2) with modulation, when $f_M = 8.0Hz$;

(d) oscillations of the liquid surface oscillations $Z(t)$ (at the boundary x_0) at high frequency powerful sinusoidal excitation of the wall 2 by actuator 8 and low frequency sinusoidal excitation of the wavemaker 3. The combinational frequencies are absent.

On the other hand, this case is more complicated, than the acoustical waves, owing to significant wave dispersion, which lead to the specific distortions of the wave oscillograms. However, it is obvious, that the latter is not significant at high technological frequency f_{tech} ($f_{tech} = mf_M$, here f_M is the modulation frequency, $m = 1,2,...$), where high-frequency damping prevails. Actually, the ratio between the time τ_{dis} of dispersion misphasing at π of two neighboring spectral components that are multiples of f_{tech} and the time τ_{vis} of viscous damping by a factor of e is given by $\tau_{dis} / \tau_{vis} \sim 16\nu\pi^4 f_{tech}^3 g^2 / 3$ and can be increased to a value greater than unity. In this case, the dispersion is not essential if the modulation frequency is fairly large.

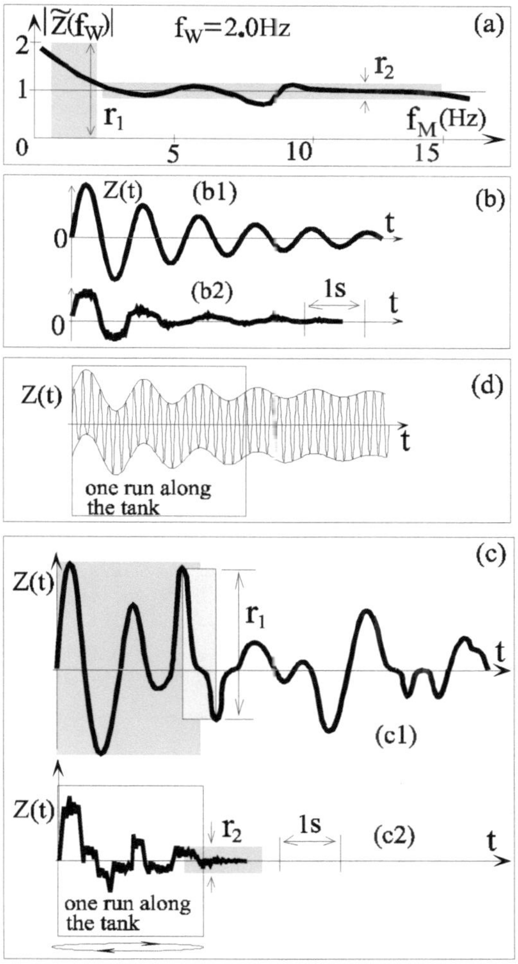

Figure 3-8: Characteristics of damping of waves in the tank 1 with modulated wall 2:

3.2.1. Balance Modulation of the Wall Reflection Coefficient

Below we consider the application of the algorithm (section 3.1.3 above, [50]) of modulation of reflection coefficient of the controlled wall (or parametric load) for the surface gravitational waves. The scheme of the experimental set-up (tank 1 with length $G_1 = 1.5m$, width $B = 0.1m$ and with water of depth $H = 0.22m$) is given in **Fig. 3-7-a** with the dispersion dependence $f(k)$ of running water waves in **Fig. 3-7-b**. The controlled wall 2 (near the "controlled" boundary x_0) is fastened by joints to the bottom of tank 1 near the left edge of tank. The wall 2 is mechanically connected with the electric actuator 8. Between left wall of tank 1 and the controlled wall 2 we support the free space (of length $G_0 = 0.03m$) due to the corrugated connection between wall 2 and lateral walls of tank 1. The hydrostatic pressure on the wall 2 is compensated by soft spring 9. The absence of waves in tank wall 2 (due to spring 9) is vertical. Taking the softness of spring 9 the own frequency of its oscillation is much smaller than the frequency of expected incident wave. The friction brake 11 is mechanically connected by joints with the left wall of tank 1. Upper position of brake 11 means that wall 2 is free. Lower position of brake 11 means that wall 2 is stopped in its current position (rigid state of the wall 2). The brake 11 is controlled by simple electric actuator (absent in **Fig. 8-a**). The sensor (float like) 4 of water surface with signal $Z(x_0,t)$ (below will be written as $Z(t)$) is spaced near the neutral vertical position (or x_0) of wall 2. To register perturbations of water pressure, the sensor 7 was fastened on the right hand surface of wall 2 near the water surface. The float like wavemaker 3 is placed near the right edge of tank 1. Wavemaker 3 is mechanically connected with electric actuator 10, and has a velocity sensor 5 (with velocity signal $U(t)$) and a sensor 6 (with signal of force $F(t)$) of the force of actuator action to the float. There were two variants to provide the soft state of the wall 2: control of actuator 8 *via* signal of sensor 7 or *via* signal of sensor 4. The second version was more successful and effective. There were three variants to provide the soft state of the wall 2: to control actuator 8 by signal of sensor 7 or by signal of sensor 4, or free shift of wall 2 without any actuators. The second version turned out to be more stable and effective. All the sensors are made on the basis of movable electronic tubes. Owing to joints, wall 2 is matched not exactly to the horizontal velocities distribution with depth in the incident wave, and this fact caused the finite coupled mass of liquid. The water mass together with the wall mass does not make the soft state of the wall 2 an ideal.

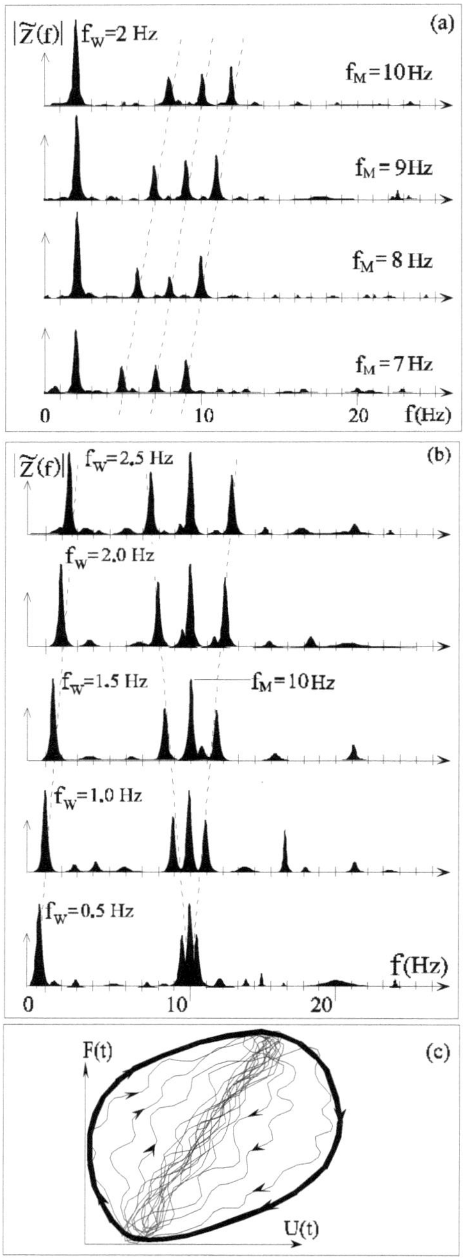

Figure 3-9: Characteristics of the modulation algorithm, used for the water surface ewaves: (a) spectra $|\hat{Z}(f)|$ of the vertical deflection $Z(t)$ (at the boundary x_0) of the liquid surface at the initial wave frequency $f_W = 2Hz$ and modulation frequencies $f_M = 7.0, 8.0, 9.0, 10.0Hz$; (b) spectra $|\hat{Z}(f)|$ of the vertical deflections $Z(t)$ (at the boundary x_0) of the liquid surface at the initial wave frequencies $f_W = 0.5, 1.0, 1.5, 2.0, 2.5Hz$ and modulation frequency $f_M = 10.0Hz$; (c) effect of the CWG on the impedance pattern of wavemaker 3, when $f_W = 2Hz$, $f_M = 8.0Hz$.

The friction brake 11 is pressing on the wall 2 in the temporal intervals $2n(2f_M)^{-1} \le t < 2(n+1)(2f_M)^{-1}$ ($n = 1,2,3,...$), where f_M is the modulation frequency. This provides the fixing of the wall 2 and its reflection coefficient $\Gamma = +1$ (in the sense of vertical displacements $Z(t)$ of water surface particles on the wall 2). In the temporal intervals $(2n+1)(2f_M)^{-1} \le t < (2n+3)(2f_M)^{-1}$ ($n = 1,2,3,...$) the friction brake 11 takes the upper position. This provides the free motion of the wall 2 (owing to the actuator 8 with sensors 7 and 4) and its reflection coefficient $\Gamma = -1$ (in the sense of vertical displacements $Z(t)$ of water surface particles on the wall 2). Thus, we obtain the balance modulation ($\Gamma = \pm 1$) of the reflection coefficient of the wall 2 with the modulation frequency f_M.

The **Fig. 3-8-a** shows the amplitude $|\tilde{Z}(f_w)|$ of the liquid surface oscillations ($\tilde{Z}(f_W)$-spectral component of the function $Z(t)$ on the frequency f_W of incident wave) at the boundary x_0 as the function of the modulation frequency f_M, when the frequency of the incident wave is $f_W = 2Hz$. The gray horizontal strip (near the amplitude of incident wave) means the consensus (absence of low frequency reflected wave) between the wall 2 and the incident wave. At frequencies f_M near zero, the wall becomes rigid and the amplitude of vertical liquid particles shift becomes double the amplitude of incident wave. At high modulation frequencies f_M, inertia of the wall 2 begins to manifest itself, reducing its consensus with the incident wave. The value r_2 / r_1 can be interpreted as a rough estimate of the reflection coefficient $|\Gamma|$ of the wall 2 at the frequency $f_W = 2Hz$.

The **Fig. 3-8-b** and **Fig. 3-8-c** show the relaxation of waves (signal $Z(t)$ in the point x_0) in the tank 1 which were previously excited by wave maker in monochromatic regime (**Fig. 3-8-b**) with wavemaker frequency $f_W = 1.0Hz$ and by pulse (**Fig. 3-8-c**): without modulation (**Fig. 3-8-b1**, **Fig. 3-8-c1**) and with modulation switched on (**Fig. 3-8-b2**, **Fig. 3-8-c2**). We noticed, that modulation provides the acceleration of relaxation. Another case is presented in **Fig. 3-8-d**: the wall 2 is excited by actuator 8 on the high frequency (independently of the wave field in water) with amplitude greater than the wave, produced by wavemaker 3, which is excited on low frequency. In **Fig. 3-8-d** only the sum of two quasimochromatic oscillations without any combinational frequencies (*i.e.* modulation effect does not appear), has been observed.

In **Fig. 3-9-a**, **Fig. 3-9-b** we observed a number of plots of modulus of Fourier spectra $|\tilde{Z}(f)|$ of signal $Z(t)$ from sensor 4 in the following two cases: (a) frequency of wavemaker is $f_W = 2Hz$, and the modulation frequency takes several positions $f_M = 7.0, 8.0, 9.0, 10.0Hz$; modulation frequency is $f_M = 10.0Hz$, and frequency of incident wave (wavemaker) takes several positions $f_W = 0.5, 1.0, 1.5, 2.0, 2.5Hz$. Thus, **Fig. 3-9-a** and **Fig. 3-9-b** illustrates the parametric conversion of incident wave frequency and the appearance of combinational frequencies in wave spectra.

Fig. 3-9-c gives the experimental diagram of the wavemaker 3 impedance. This figure shows, how the phase trajectory (like elliptic trajectory, *i.e.* waves in the tank 1 are stationary, or phase difference between signals $U(t)$ and $F(t)$ is about $\sim \pi/2$) was transformed into the linear one, when the modulation became switched on. We can interpret this fact as the absence (suppression) of reflected waves at low frequency $f_W = 2Hz$ of incident wave, radiated by wavemaker 3 (when $f_M = 8.0Hz$). The above algorithm can be classified as {LS/LT/CA/M} (see **Fig. 1-3**).

3.2.2. Cyclical Wave-Bolt

Above, we described an active damping system based on the rigidity modulation of the controlled wall. Such an approach with balanced (bipolar) modulation of the reflection coefficient ($\Gamma = \pm 1$) does not reduce the operating volume of the basin and ensures an almost instantaneous response to the incident wave. However, such a system has the following drawbacks:

(1) requires taking special measures to provide for a soft state of the controlled wall (use of an electric drive and a wave displacement pickup);

(2) requires taking special measures to provide for hermetic sealing of the free air space behind the controlled wall (use of a corrugated film to minimize the resistance of the controlled wall to motion without friction against the flume wall);

(3) requires taking special measures to provide for hydrostatic equilibrium of the controlled wall (use of a rubber rope to compensate for hydrostatic pressure, sufficient strength (thickness) of the controlled wall);

(4) is fundamentally unable to provide for complete matching of a plane controlled wall with the horizontal velocity distribution of particles in an arbitrary incident wave;

(5) is fundamentally inefficient for bodies with finite wave dimensions (not greater than incident wavelength) in three-dimensional problems.

Taking this into account, we consider a half-balance method of reflection damping, in which the reflection coefficient of the protection walls on the way of the incident wave is switched between the values $\Gamma = 0$ and $|\Gamma| = 1$. This approach applies to three-dimensional problems and does not require solving the hydrostatic problem, since the vertical liquid-air interface is absent here. Thus far, it was implied that the reflected wave had already been formed at some frequency and must be suppressed or absorbed. However, the situation is possible where the reflected wave is excluded since the incident wave is always behind the protected surface running away from the former with velocity $V > c$, where c is the propagation velocity of the incident wave. Implementation of this idea requires the use of a technological intermediary in the form of a CWB [63], [64], [62], [59], [61] to match the runaway condition with the condition of immobility of the protected surface (basin wall). For this, on the path of the incident wave, we mount an echelon of films 12, at equal distances with spatial period D (**Fig. 3-10-a**), in parallel to the incident wave front, and a special device for control of their tension. The vertical loose film represents a transparent wall ($\Gamma = 0$), and the vertical tight film represents an opaque wall ($|\Gamma| = 1$). At the initial time, all the films are tightened simultaneously (become opaque) and are loosened alternately (become transparent), beginning with the Film that is next to the incident wave. Thus, the film-loosening wave (**Fig. 3-10-b**) runs away from the incident wave, and its velocity V is greater than the maximum velocity $c_{\max}$ of wave perturbations in the flume (with allowance for the wave dispersion). As the loosening wave reaches the deepest CWB layer (*i.e.*, the film that is next to the basin wall), all the films are stretched simultaneously. Then the process is repeated periodically with frequency f_B. Two neighboring films in tense state form the so-called virtual resonator (VR), whose minimum eigenfrequency is inversely proportional to its dimension D, which is fairly large. Viscous damping of gravity surface waves is of fundamental importance for such oscillations of the liquid. If the tension of two neighboring films is rapid, then the current distribution of the wave field between them transforms to initial conditions for the subsequent oscillations of the liquid in the VR. Damping of these oscillations over a time much smaller than the modulation period $1/f_B$ ensures effective suppression of reflections. The limiting reflection coefficient $|\Gamma| \approx 1/N$, where N is the number of films in the CWB, at the incident wave frequency is due to the front film of the CWB, which cannot but reflect the incident wave at the beginning of each period $T_B = 1/f_B$ during the time interval $T_0 = D/V$; it is taken into account that the waves have time to get damped in the remaining resonators. Reflections from the wall with which the incident wave does not collide are absent [63], [64], [62], [59], [61]. **Fig. 3-10-b** shows the spatial-temporal diagram (film-loosening wave) of the tensing forces $R_1(t), R_2(t), \ldots, R_{N-1}(t), R_N(t)$ switching "on" (flooded black graphs) and "off", applied to the films in the points $x = D, x = 2D, \ldots,\ x = (N-1)D, x = ND$ respectively. We noticed, that the wave (with velocity $V = 1.5 m/s$) of debilitation of films always outstrips the incident wave front (with velocity $c_{\max} = 1.45 m/s$). Radiation from the inner side of the spatial tension jump due to the loosening wave is also excluded, since the loosening operation does not require mechanical work. Since only a switching from tight to loose states, not vice versa, takes place in each cycle, the "superluminous" tightening wave that could create the self-radiation of the CWB is absent here. Radiation only occurs at the time of simultaneous stretching of all the films, but it takes place only at higher frequencies and does not contribute to the reflection at the incident wave frequency.

The films in tension represent VR walls. The stretching time τ_c of the films is also the characteristic scale of their high-frequency horizontal oscillations, which mainly correspond to the lower oscillation mode of the tense ($\tau_c << T_0$); the time τ_c is determined by the tension of each film $|R_n|$. It is desirable to use the largest tension $|R_n|$ allowed by the tightening device and film strength. Two goals are reached simultaneously in this case: the maximum separation of VRs from each other and the maximum absorption of the film oscillation energy by the liquid (see the text below). Note that the eigenfrequencies of the tense film are outside the sensitivity limits of the floating sensor 4 (in the point $x_0 > L$, with output signal $Z(t)$) of liquid surface oscillations (**Fig. 3-10-a**). The film in tension is an equivalent of a tense string, and its shape before tension transforms to the initial conditions of string oscillations. Both ends of such a string are fixed, and this boundary-value problem does not have a zero eigenfrequency, including the case of a string with inhomogeneous parameters (in particular, the string with an inhomogeneous distribution of mass). The minimum eigenfrequency $f_F \sim 1/\tau_c$ of oscillations of a tense film is much greater than the frequency of the incident wave, $f_F >> 1/\tau_W$ and than the lower eigenfrequency of the virtual resonator, $f_F >> f_{VR} = \sqrt{g/(\pi D)}$. Hence, oscillations of a tense film do not contribute to the reflection at the incident wave frequency. The front film of the CWB has a minimum time $T_0 = D/V$ in tense state, but even this time is sufficient for the provision of $n_F \sim T_0/\tau_c > 3$ $n_F \sim T/\tau_c' > 3$ (see the text below) oscillations of its lower mode. With respect to virtual resonators, the films oscillating in tension represent a source of exciting oscillations of the liquid there. Eventually, film tension work transforms to thermal energy of the liquid, mainly because of the high-frequency viscous dissipation. The basis for the experimental setup is a flume 1 of length $G_2 = 1.5m$ and width $B = 0.1m$, filled by water of depth $H = 0.22m$ (**Fig. 3-10-a**). In the right-hand part, the flume has a floating wavemaker 3, connected to the electric drive and equipped by a velocity pickup 6 and a liquid action pickup 5. A floating sensor 4 is mounted at the front edge of the CWB to measure the vertical displacement $Z(t)$ of the liquid surface. The CWB being

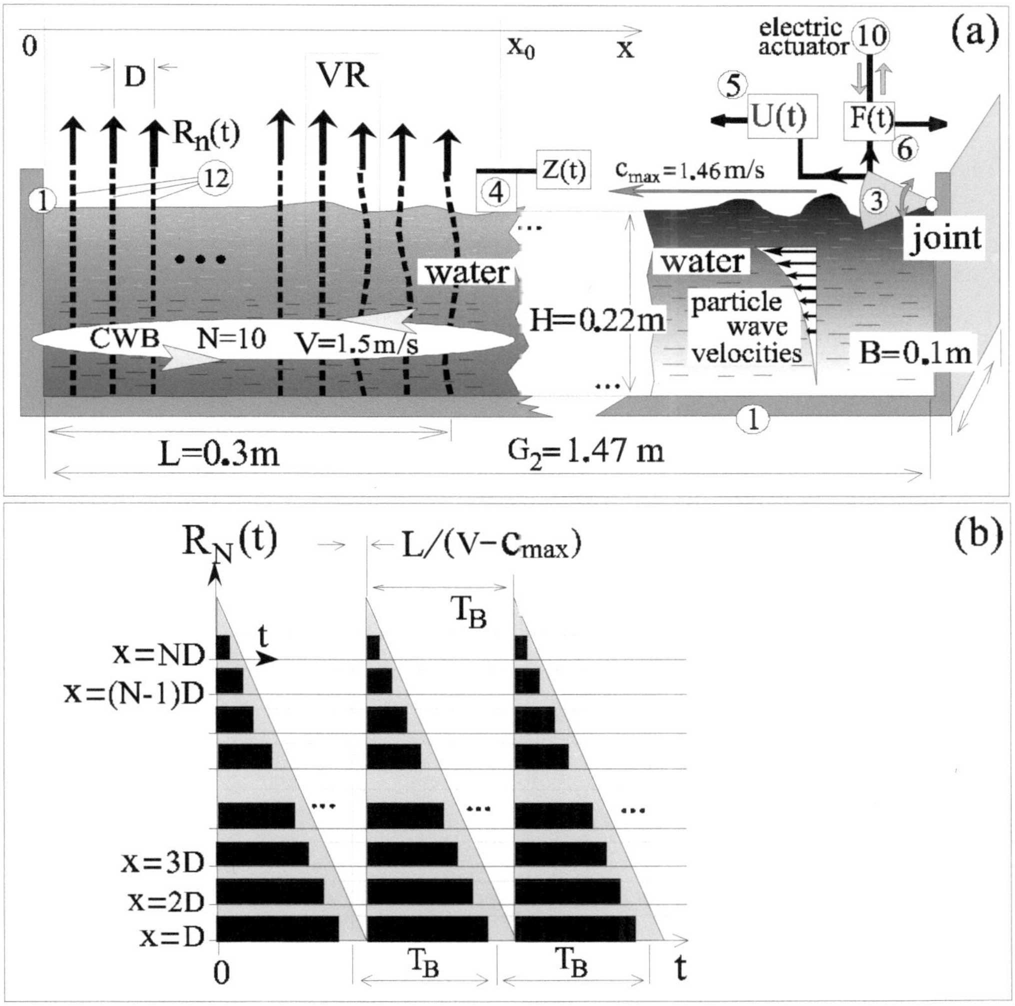

Figure 3-10: The experimental setup for the investigation of the Running Cyclical Wave Bolt (CWB-1) (a); spatial-temporal diagram (film-loosening wave) of the tensing forces $R_1(t)$, $R_2(t)$,..., $R_{N-1}(t)$, $R_N(t)$ switching "on" (flooded black graphs) and "off", applied to the films in the points $x = D, x = 2D$,..., $x = (N-1)D$, $x = ND$ respectively (b).

investigated represents an echelon of $N = 10$ flexible light nonstretchable lavsan films 12, having a spatial period $D = 0.03m$ and occupying an area of length $L = ND = 0.3m$, which corresponds to about one-tenth of the wavelength of the lower mode of the flume. The chosen model of CWB has the following main parameters: $T = 0.02s$ -clock interval of change of states of the neighboring films, $V = D/T_0 = 1.5m/s$ -runaway velocity of the film-loosening jump, $A_{x,z} \sim 10^{-2}m$ -characteristic amplitudes of displacement of particles in the incident wave along x (and z), $\tau_W \sim 2.5s$ -temporal scale of the incident (for example, seiche) wave, $T_B \sim 0.8s$ -period of the CWB, $f_B = 1/T_B \sim 5.5Hz$ -frequency of the CWB, and $A_{HF}/D \sim 4A_x HD^{-2}T_B\tau_{WD}^{-1} = 0.4$ -maximum relative vertical amplitude of displacement of particles in high-frequency waves in closing up the VR. Then, taking into account the cinematic viscosity $\nu = 10^{-6}m^2/\sec$, the water density $\rho = 10^3 kg/m^3$, the acceleration due to gravity $g = 9.8m/s^2$, the characteristic capillary scale $\varepsilon = \sqrt{2\chi/(\rho g)} = 3.86 \times 10^{-3}m$, the water surface tension coefficient $\chi = 7.3 \times 10^{-2} N/m$ for $20^o C$, the high-frequency damping factor $\sim \exp(-\gamma t)$ (where $\gamma = \alpha f^4$ and $\alpha = 32\pi^4 \nu g^{-2}$), and the maximum velocity of the wave perturbation $c_{\max} = \sqrt{gH} = 1.46m/\sec$ (in the shallow-water approximation), we formulate the conditions for efficiency of the proposed CWB model:

(1) the film tension reached in this experiment, $|R_n| = 50$ units SI (newtons), determined the stretching time: $(BH^2 A_x \rho/|R_n|)^{1/2} = 0.0067\sec$;

(2) the small reflection from the "transparent" loose film is reached by the small weight and flexibility of the films;

(3) the small transparency of the "opaque" state is determined by the high tension of the films. Tension of four or five front films, where the initial amplitude of the relaxed waves is at a maximum, is especially important. Oscillations in each VR must be damped independently of the remaining VRs;

(4) the leak of the liquid between the neighboring VRs was small, that is, the volume of the liquid in the VR was constant on the temporal scale τ_W so that $h/B < 0.016$ ($h \sim 10^{-3}m$ is the gap between the film and the flume walls). Fulfillment of this condition guarantees the absence of zero eigenfrequencies in each VR;

(5) the thickness of the boundary layer was much smaller than the distance between the films: $D^{1/2}H^{-1/4}A_z^{1/2} = 2.52 \times 10^{-2}(m^{3/4}) >> 1.6 \times 10^{-3}(m^{3/4})$, *i.e.*, "creeping" flow between the films was absent;

(6) the damping of oscillations had an oscillatory nature at the lower VR mode. For this, the natural damping factor in the waveguiding medium must be smaller than the lower eigenfrequency f_{VR} of oscillations in the VR. In the opposite limiting case, the large damping factor will result in "freezing" of oscillations at technological frequencies, the absence of field operation, and, therefore, the absence of absorption: $\gamma < f_{VR}$, $D > 2^{5/3}\nu^{2/3}g^{1/3} = 4.7 \times 10^{-5}m$;

(7) effective absorption was reached at the oscillation frequency f_F of the tense film: $[\gamma(f_F)L/V] >> 1$;

(8) capillary effects for the lower VR mode were small: $\{D, A_{x,z}\} >> \varepsilon = 3.86 \times 10^{-3}m$;

(9) overlappings were absent, *i.e.*, the longitudinal displacement of particles in the wave was smaller than the distance between the films: $A_x T/(\tau_W D) = 1.3 \times 10^{-2} < 1$. The work Q_F performed in rapid tension and stretching of the film (contraction of its length by $\Delta z \sim 16A_x^2/H$) over the time $\tau_c \sim 0.0067s$ must be estimated roughly as $Q_F \sim |R_n|\Delta z \approx 2BH\rho A_x^3/(3\tau_c^2) \approx 3.6 \times 10^{-3}$ units of SI (Joules). This work is considerably greater than the incident wave energy $Q_W \sim \rho g B A_z \tau_c \tau_W/(8\pi) \approx 2.8 \times 10^{-2} J$ fed over the time τ_c and transforms to thermal energy owing to the viscous damping of film oscillations.

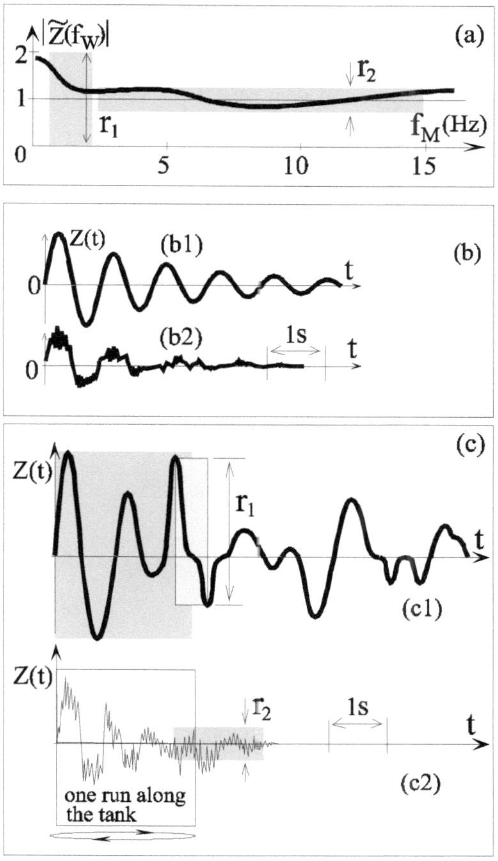

Figure 3-11: Characteristics of damping of waves in the tank 1 with CWB:

(a) modulus $|\tilde{Z}(f_w)|$ of spectrum amplitude of the liquid surface oscillations $Z(t)$ (at the boundary x_0) on the incident wave frequency (given by wavemaker 3, $f_W = 1.0Hz$) at various CWB frequencies f_B;

(b) relaxation of the liquid surface oscillations $Z(t)$ (at the boundary x_0) after sinusoidal source (with wavemaker frequency $f_W = 1.0Hz$) switching off: (b1) with CWB switched off; (b2) with CWB switched on, when $f_B = 5.0Hz$;

(c) relaxation of the liquid surface oscillations $Z(t)$ (at the boundary x_0) after pulse source switching off: (c1) with CWB switched off ; (c2) with CWB switched on, when $f_B = 5.0Hz$.

The **Fig. 3–11–a** shows the amplitude $|\tilde{Z}(f_w)|$ of the liquid surface oscillations ($\tilde{Z}(f_W)$-spectral component of the function $Z(t)$ at the frequency f_W of incident wave) at the boundary x_0 as the function of the CWB frequency f_B, when the frequency of the incident wave is $f_W = 1.0Hz$.

The gray horizontal strip (near the amplitude of incident wave) means the consensus (absence of low frequency reflected wave) between the CWB and incident wave. The value r_2 / r_1 can be interpreted as a rough estimate of the reflection coefficient $|\Gamma| \sim 0.2$ of CWB on the frequency $f_B = 8.0Hz$.

The **Fig. 3-11-b** and **Fig. 3-11-c** show the relaxation of waves (signal $Z(t)$ in the point x_0) in the tank 1 which were previously excited by wave maker in monochromatic regime (**Fig. 3-11-b**) with wavemaker frequency $f_W = 1.0Hz$

and by pulse (**Fig. 3-11-c**): with CWB switched off (**Fig. 3-11-b1**, **Fig. 3-11-c1**) and with CWB switched on (**Fig. 3-11-b2**, **Fig. 3-11-c2**) with frequency $f_B = 5.0Hz$. We noted, that modulation provides the acceleration of relaxation.

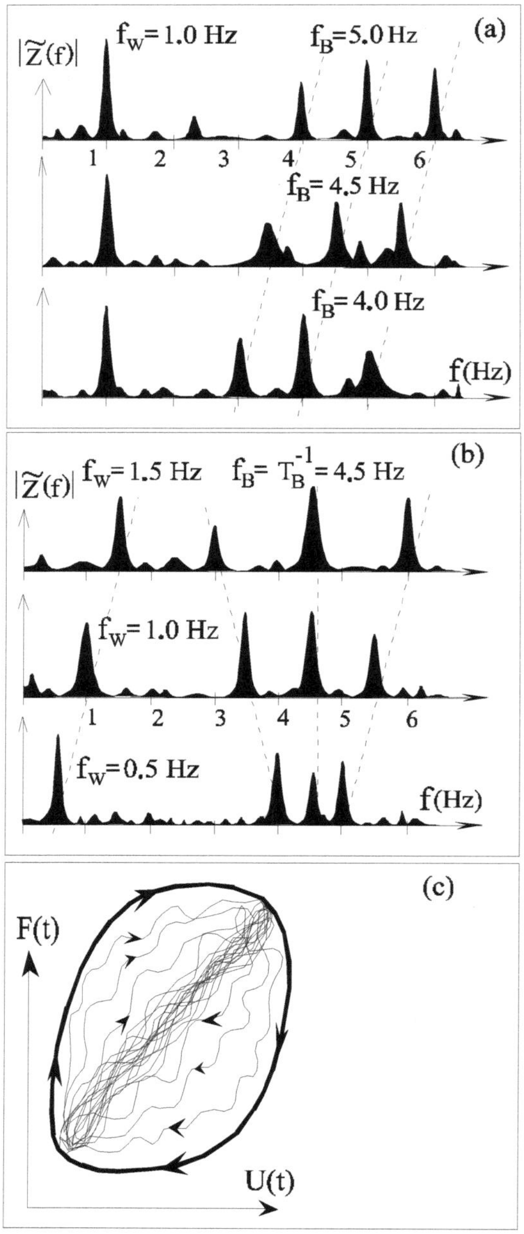

Figure 3-12: Spectra $|\tilde{Z}(f)|$ of the vertical deflection $Z(t)$ of the liquid surface at the initial wave frequency $f_W = 1.0Hz$ and CWB frequencies $f_B = 4.0, 4.5, 5.0Hz$ (a); spectra $|\tilde{Z}(f)|$ of the vertical deflections $Z(t)$ of the liquid surface at the initial wave frequencies $f_W = 0.5, 1.0, 1.5Hz$ and CWB frequency $f_B = 4.5Hz$ (b); effect of the CWB on the impedance pattern of wavemaker 3.

In **Fig. 3-12-a** and **Fig. 3-12-b** we observed a number of plots of modulus of Fourier spectra $|\tilde{Z}(f)|$ of signal $Z(t)$ from sensor 4 in the following two cases: (a) frequency of wavemaker is $f_W = 1.0Hz$, and CWB frequency takes several positions $f_B = 4.0, 4.5, 5.0Hz$; CWB frequency is $f_B = 4.5Hz$, and frequency of incident wave (wavemaker) takes several positions $f_W = 0.5, 1.0, 1.5Hz$. Thus, **Fig. 3-12-a** and **Fig. 3-12-b** illustrate the parametric conversion of incident wave frequency and the appearance of combinational frequencies in wave spectra.

Fig. 3-12-c gives the experimental diagram of the wavemaker 3 impedance. This figure shows, how the phase trajectory (like elliptic trajectory, *i.e.* waves in the tank 1 are stationary, or phase difference between signals $U(t)$ and $F(t)$ is about $\sim \pi/2$) was transformed into the linear one, when the CWB became switched on. We can interpret this fact as the absence (suppression) of reflected waves at low frequency $f_W = 2Hz$ of incident wave, radiated by wavemaker 3 (when $f_B = 5.0Hz$).

In this paper, we studied an approach to the problem of active quenching of reflected waves based on parameter modulation of the waveguiding medium in a thin layer on the path of the incident wave and performed without measuring of wave fields. The key operation in this approach is the conversion of the low-frequency incident (or reflected) wave to highfrequency waves for which the waveguiding medium is absorbing (opaque). We developed the structure of a cyclic wave gate as a technological intermediary between the protected surface and the incident wave. **Fig. 6** shows the evolution of the impedance trajectory of the wave generator $F(t)$, $U(t)$, where $F(t)$ is the signal from the liquid action pickup 5 and $U(t)$ is the signal from the liquid velocity pickup 4. With the CWB switched off and nonresonant excitation of wave generator 2, the trajectory had a quasi-elliptic form. Switching on the CWB leads to degeneracy of the $F(t)$, $U(t)$ trajectory to a linear one, which is indicative of the decrease in the Q-factor of flume 1 as a resonator. We note the following practical advantages of the CWB compared with the quenching system studied in [50]:

(1) the CWB does not require the creation (and the use of electric drives and wave sensors) of a "soft" boundary, which, in principle, excludes the self-excitation problem that is conventional for active systems;

(2) unlike the hinged controlled wall, the free film of the CWB is matched with the arbitrary vertical distribution of the horizontal velocity of particles in the incident wave;

(3) if the films are flexible and light, then the same CWB is also efficient for quenching internal gravity waves in a liquid;

(4) the CWB can be used in three-dimensional boundary-value problems in the form of a parametric coating. The coating represents a three-dimensional foam-like structure, whose cavities (VRs) are separated by walls of controlled transparency. The surface separating the region of transparent (outer side) and opaque (inner side) walls moves as the film-loosening wave described above. The above algorithm can be classified as {LS/LT/CA/M} (see **Fig. 1-3**).

In this section, we studied an approach to the problem of active damping of reflected waves based on parameter modulation of the waveguiding medium in a thin layer on the path of the incident wave and performed without measuring wave fields. The key operation in this approach is the conversion of the low-frequency incident (or reflected) wave to highfrequency waves for which the waveguiding medium is absorbing (opaque). We developed the structure of a cyclic wave bolt as a technological intermediary between the protected surface (basin wall) and the incident wave. Owing to the CWB, the incident wave perceives the actual immobile wall as running away with a velocity greater than the velocity of the wave and, therefore, the wall does not create reflection in the frequency range of the incident wave. This system is based on the local action principle and actually perceives the incident wave as a time-constant field. This makes hydrostatic problems very important, especially, the problem of keeping the volume of liquid constant in a virtual resonator. The obtained solution does not reduce the combination of circuits with constant parameters that is conventional for the known active methods, and ensures quenching in the absence of information on the length or period of the wave being quenched. The thickness of the active absorbing layer can be much smaller than the length of the incident wave. The use of a high spatial-temporal resolution for parameter control of the waveguiding medium in the active region of the CWB and the opacity of this medium for high-frequency waves make it unnecessary to measure the wave field and generate a cancellation wave. Based on the results obtained in the experiments with gravity surface waves in a liquid, one can develop applications of the CWB principle for waves of different physical nature, in particular, sound or electromagnetic waves.

3.3. ELECTROMAGNETIC WAVES

In this section, we investigate the superwideband (temporal) representation of electromagnetic boundary problem. Using the modeling elements the fast electronic switches, controlled by fiber optic cables, the metallic needles, we will consider the formation of the radio-wave-absorbing (nonreflecting) coating. This parametric coating is organized and operated by control algorithm of cyclical wave-bolt (CWB, see above the Sections 3.1.4, 3.1.5, 3.1.6, 3.1.7). The parametric version of "black body" is considered analytically. It is shown that absorbing coating can be formed from metal needles connected by semiconductor electronic switches. When the electronic switches are much smaller than the needles and exhibit a sufficient high-speed performance and the needles are much smaller than the coating thickness, this thickness can be made much smaller than the wavelength.

3.3.1. Solution of the Absorption Problem

The linear dissipation mechanism that ensures efficient (~90%) absorption of the wave (oscillator) energy with time period $\sim\tau$ necessitates several periods. Attenuation is observed if there is enough time for a wave to complete its work against dissipative forces. Constant parameter systems can exhibit efficient absorption of waves with period $\sim\tau$ in a coating of thickness $L \ll c\tau$ only in resonant oscillators with lumped parameters.

3.3.2. Virtual Resonator

Consider a plane smooth (with characteristic temporal $\sim\tau$ and spatial scales $\sim c\tau \gg L$) incident electromagnetic wave that is linearly polarized and propagates to the left in free space (**Fig. 3-13-a1**). Let $E(x+ct)$ be the electric-field intensity of this wave. Assume that the wave consists of nonoverlapping short pulses (wave fragments or plane wave packets) of duration $\tau_0 \ll L/c$ and length $\ell_0 = c\tau_0 \ll L$ that travel immediately after each other and have parallel fronts: $E(x+ct) = \sum_n E_n(x+ct)$, where $E_n(\xi) = E(\xi)\Pi(\xi + n\ell_0)$, $\Pi(\xi) = I(\xi) - I(\xi - \ell_0)$, $I = 0$ when $\xi < 0$, $I = 1$ when $\xi \geq 0$ and I are the unit functions (**Fig. 3-13-a2**). Each fragment $E_n(x+ct)$ may propagate independently of the remaining fragments without dispersion, that is, retaining its shape. The combination of all fragments forms initial smooth incident wave $E(x+ct)$. Let us choose the decomposition of wave E such that, at certain instant $t = t^*$ when the ends of wave fragment $E_{n^*}(x+ct)$ with number $n = n^*$ coincide with the ends of the segment $x \in [0, \ell_0]$, thin impenetrable conducting walls appear at the points $x = 0$ and $x = \ell_0 \ll L$ (**Figs. 3-13-a3, 3-13-a4, 3-13-a5**). This operation (like instant plating of boundary surface by metal) gives reflection, but does not produce electromagnetic radiation, unlike the acoustical case (see controlled $T \Leftrightarrow L$ walls in Section 3.1.4). The specific structure of walls exhibiting controlled transparency is analyzed below. When $t > t^*$, the wave fragments with the numbers $n > n^*$ travel in this structure to the left, as before, and form a smooth wave with a stepwise trailing edge. The wave fragments with the numbers $n < n^*$ are reflected from the boundary (wall) $x = \ell_0$, travel to the right, and form a smooth reflected wave with a stepwise leading edge. When $t > t^*$, the wave packet with the number $n = n^*$ is locked in the resonator with the walls $x = 0$ and $x = \ell_0$, where it experiences multireflections and loses $A^2 \times 100\%$ of its energy in each time interval $\tau_0 = \ell_0 / c$ ($0 < A < 1$ is the amplitude absorption coefficient of the resonator walls $x = 0$ and, which is determined, for example, by the finite conductivity of the walls, $\mathrm{Im}\, A = 0$).

3.3.3. Oscillatory Attenuation of the Field

Multireflections of wave fragment $E_{n^*}(x+ct)$ of length ℓ_0 in a resonator of the same length correspond to oscillations at the resonator's eigenfrequencies $n\pi c/\ell_0$ ($n = 1, 2, \ldots$), which do not contain zero frequency. This is due to the absence of free charges in metallic cavity for instance. So VR with metallic walls do not have zero eigenfrequencies, unlike VR, formed by $T \Leftrightarrow E$ walls in Sections 3.1.3, and 3.1.4. for longitudinal acoustic waves. The instantaneous electric-field distribution of wave packet $E_{n^*}(x+ct^*)$ at t^* can also be interpreted as the distribution involved in the initial conditions for resonator oscillations.

Thus, in the presence of the resonator walls, the energy of short wave fragment $E_{n^*}(x+ct)$ of smooth incident wave $E(x+ct)$ is transformed into electromagnetic oscillations whose energy is concentrated near the frequencies

$$\omega_n = n\pi c/\ell_0 \quad (n = 1, 2, \ldots) \tag{3-9}$$

During the time interval when the resonator exists $t-t^*>0$, spectral power $|\tilde{u}(t,\omega_n)|^2$ of the mode of frequency ω_n, which was initially locked, is reduced to the quantity $|\tilde{u}(t,\omega_n)|^2 \le |\tilde{u}(t^*,\omega_n)|^2[1-A(\omega_n)]^{-2ct/\ell_0}$, when $0<A(\omega_n)<<1$. Here, $\tilde{u}(t,\omega_n)=\int_0^{\ell_0} u(x,t)\exp(-i\omega_n x/c)dx$ and $u(x,t)$ is the electric field in the resonator. During the aforementioned time interval, wave energy is transformed into heat energy. It is shown below that, the lowest resonator eigenfrequency $\omega_1=\pi c/\ell_0$, the absorption coefficient is lower than at each of the remaining eigenfrequencies $A(\omega_1)<A(\omega_n)$.

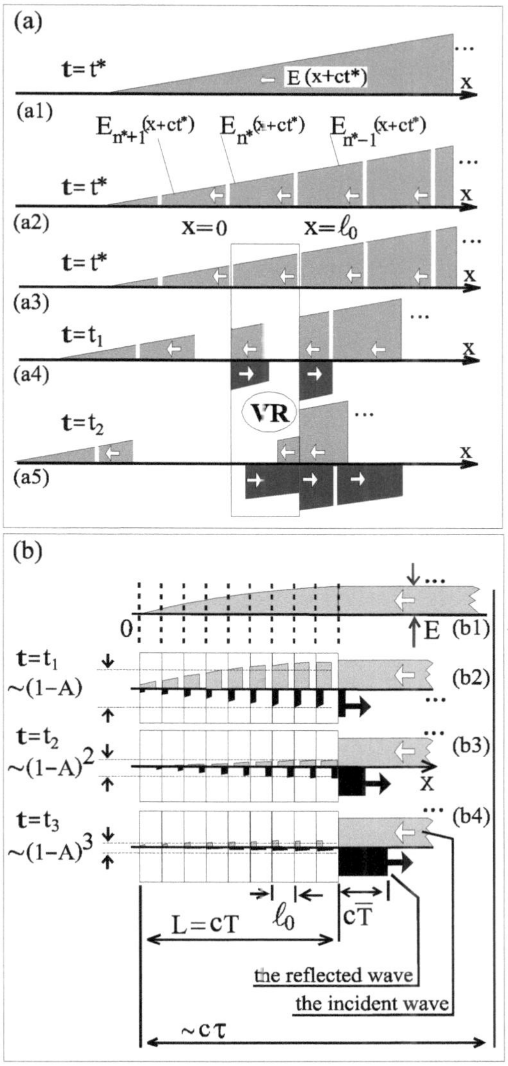

Figure 3-13: Typical scenario (a) of frequency conversion of smooth (a1) incident wave (virtual splitting of initial incident wave at the instant $t=t^*$ into a set of wavelets (a2)) with appearance of two nontransparent walls in the points $x=0$ and $x=\ell_0$ instantly switched on at the moment $t=t_1$ (a3-a5); electric field spatial distributions at the moments $t=t^*$ (a3) $t=t_1$ (a4) and $t=t_2$ (a5), where $t_1<t_2<t_1+(\ell_0/c)$, $t_1+(\ell_0/c)<t_3<t_1+2(\ell_0/c)$. The sequence (b) of the phases (down-steps) of CWB action: smooth (b1) incident wave; the walls of CWB are switched on (*i.e.* opaque since the moment $t=t^*$), wave fragments (gray run to the left, black run to the right) are shown at the instants $t=t_1>t^*$ (b2), $t=t_2>t_1$, (b3) $t=t_3>t_2$ (b4) with characteristic factors $\sim(1-A)$ ((b2), one reflection in VR), $\sim(1-A)^2$ ((b3), two reflections in VR)), $\sim(1-A)^3$ ((b4), three reflections in VR) of dissipative suppression of waves in VRs respectively.

Assume that the absorption coefficient of a certain hypothetic wall is frequency-independent and equals to $A(\omega_n) = A(\omega_1)$. Each reflection from the resonator's walls then reduces the amplitude of wave fragment $E_{n*}(x+ct)$ by factor $A(\omega_1)$. Recall that this fragment, which was initially locked, retains its shape and duration. The maximum value $u_0(t) = \max_{0 \le x < \ell_0} u(x,t)$ of the resonator field observed between real walls satisfies the following relationship at the moment $t \ge t^*$ [84]:

$$u_0(t) \le u_0(t^*)[1 - A(\omega_1)]^{-c(t-t^*)/\ell_0} . \qquad \textbf{(3-10)}$$

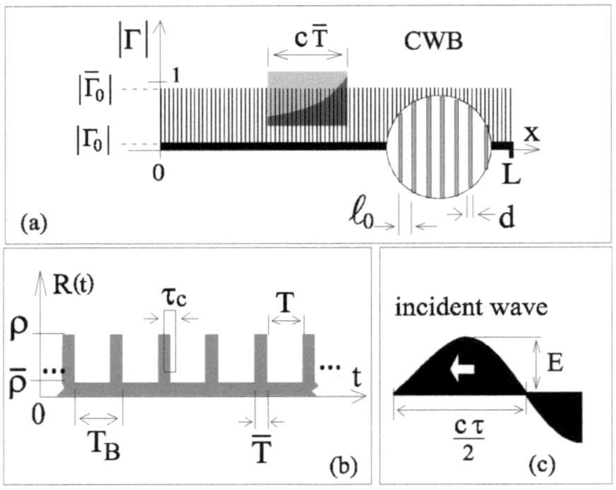

Figure 3-14: Incident wave (c) and spatial (a) and temporal (b) structure of CWB: (a) $|\Gamma|$−modulus of the wall's reflection coefficient; $|\bar{\Gamma}_0|, |\Gamma_0|$ −modulus of reflection coefficient of the CWB's wall in the opaque (our goal is $|\bar{\Gamma}_0| = 1$) and transparent (our goal is $|\Gamma_0| = 0$) states respectively, ℓ_0 -dimension of one virtual resonator (VR), d -width of the wall of VR, $\bar{T}$ −temporal scale of opaque state of VR's walls (temporal scale of relaxation of oscillation in VR); c −light speed; (b) $R(t)$ -temporal distribution of resistance of electronic switches in opaque state of wall $R = \rho$ (opened switches) and $R = \bar{\rho}$ (closed switches) in transparent state, T_B −temporal period of CWB, τ_c −temporal scale of transition from the transparent state to opaque one (and back), *i.e.* switching temporal scale; (c) incident wave with electric amplitude E, temporal scale τ, spatial scale $\lambda = c\tau$.

3.3.4. Finite Spatial-Temporal Resolution

In the foregoing, it has been assumed that the reflection, transparency, or absorption coefficient of an arbitrarily thin wall is switched during an arbitrarily short time interval. Now, let us take into account that the walls are of certain nonzero thickness $r_c > 0$ and their state is switched during certain nonzero time interval $\tau_c > 0$. Thus, the process of oscillatory attenuation involves only a part of the energy of wave fragment $E_{n*}(x + ct^*)$ rather than the total energy. This part of the energy, which is locked in the resonator, is determined by the expression

$$\delta = \left[\int_{\tau_c}^{L-2\tau_c - r_c} E_{n*}^2(x,t^*)dx \right] \left[\int_0^{L_0} E_{n*}^2(x,t^*)dx \right]^{-1} = 1 - (2c\tau_c + d)\ell_0^{-1} < 1,$$

and corresponds to the spatial interval whose points do not have enough time to react to finite time $\tau_c > 0$ of switching of the walls' reflection coefficient and to finite thickness of the walls $d > 0$. To simplify the further analysis, let us assume that the remaining part of wave fragment E_{n*} with the relative energy $1 - \delta = (2c\tau_c + d)/\ell_0 << 1$ does not participate in the oscillatory attenuation of the resonator field and is completely converted into the waves of frequency $\sim 2\pi/\tau$ that are reflected from the wall $x = L_0$ and transmitted through the wall $x = L_0$. Let the conditions for a high spatial–temporal resolution of the elements involved in the control of the boundary value problem,

$$d << \ell_0, \ \tau_c << \ell_0 / c \quad \textbf{(3-11)}$$

be fulfilled. Then, most of the energy of wave fragment $E_{n*}(x+ct*)$ participates in the process of oscillatory attenuation of the resonator field, which was described in section 3-3-3.

3.3.5. The Dissipative Mechanism of Attenuation

In the foregoing, it has been assumed that the wall's opacity and wave absorption in the walls are due to the switching of electric conductance σ of the spatial segments $|x| \le r_c / 2$ and $|x - \ell_0| \le d / c$ during the time interval $t* \le t < t* + \tau_c$ from $\sigma = 0$ to nonzero conductance $\sigma > 0$. Amplitude absorption coefficient A of such a wall is determined by the relationship [85]

$$A = (8\omega_1 \varepsilon_0 / \sigma)^{1/4} > 0 \quad \textbf{(3-12)}$$

At the lowest resonator frequency $\omega_1 = \pi c / L_0$, this coefficient equals to

$$A(\omega_1) = (8\pi c \varepsilon_0)^{1/4} / (\ell_0 \sigma)^{1/4} \quad \textbf{(3-13)}$$

if the wall's thickness

$$d > \Delta \quad \textbf{(3-14)}$$

exceeds the skin-layer thickness

$$\Delta = \sqrt{2 / \sigma \omega_1 \mu_0},$$

where ε_0 and μ_0 are the permittivity and permeability of free space, respectively. Condition (3-14) ensures the opacity of the walls. This is necessary for isolation of wave fragment $E_{n*}(x+ct)$ and transformation of its energy into the energy of oscillations at resonator eigenfrequencies (3-9). The higher the wall conductance, the better isolation. The condition for the nonzero absorption coefficient $A > 0$ provides for efficient attenuation of resonator oscillations. This condition is not fulfilled in the presence of infinite conductance. It can be easily checked that values ℓ_0, d and σ satisfy the conditions (3-14) and $d << \ell_0$ can always be found.

3.3.6. A Cyclic Wave Bolt

Consider a wave with period τ. The portion of the wave energy concentrated in a wave fragment with the spatial scale $\sim L << \tau c$ can be absorbed during the time interval $\sim \ell_0 / c << \tau = L / c$ if the energy of this wave fragment is transformed into the energy of waves of frequencies $\sim \pi c / \ell_0$.

Let us analyze an elementary 1D version of a cyclic wave bolt for a scalar acoustic field. The path of the incident wave (**Fig. 3-14-c**) with the characteristic temporal and spatial scales $\tau = 2\pi / \omega$ and $\lambda = c\tau$ (is the incident wave frequency) contains a sequence of $N >> 1$ walls of thickness $d << \ell_0$ located at the points $x_n = n\ell_0$ ($\ell_0 = L / (N-1)$, $n = 1, 2, ..., N$). The total thickness of this structure is L (**Fig. 3-14-a**).

At each instant, all of the walls of the cyclic wave gate have the same reflection factors, $\Gamma(t)$. Reflection factor $\Gamma(t)$ has only two stationary states:

(1) the transparent state with $\Gamma = \Gamma_0$ (when $0 < |\Gamma_0| << 1$) observed in the time intervals $t \in \hat{T}_m = \{mT_B - T \le t < mT_B\}$ of duration $T < T_B$ and

(2) the opaque state with $\Gamma = \Gamma_0$ (when $0 < 1 - |\Gamma_0| << 1$) observed in the time intervals $t \in \hat{\hat{T}}_m = \{(m+1)T_B \le t < (m+1)T_B + \overline{T}\}$ of duration $\overline{T} << T$, where $m = 1, 2, ..., N$. The second state may be due to the walls' conductivity.

Switching from the transparent to opaque state of the walls occurs over the time interval $\tau_c << \ell_0 / c$ with the period $T_B = T + \overline{T} << \tau$. During interval $\hat{T}_m$, the gate is occupied by the incident wave in the absence of reflections. After that, during interval $\hat{\hat{T}}_m$, the gate is closed. When $t \in \hat{\hat{T}}_m$, the gate reflects the wave incident from the outside and almost completely absorbs the wave that entered the gate in preceding transparent-state interval $t \in \hat{T}_m$ and is locked there. At the instants of switching to the opaque state, each wall (except the rightmost one shown in **Fig. 3-14-b**) very quickly, during the interval $\tau_c << \ell_0 / c$, splits the incoming wave into the wave incident from the right (and reflected to the right, **Fig. 3-13-b**) and the outgoing wave traveling to the left. Both of the waves formed when the opaque walls of the gate are switched on (except the exterior right wall $x = L$) and they become trapped in the interior of N small-size virtual resonators (the space between neighboring reflecting walls). The instantaneous spatial distribution of the field becomes an initial condition for the subsequent field oscillations inside a virtual resonator. These oscillations occur at the eigenfrequencies $\omega_n = n\pi c / \ell_0$, where $n = 1, 2, ..., N$; that is, the zero eigenfrequency is absent. During time interval $\overline{T}$, the wave travels $\overline{T}c / \ell_0 > 1$ times along the virtual resonator. Therefore, when next transparent-state interval $\hat{T}_{m+1}$ begins, the amplitude of high-frequency (HF) oscillations inside the virtual resonator is attenuated to a level that is not higher than

$$|\Gamma_1| \le [1 - A(\omega_1)]^{-c\overline{T}/\ell_0} \qquad \textbf{(3-15)}$$

measured with respect to the initial amplitude. This follows from relationship (3-3). In (3-10), $A(\omega_1)$ is the absorption coefficient of the wall at the lowest eigenfrequency $\omega_1 = \pi c / \ell_0$ of the virtual resonator. Quantity $|\Gamma_1|$ also determines the maximum contribution of the field that has not decayed in all of the virtual resonators to the absolute value of the reflection coefficient $|\Gamma_B|$ of the cyclic wave bolt at the low frequency of the incident wave. Obviously, when $\ell_0 \to 0$ (at $d << \ell_0$), the resonator oscillations may decay even in spite of small absorption coefficient $A(\omega_1) << 1$ and small life time of the virtual resonator.

The (3-10) yields the estimate of the maximum contribution

$$|\Gamma_2| \le (2c\tau_c + d) / \ell_0 << 1 \qquad \textbf{(3-16)}$$

and quantity $|\Gamma_B|$ for the waves that are not involved in oscillatory attenuation of the resonator field and are due to the semitransparent states of the resonator walls observed during finite switching time τ_c. The maximum contribution of pulse reflection from the exterior gate boundary $x = L$ to $|\Gamma_B|$ can be majored as

$$|\Gamma_3| \le \overline{T} / T_B << 1. \qquad \textbf{(3-17)}$$

Thus, the main contribution to the absolute value of the total reflection coefficient of the cyclic wave gate, $|\Gamma_B|$, observed at the low frequency of the incident wave in each cycle is provided by the following waves: the wave completely reflected from the right end of the gate during short time $\overline{T}$ (contribution $|\Gamma_3|$), the waves that have not completely decayed during life time $\overline{T}$ of virtual resonators (contribution $|\Gamma_1|$), the waves transmitted through the resonator walls in the process of switching (contribution $|\Gamma_2|$), and the waves caused by transient charge-redistribution processes initiated by switching of the walls' parameters (contribution $|\Gamma_4|$). When the conditions

$$\left(1-\left|\overline{\Gamma}_0\right|\right),\ \left|\Gamma_0\right| << \left|\Gamma_1\right| + \left|\Gamma_2\right| + \left|\Gamma_3\right| + \left|\Gamma_4\right| << 1$$

are fulfilled and all of the above contributions are small,

$$\left|\Gamma_1\right|,\ \left|\Gamma_2\right|,\ \left|\Gamma_3\right|,\ \left|\Gamma_4\right| << 1 \qquad \textbf{(3-18)}$$

and $\left|\Gamma_B\right|$ can be majorized as (3–18). Note that the gate efficiently operates when, in addition to (3-13) and (3-15)-(3-17), the relationships $T \le L/c$ and (3-18) are satisfied. The above algorithm can be classified as {LS/LT/CA/M} (see **Fig. 1-3**).

3.3.7. Broadbandness of Absorption

The gate absorbs an incident wave with a certain efficiency, which does not depend on any quasi-resonance relationship between the parameters of the gate and incident wave. This is a fundamental difference of the coating described above from all constant-parameter coatings except the classical thick low-absorbing coating. A cyclic wave gate is a nonresonant absorbing parametric microstructure; it periodically (with the period $T_B << \tau$) converts most energy $\left(1-\left|\Gamma_2\right|\right)$ of the low-frequency (LF) incident waves to the waves of high-frequency $\sim \pi c/\ell_0$. These efficiently decay in small-size elements (virtual resonators) of the microstructure during the relatively short lifetime of resonators $\overline{T} << T_B$. In the interior of the virtual resonators, only a very weak field component (on the order of $\sim \overline{T}T_B^{-1}\ell_0(c\tau)^{-1} << 1$ with respect to the amplitude of the incident wave) is observed at the incident-wave frequency. As the gate's microstructure becomes more diminutive and the speed of its performance grows, the part of the incident-wave energy transformed by the gate into HF waves increases and the absorption efficiency is enhanced. The LF reflection from the front end of the gate is reduced during time $\overline{T}$. This asymptotic construction is valid only because the exponential factor (**Fig. 3-14-a**) of oscillation attenuation inside virtual resonators prevails over the linear factors of an incident-wave splitting into fragments. It was mentioned in the Introduction that the frequent nonuniformity of absorption is due to an interaction between absorbing elements at the incidentwave frequency, which is manifested in counter propagating waves. The absorbing elements of the proposed microstructure are virtual resonators that practically do not interact at the incident wave frequency, since, during relatively short time intervals $\hat{\overline{T}}_m$ of duration $\overline{T}$, virtual resonators, where the wave is locked, are separated from each other by opaque walls. At the beginning of each relatively long time interval $\hat{T}_m$ of duration $T >> \overline{T}$, the resonator walls become transparent. However, the field inside them has already decayed. The wallreflection coefficients are switched during time τ_c, which is much shorter than the time ℓ_0/c (see (3-11)) of wave propagation between absorbing elements (resonator walls). Since the absorbing elements of an efficient gate interact only weakly (that is, $\left|\Gamma_B\right| << 1$), the observed absorption is nonuniform. This is characterized by the frequency scale $\sim 2\pi c/L >> 2\pi/\tau$ and the relative intensity of order $\sim\left|\Gamma_B\right|$, when $\tau_c << \ell_0/c$. In other words, the lower the quantity $L/c\tau = L/\lambda << 1$ providing a given density of reflection coefficient $\left|\Gamma_B\right|$, the smoother its frequency dependence.

3.3.8. Controlled Transparency of the Walls of Virtual Resonators

The following design versions can be used to create plane walls with controlled transparency for a normally incident plane electromagnetic wave.

3.3.8.1. A Reflecting-Wall-Wire-Grid Structure

Consider a grid that is formed of perfectly conducting threads (wires) of thickness $d << \ell_1$ parallel to the incident electric-field vector and has period ℓ_1. At frequency ω, this grid has the reflection coefficient [86]

$$\overline{\Gamma}_0(\omega) = (1+i\gamma\omega)^{-1}, \qquad \textbf{(3-19)}$$

which approaches unity, $|\Gamma_0| \to 1$, when $\omega\gamma << 1$. Here, $\gamma = (\ell_1 / \pi c)\ln(\ell_1 / \pi d)$. The limit $|\Gamma_0| \to 1$ means that, at the distance $x >> \ell_1^2 / \lambda$ from the grid, all of the currents $J_0 = j_0 \ell_1 = E\ell_1 \rho_{em}^{-1}$, excited by the incident wave in each wire (**Fig. 3-15-c**) of the grid produce a field that is equivalent to the field reflected from a metal plane (**Fig. 3-15-b**) with the linear density of the surface current $j_0 = E / \rho_{em}$, where E is the amplitude of the incident-wave electric field and $\rho_{em} = \sqrt{\mu_0 / \varepsilon_0} = 120\pi\ \Omega$ is the impedance of free space. At the spatial frequencies $\sim 2\pi / \ell_1$, the reactive field of discrete currents decays as $\sim \exp\{-x[(2\pi / \ell_1)^2 - (2\pi / \lambda)^2]^{1/2}\}$ at distance x from the grid.

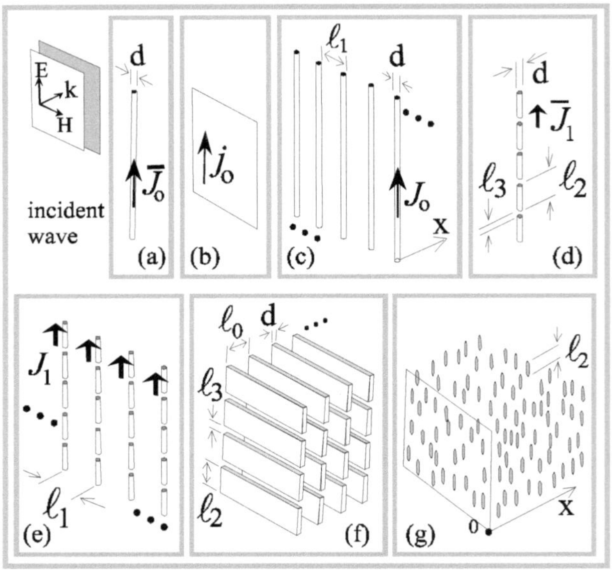

Figure 3-15: Several models of electromagnetic boundary problems with incident wave: solitary metal wire (a); infinite metal sheet (b); array (grid) of wires (c); solitary splitted metal wire (array of needles) (d); array (grid) of splitted metal wires (e); artificial media of metal strips (f); half-space filled with metallic needles (g).

3.3.8.2. A Transparent-Wall–Needle-Grid Structure

Owing to interaction between solid threads, absolute value $|J_0|$ of the current in each wire of the grid [86] is much greater than the current induced in an isolated infinite thread (**Fig. 3-15-a**) of diameter d by a normally incident plane linearly polarized wave $|\bar{J}_0| \approx E / \ln(2.36 \times \lambda / d)$ [72], that is $|J_0| >> |\bar{J}_0| \approx E / \ln(2.36 \times \lambda / d)$. In order to reduce interaction between wires, let us split solitary wire (**Fig. 3-15-d**) and each wire in grid (**Fig. 3-15-e**) into the needles of the length $\ell_2 << \lambda$ with isolating gaps of thickness ℓ_3 such that $d << \ell_3 << \ell_2$. The conduction currents in needles produce a reflected wave. The longing of these currents to zero means the longing of reflection to zero as well. The incident wave induces conduction current J_1 in a needle, that is the recharging current of electric capacitances $\bar{C}$ between neighboring needles (**Fig. 3-15-e**). This capacitance $\bar{C}$ differs weakly from transitive capacitance $C_S \approx \bar{C}$ of the electronic switch shaped as disk with thickness ℓ_3 and radius r_S (**Fig. 3-16-f**). As needle length ℓ_2 decreases, the electrostatic charges at the needle end more intensely and counteract the incident field and reduce the current. Below we can neglect the dipole-like field E_{dip} of needle's edges, if satisfy the condition $E_{dip} / E \approx (\varepsilon_S / 2) r_S^2 \ell_2^{-2} << 1$, where $\varepsilon_S \sim 15$ is a relative dielectric constant of electronic switch's semiconductor material. On the frequency ω of incident wave, the radiation resistance $\sim (2\pi\rho_{em} / 3)(\ell_2 / \lambda)^2 << \rho_{em}$ of solitary needle with length ℓ_2 is very small. Current J_1 is characterized by a step distribution over the spatial period $\ell_2 + \ell_3$. A normal plane reflected wave is produced by only one spatial harmonic of the conduction current, which corresponds to the zero spatial frequency. The current spatial harmonics at nonzero frequencies produce evanescent reactive fields, which follow from the relationship $(\ell_2+\ell_3)<<\lambda$. Current J_1 characterizes the process of switching of quasi-static charge distributions in a plane grid of noninteracting threads during the time $\sim\tau=2\pi/\omega$. Static charges assuredly exceed the charges accumulated at the needle ends during certain finite time $\sim\tau$. Therefore, in the first-order approximation, the absolute value of the reflection coefficient of a fragmented (transparent) grid can be majorized by the ratio of the aforementioned currents $|\Gamma_0|\approx|J_1|/|J_0|$ or

$$|\Gamma_0|\approx(\ell_2+\ell_3)\ell_1^{-1}\rho_{em}[(\omega\bar{C})^{-2}+(\omega\bar{L})^2]^{-1/2}, \qquad \textbf{(3-20)}$$

hence it follows that $|\Gamma_0|\to 0$ as $\bar{C}\omega\to 0$ (where $\bar{C}\approx C_S=\varepsilon_0\varepsilon_S\pi r_S^2\ell_3^{-1}$ is the capacitance of electronic switch, $\bar{L}\approx(3\pi)^{-1}\mu_0\ell_2\{\ln(2\ell_2/d)-(11/16)\}$ is the inductance of one needle [72], $\omega=2\pi/\tau$ –frequency of incident wave). Strictly speaking, besides capacitance $\bar{C}$, one must take into account the own capacitance $\bar{\bar{C}}\approx\pi\varepsilon_0\ell_2\ln^{-1}(2\ell_2/d)$ [72] of metallic needle. To simplify the further description, we assume $\bar{C}>>\bar{\bar{C}}$ or $\ell_3\ell_2 r_S^{-2}\varepsilon_S^{-1}\ln^{-1}(2\ell_2/d)<<1$. On the frequency ω, the impedance of spatial period $\ell_2+\ell_3$ of needle structure has a character of capacitance, *i.e.* $(\bar{L}\bar{C})^{-1/2}>>\omega$. The transparency coefficient of a medium formed from thin metal strips that are unboundedly fragmented can be obtained using the results of classical study on the theory of lens antennas (see below the section 3.3.8.4.).

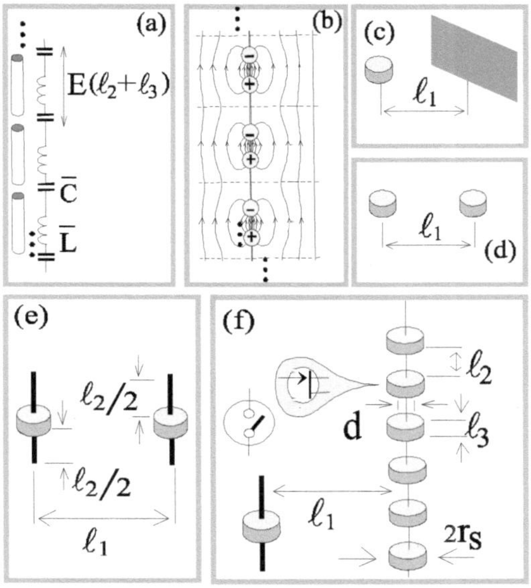

Figure 3-16: The interaction between needles: local capacitances $\bar{C}$ and inductances $\bar{L}$ of the needles, which are placed on the line (a); electric field, induced by the incident wave near the line of needles (b); local capacitance near the infinite metal sheet (mirror) (c); two interacting capacitances (d); interaction between capacitance, equipped with needles (e); interaction between one local capacitance and line of capacitances (f) (the chains "needle-switch").

3.3.8.3. Interaction Between Needles

The estimate (3-15) $|\Gamma_0|$ includes an interaction only between the ends of adjacent needles but neglects the interaction between lines of needles at the distances $\geq\ell_1>>\ell_2>>\ell_3$ (see **Fig. 3-18**) from each other. Each switch (with capacitance C_S in free space) is placed among another line of needles and switches. Due to these neighboring capacitors and needles, effective capacitance $\bar{C}_S$ always includes additional induced capacitance $\Delta C_S=(\bar{C}_S-C_S)>0$. Any metallic body, interacting with the switch (and capacitor), can be considered as some "metallic mirror". The electrostatic source (switch) and its image in mirror participate in frequentative interaction, and this induces an additional capacitance $\Delta C_S > 0$. Under assumption about smallness of $\Delta C_S / C_S) << 1$ we consider one-shot interaction. The interaction is increasing with greater dipole moment of a source, with smaller distance source-mirror, with greater dimension of mirror. The reflection of capacitor (**Fig. 3-16-c**) in the mirror (infinite metallic sheet), parallel with capacitor's axis, causes the relative increment $(\Delta C_S / C_S)_0 \approx 2^{-5} r_S^2 \ell_3 \ell_1^{-3} << 1$ of his capacitance C_S, where ℓ_1 -distance between capacitor and mirror. Placing the same else capacitor (**Fig. 3-16-d**) instead of the metallic plane, we will obtain $(\Delta C_S / C_S)_1 \approx 2^{-4} r_S^4 \ell_3^2 \ell_1^{-6} << 1$. For two capacitors equipped with pair of metallic needles (**Fig. 3-16-e**) we will obtain $(\Delta C_S / C_S)_2 \approx 2^{-4} r_S^4 \ell_2^2 \ell_1^{-6} << 1$. Note that capacitance $\overline{C}_S$, which was determined in electrostatics (*i.e.* at $\omega = 0$), is always greater, than capacitance $\overline{C}_S(\omega) = \mathrm{Re}\{J(\omega)/[i\omega U(\omega)]\}$, defined on any finite frequency $\omega \neq 0$, where $J(\omega)$ and $U(\omega)$, are the amplitudes of current and differences of potentials on capacitors contacts. This is due to the fact, that electrostatics gives infinite time for accumulation of the charges (and $\overline{C}_S(\omega) \le \overline{C}_S(0)$), unlike any $\omega \neq 0$. So we can consider the estimate $\overline{C}_S = \overline{C}_S(0)$ as upper. After summarizing all increments of capacitance, induced by all elements of one line (**Figs. 3-15-d, 3-16-a, 3-16-b, 3-16-f**) of needles with switches (capacitors) we obtain $(\Delta C_S / C_S)_3 \le 2^{-4} r_S^4 \ell_2^2 \sum_{n=-\infty}^{+\infty} [r^2 + n^2(\ell_2 + \ell_3)^2]^{-3} << 1$, when $(\ell_2 + \ell_3)/\ell_1 << 1$. For infinite array of needles with capacitors (switches) (**Fig. 3-15-e**) at $r = |n|\ell_1$ $r = |n|\ell_\#$ $(n = \pm 1, \pm 2, \ldots)$ we obtain $(\Delta C_S / C_S)_\Sigma \le 3\pi 2^{-6} r_S^4 \ell_2 \ell_1^{-5} \zeta(5) << 1$, where $\zeta(5) \approx 1.04$–Dzeta-function of argument 5. So we can state that, $\overline{C}_s \approx C_s$ and estimation (3-15) $|\Gamma_0|$ is sufficiently accurate. Thus, neglecting interaction between needles, we assume that $\overline{C} = 4\pi\varepsilon_0 \varepsilon r_S^2 \ell_3^{-1} / 4$, where $\varepsilon \sim 15$ is the relative permittivity of the semiconductor material of a disk-switch placed between the ends of two adjacent needles, ℓ_3 is the thickness of the disk, and r_S is the radius of the disk.

3.3.8.4. Another Version of Radio-Transparent Metal-Air Structures

Above we investigated the effect of reflection reduction *via* splitting of continuous wire grid in cross direction, and electric currents induced by incident wave. Now we will illustrate this effect by two examples of semiinfinite metal-air structures (from theory of lens-like VHF antennas).

(a) The transparency of structure, formed by metal strips. The exact general solution for reflection coefficient of this structure was obtained in [87]. Infinite metal sheet (of thickness d) is cut into a periodical sequence (with period $\ell_2 + \ell_3$) of strips of width ℓ_2 and gap ℓ_3 between strips. These sheets are cut which are parallel to each other, and they are spaced as semi-infinite sequence with longitudinal period ℓ_0 (**Fig. 3-15-f**). Incident wave has linear polarization, and electric vector is perpendicular for the slots between strips on each sheet. We assume that metal strips have zero electric resistance. In the case $d << \ell_3 << \ell_2 << \ell_0 << \lambda$ of our interest, we obtain from [87] the following modulus of reflection coefficient:

$$|\Gamma_{st}| \cong \left(\frac{\ell_2+\ell_3}{\pi\ell_0}\right)\ln\left[\frac{2(\ell_2+\ell_3)}{\pi\lambda}\right] + 8\left\{\left(\frac{\ell_2+\ell_3}{\lambda}\right)\ln\left[\frac{2(\ell_2+\ell_3)}{\pi\ell_3}\right]\right\}^2 << 1. \qquad (3\text{–}21)$$

For instance, if $(\ell_2 + \ell_3)/\ell_3 = 10.0$, $(\ell_2 + \ell_3)/\ell_0 = 0.1$ and $(\ell_2 + \ell_3)/\lambda = 0.05$, we obtain $|\Gamma_{st}| \cong 0.127$. So we can state, that above structure is transparent for VHF waves and practically opaque visually. The ratio "(metal)/(air+metal)" in volume for this structure (or an artificial media) is $\nu \approx \ell_2 \ell_0^{-1} (\ell_2 + \ell_3)^{-1} d << 1$. Electronic control of refraction coefficient of metal strip media was described in [88]. There the lens like antenna (with electronic beam scanning) was represented by the system of parallel metal strips of a finite length. The strips, spaced on same line, were loaded consecutively by VHF diodes. The voltage, applied to a line of strips, unites several strips as one long strip. If the voltage on diodes is switched off, several strips of one line were acting as group of isolated metal strips. This operation allows to control refraction coefficient of antenna and its focusing.

(b) The half-space, filled with metal needles. As one example of transparent metal-air structures, we consider the classic case of artificial media: the half-space, filled casually, with a lot of metal needles with average concentration n_ε of needles per unit of volume (**Fig. 3-15-g**). Needles have the identical length ℓ_2 (with identical diameters $d << \ell_2$) and are oriented along electric field of incident wave. We assume, that the above metal-air structures have a small average volume concentration $n_\varepsilon \approx 4\ell_2^{-2}\ell_0^{-1}$ and ratio $v \approx 4\pi d^2 \ell_2^{-1}\ell_0^{-1}$ of volumes (metal)/(air+metal). This causes artificial relative dielectric constant $\varepsilon \approx 1 + 2\pi n_\varepsilon \ell_3^2 d$ and modulus of reflection coefficient $\sim 4\pi\ell_0^{-1}\ell_2^{-2}\ell_3^2 d$ of half-space. However, this estimate does not include the interaction between the adjacent ends of needles. This interaction was included in estimate (3-21) and for metal strips system, described above.

3.3.8.5. Switching of the Reflection Coefficient of the Grid

To provide fast state switching, let us connect the adjacent ends of conductors by electronic switches. The typical temporal scale of electric resistance $R(t)$ is the minimum temporal scale of the cyclical wave bolt operation. The switch resistance is a single parameter controlled in the binary mode: in the open state, $R(t) = \overline{\rho}$ for $t \in \hat{\overline{T}}_m$ and, in the closed state, $R(t) = \rho$ for $t \in \hat{T}_m$ (**Fig. 3-14-b**). It is assumed that $\rho >> \overline{\rho}$. Ideally, $\rho = \infty$ and $\overline{\rho} = 0$. In the intervals $t \in \hat{\overline{T}}_m$, the system of metal segments is brought together by electronic switch currents and almost completely reflects the incident wave. In the intervals $t \in \hat{T}_m$, when switches are closed, the system disintegrates and becomes transparent for the incident wave. Time τ_s of change $\rho \Leftrightarrow \overline{\rho}$ is the smallest temporal scale of the microstructure considered.

3.3.9. Oscillation Attenuation in Virtual Resonators

During time intervals $\hat{\overline{T}}_m$, each pair of neighboring wire grids (with open switches) spaced by distance ℓ_0 forms a virtual resonator with the initial conditions determined by the current field distribution at the instant when the resonator is formed. This resonator is similar to a cavity in metal and does not have the zero eigenfrequency. Resonator oscillations at frequencies $n\pi c / \ell_0$ ($n = 1, 2, ...$) are attenuated according to the dissipative and filtering mechanisms. Their contribution $|\Gamma_1|$ to reflection coefficient $|\Gamma_B|$ of the cyclic wave gate at the incident-wave frequency is due to incomplete decay of resonator oscillations during time interval $\overline{T}$.

3.3.9.1. The Dissipative Attenuation Mechanism

The dissipative mechanism of attenuation involves conversion of the energy of virtual-resonator oscillation into heat released in open electronic switches with finite resistance $\overline{\rho}$ (opened switch) and in metal needles with length ℓ_2 and resistance ρ_0 during "opaque" temporal intervals $\hat{\overline{T}}_m$ of CWBs cycle (see section 3.3.6.). We consider weak absorption of linearly polarized planar incident wave (with density $|\mathbf{E}|^2 (2\rho_{em})^{-1}$ of power stream) by infinite planar grid of metallic needles, parallel to the wave's front (**Fig. 3-15-e**, and **Fig. 3-16-f**). The chains "needle-switch" are assumed parallel with the electric field $\mathbf{E}$ of incident wave. We will assume also that active resistance ρ_0 of needle is much lesser than the resistance $\overline{\rho}$ of electronic switch (opened), and voltage on $\overline{\rho}$ is much lesser than "voltage" (**Fig. 3-16-a**) in the incident wave on spatial period of array, *i.e.* . The $(\ell_2\ell_3)^{-1}$ switches are spaced per unit of array's area. Each switch dissipates power $\overline{\rho}|J_0|^2 / 2$. Dissipation causes the absorption coefficient (in power) $A^2 \approx \rho_{em}^{-1}\overline{\rho}\ell_3\ell_2^{-1} << 1$ (see (3-10)). Thus, when $\overline{\rho} >> \rho_0$, component $|\Gamma_{1(1)}|$ of the reflection coefficient at the incident-wave frequency that is due to incomplete decay of virtual-resonator oscillations during time interval $\overline{T}$ is determined by the relationship

$$|\Gamma_{1(1)}| = \exp[-(\overline{\rho} / \rho_{em})(\ell_1 / \ell_2)(c\overline{T} / 2\ell_0)] << 1. \qquad \textbf{(3-22)}$$

We must ensure the condition (3-22), taking $(\overline{\rho} / \rho_{em})(\ell_1 / \ell_2) << 1$, to suppress the necessary factor $(c\overline{T} / 2\ell_0) >> 1$ (multiple run in VR).

3.3.9.2. The Filtering Mechanism of Attenuation

The filtering mechanism is based on the transparency of a perfectly conducting metal grid with period ℓ_1 at the frequencies $\omega << 1/\gamma$ and its opacity at the frequencies $\omega >> 1/\gamma$ ($\gamma = (\ell_1/\pi c)\ln(\ell_1/\pi d)$, see section 3.3.7.1). The grid is characterized by reflection coefficient (3-19). This grid can be regarded as a mask in a frequency band characterized by the transmission coefficient $|\bar{\Gamma}_0|^2 = 1-|\Gamma_0|^2 = (\omega\gamma)^2[1+(\omega\gamma)^2]^{-1}$, when $c^{-1}\ell_1\ln(\ell_1/\pi r_c) << L_0/c$ and it is assumed that the resonator's loss (like $A^2 << 1$ above) is determined by the HF transparency of the resonator walls, and we obtain

$$|\Gamma_{1(2)}| = \exp[-2^{-3}\pi^{-1}\ell_1\ell_0^{-2}c\bar{T}\ln(\ell_1/2d\pi)] << 1. \qquad \textbf{(3-23)}$$

Multiple occurrences of wave trains in a virtual resonator are connected with multiple crossing of wire grids acting as a HF filter and cannot give significant contribution $|\Gamma_{1(2)}|$ to the reflection coefficient at the frequency of the incident wave. **Fig. 3-17-a** shows the conversion of any initial wave distribution with nonzero spatial mean value in "central" VR (numbered by n) into a sequence of distributions in adjacent VRs: to the left VRs, numbered by $n-1$, $n-2$ ..., and to the right VRs, numbered by $n+1$, $n+2$ ...) with added oscillation and zero spatial mean value. This means, that the wave, crossing the grid-like wall of VR, cannot cause the contribution to $|\Gamma_B|$ on the low frequency of incident wave.

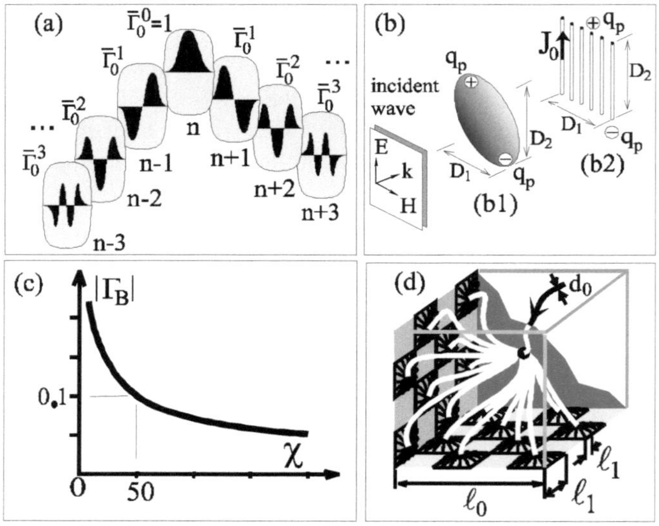

Figure 3-17: Conversion of space-time shape of wave pulses (a), leaving the n-th VR on the both sides and passing grid like walls with transparency coefficients $\bar{\Gamma}$ at the opaque intervals $\bar{\bar{T}}_m$. Accumulation of charges (b) at the poles of metallic body (b1) and wire array (b2). Modulus of reflection coefficient $|\Gamma_B|$ as function of χ (c). Cable structure (d) of cubic virtual resonator.

3.3.9.3. The Resonance Mechanism of Absorption

The fact that a virtual resonator has distinct resonance frequencies can be employed to intensify attenuation. For this let us place a resonance-absorbing optimally loaded electric dipole of a small dimension, $\ell << L_0$, at a point where the fundamental mode field reaches its maximum (in order to eliminate the electrostatic interaction with the walls of the virtual resonators see section 2.3.3.). This situation can be attributed to a particular case of the dissipative mechanism of attenuation. As an oscillatory circuit, the electric dipole (a system of such dipoles is transparent for an LF incident wave) can be characterized by radiation capacitance, inductance, and resistance. This resonance mechanism of absorption can ensure almost complete decay of the virtual-resonator field during the time interval approximately equal to $\sim L_0/\ell >> 1$ periods of the virtual-resonator fundamental mode.

3.3.10. Transient Processes

When electronic switches are simultaneously closed or opened, charges on the metal elements of the microstructure are redistributed. Electronic switches cannot perform work on the field. The energy of transient processes in the system is a part of the incident wave energy (**Fig. 3-17-b**).

Switch closing is accompanied by charge and current oscillations at very high eigenfrequencies $\sim (\pi c / \ell_2) \gg 2\pi / \tau$ of needles. These oscillations experience radiation relaxation during a time interval that is much shorter than T. The charge that has been accumulated by the instant of switch closing is retained. The initial energy $\sim \overline{L} J_0^2 / 2$ of these oscillations is concentrated in the inductance $\overline{L}$ of the needle that carries current J_0 (see section 3.3.8.1.) before the switches are closed. These oscillations cause radiation due to which the energy is fully relaxed during the time $\sim (3/2)\pi^{-2} c^{-1} \ell_2 [\ln(2\ell_2 / d) - 1] \ll T$.

Simultaneous switch opening (the transition $\hat{T}_m \to \hat{\overline{T}}_m$, see section 3.3.6.) is equivalent to instantaneous formation of a solid metal grid or metal plane. The magnetic field on the boundary does not have discontinuities in time or along the normal to the boundary. A fragment of a plane incident wave observed between two metal planes is multiply reflected by them and produces a nonzero time average circulation of the pulsating current (an analog of the magnetic current) on a contour. The contour plane is parallel to the electric field vector and the incident wave vector. The time-average circulation of the induced current corresponds to the incident-wave field observed at the instant of switch opening. The amplitude of this current (caused by initial charge $\sim \ell_2 E \overline{C}$) exponentially decreases in time as resonator oscillations decay. Switch opening causes discharging of switch capacitances $\overline{C}$, which slightly increases ($\Delta J_0 / J_0 \sim \varepsilon \pi r_S^2 \ell_2 (c \overline{T} \ell_3 \ell_0)^{-1} \ll 1$ with $\varepsilon \sim 15$) the average grid current J_0 during interval $\overline{T}$. The field of a plane linearly polarized incident wave induces currents on the exterior surface of the structure (as in the problem of diffraction from a metal object) and causes separation and a certain accumulation of charges at the edges (poles) of the microstructure that limit it in the direction of the incident electric-field vector. The electrostatic energy of pole charges is retained when switches are rapidly opened and can be converted into the energy of the scattered field at the low frequency of the incident wave during the subsequent temporal period of the microstructure. The time of pole charge increase cannot exceed the characteristic time interval $\tau / 2$, during which the incident electric field retains its sign. We accept, that the total current of wire array $J_\Sigma \sim E \rho_{em}^{-1} D_1$ accumulates polar charges $q_P \sim \overline{\tau} J_\Sigma$, where $\overline{\tau} \sim \overline{T} T_B^{-1} \tau / 2$ is effective time of accumulation. Assume that the energy $W_P \sim (4\pi \varepsilon_0 D_2)^{-1} q_P^2$ of pole charges is completely converted into the energy of the scattered wave at the low frequency of the incident wave and the rate of pole-charge accumulation is independent of the charge field, i.e. $\lambda D_2^{-2} D_1 \overline{T} T_B^{-1} / 8\pi \ll 1$. The contribution of pole charges to the reflection coefficient can then be roughly majorized as $|\Gamma_4| \leq (W_P / W_0)^{1/2}$ or

$$|\Gamma_4| \leq (4\pi)^{-1/2} \lambda^{1/2} D_1^{1/2} D_2^{-1} T_B^{-1} \overline{T} \ll 1, \tag{3-24}$$

where D_1 and D_2 are the transverse and longitudinal (relative to the incident electric-field vector, **Fig. 3-17-b**) external measurements of the microstructure, respectively, $W_0 \sim \tau D_1 D_2 E^2 (4\rho_{em})^{-1}$ is the incident wave energy, injected *via* cross section $D_1 D_2$ during the time $\tau / 2$. $|\Gamma_4|$ can be sufficiently small, if $\lambda^{1/2} D_1^{1/2} D_2^{-1} \ll 1$ and (or) $T_B^{-1} \overline{T} \ll 1$.

3.3.11. The Reflection Coefficient of the Cyclical Wave Bolt

Let us sum up estimates (3-15)-(3-18) and (3-20), (3-22)-(3-24) of the factors substantially contributing to reflection coefficient $|\Gamma_B|$ at $T = L / c$. To simplify the analysis, we assume below that the wave is attenuated during one pass along a virtual resonator only slightly. Taking into account only contributions $|\Gamma_1|$, $|\Gamma_2|$, $|\Gamma_3|$, and $|\Gamma_4|$ to $|\Gamma_B|$ in the case when $\overline{T} T_B^{-1} = \ln(\chi) / \chi \ll 1$, we find that the reflection coefficient of the gate reaches its minimum $\hat{\Gamma} = \min |\Gamma_B| = (1 + \ln \chi) / \chi \ll 1$ (**see Fig. 3-17-c**). For the dissipative mechanism of attenuation, $\chi = v\alpha \gg 1$,

$\nu = 2^{-1}\bar{\rho}\rho_{em}^{-1}\ell_1 L_0^{-1} N >> 1$, where χ and ν are the parameters characterizing the degree of the structure miniaturization. For the filtering mechanism of attenuation, $\nu = (8\pi)^{-1}\ell_1 L_0^{-1} N \ln(\ell_1 / \pi r_c) >> 1$ where $\alpha = [1+(4\pi)^{-1/2}\lambda^{1/2} D_1^{1/2} D_2^{-1}]^{-1}$ (both for dissipative and filtering mechanisms of attenuation). Regardless of the specific attenuation mechanism, the quantity $\hat{\Gamma} = 0.1$ can be reached at χ that is no less than the value $\chi = 50$ (**Fig. 3-17-c**).

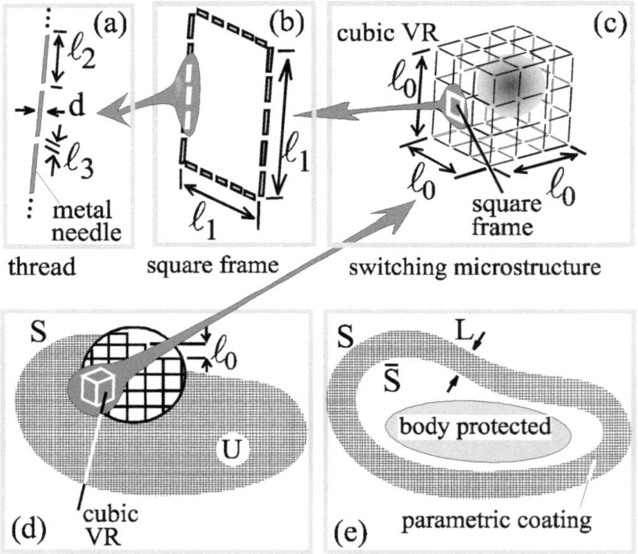

Figure 3–18: Geometrical hierarchy of metal air microstructure of absorbing coating.

As parameter χ grows, the wave bolt's microstructure becomes more miniature and exhibits a higher performance speed. The range $\lambda_1 < \lambda < \lambda_2$ of incident wavelengths for waves with permissible low coefficient $|\Gamma_B|$ is limited from above by component $|\Gamma_4|$, which increases as $\sim \lambda^{1/2}$. This range is limited from below by reflections that occur in the intervals $t \in \hat{\bar{T}}_m$ and grow as $\sim \lambda^{-1}$. Note that quantity $|\Gamma_B|$ increases at the ends of the operating frequency band more smoothly as the values $\bar{T}T_B^{-1}$, $c\tau_c / \ell_0$, ℓ_3 / ℓ_2, ℓ_2 / ℓ_1, ℓ_1 / ℓ_0 and ℓ_0 / λ decrease. Applying a control signal (**Fig. 3-14-b**) to each electronic switch of a reasonably designed system (**Fig. 3-17-d**) requires a small relative volume ($\sim [(3/4)\pi d_0^2 \ell_0^{-1} + 3\pi d_0^2 \ell_2^{-1} \ell_0^{-1}] << 1$) of a control fiberoptic cable with diameter $\sim d_0$. When switches with gate options and thin fiber-optic control cables (having equal lengths but containing folds for simultaneous reaction of spaced switches) are employed, scattering of the field from the cables can be neglected. At present, there is no problem in producing a laser-pulse duration of $\sim 10^{-12} s$ and a performance speed of $150 Gbit/s$ for switches and microelectronic photodetectors (*i.e.*, devices transforming the light intensity to a logical level) [45]. Microelectronic elements transforming the average level of the optical fiber illumination to a constant feed voltage can be applied to galvanically isolate microelectronic switches.

3.3.12. A 3D Microstructure

A 3D microstructure (**Fig. 3-18-d**) consists of 3D virtual resonators (for example, of a cubic shape, **Fig. 3-18-c**) and occupies spatial region U bounded by closed surface S. Needles of diameter d and length $\ell_2 >> d$ and electronic switches placed in gaps of the width $\ell_3 << \ell_2$ form controlled-conductance frames (**Fig. 3–18–b**) and threads (**Fig. 3-18-a**), the threads form a plane grid (verge of the cubic resonator) with square cells (**Fig. 3-18-b**) of the width $\ell_1 >> \ell_2$, the grids form a cube (the cube is inwardly empty) with the side $\ell_0 >> \ell_1$ (**Figs. 3-17-d** and **3-18-c**), and the cubes or virtual resonators densely fill region U (**Fig. 3-18-d**). When the electric-field vector is arbitrarily oriented with respect to wires, the small reflection coefficient of virtual-resonator walls may change by a factor $\sqrt{2}$ of within transparent-state time intervals $\hat{T}_m$; the small wave transmission coefficient may simultaneously change by a factor of within opaque-state time intervals $\hat{\hat{T}}_m$. The ratio "(metal)/(air+metal)" in volume for this structure is $\nu \approx 6d^2\ell_1^{-1}\ell_0^{-1} << 1$. The spatial-temporal parameters of the microstructure are characterized by the following hierarchy

$$\tau >> T_B >> \overline{T} >> \ell_0 / c >> \ell_1 / c >> \ell_2 / c >> \ell_3 / c >> \tau_c \,. \qquad \textbf{(3-25)}$$

Now, let us make a cavity with surface in the interior of such that surface that is separated by a microstructure layer of thickness no less than L (**Fig. 3-18-e**). Assume that reflection from a 1D layer of thickness L can be neglected. Then, if a body with arbitrary scattering characteristics is placed in this cavity, the field observed in the exterior of S will not change. Thus, we obtain an absorbing parametric coating whose efficiency is independent of the direction of the incident wave propagation. Anisotropy arises because interaction between coating elements (manifested, for example, as the phase difference of the waves reflected from the interior and exterior layers of an interference coating) depends on the direction of an incident wave. However, the elements of the proposed coating (virtual resonators) practically do not interact: during the time intervals $t \in \hat{\hat{T}}_m$, when the resonator field is nonzero, they are isolated from each other by walls; in the intervals $t \in \hat{T}_m$, when they are open, the resonator field has already almost completely decayed. When elements do not interact, the dependence of such interaction on the incident-wave direction is of no importance.

3.3.13. Scattering of an Incident Wave by the Microstructure

Let us express the electric and magnetic fields in free space in terms of vector potential $\zeta = \{\zeta_1, \zeta_2, \zeta_3\}$ (or $\zeta = a_1\zeta_1 + a_2\zeta_2 + a_3\zeta_3$) according to the wellknown relationships $\mathbf{E} = -\zeta'_t$, $\mathbf{H} = curl(\zeta)$. Cartesian components $\zeta_i = \zeta_i(\mathbf{r},t) = \zeta_i(x_1,x_2,x_3,t)$ (oriented along the x_j axes with unit vectors $\mathbf{a}_j$, $j = 1,2,3$) of the vector potential satisfy the scalar wave equation $(\zeta_i)''_{tt}c^{-2} = (\zeta_i)''_{x_1x_1} + (\zeta_i)''_{x_2x_2} + (\zeta_i)''_{x_3x_3}$. The effect of the orientation of metal needles in closed grids on the field polarization is insignificant inside virtual resonators, since the field decays during time interval $\overline{T}$. Outside surface S, polarization effects are also insignificant, since the reflected field is of the relative order $|\Gamma_B| << 1$. At instant $t > t_0$, field $\zeta_i(\mathbf{r},t)$ that propagates in free space and satisfies the wave equation is determined by values $\zeta_i(\mathbf{r}_0,t_0)$ and $[\zeta_i(\mathbf{r}_0,t_0)]'_t$, at instant t_0 on the surface $|\mathbf{r}-\mathbf{r}_0| = c(t-t_0)$: $\zeta_i(\mathbf{r},t) = \hat{P}[t_0,t;\Phi_0^i,\Psi_0^i]$, where $\hat{P}[t_0,t;\Phi_0^i,\Psi_0^i] = (\partial/\partial t)\hat{F}[\Phi_0^i] + \hat{F}[\Psi_0^i]$, $\Phi_0^i = \zeta_i(\mathbf{r},t_0)$, $\Psi_0^i = [\zeta_i(\mathbf{r},t_0)]'_t$, where operator $\hat{F}[\psi]$ acts on function $\psi(\mathbf{r}_0)$ according to the Poisson formula [82], [83] $\hat{F}[\psi] = \{4\pi c^2(t-t_0)\}^{-1} \iint_{|\mathbf{r}-\mathbf{r}_0|=c(t-t_0)} \psi(\mathbf{r}_0)dS(\mathbf{r}_0)$. However, the same result for the field at t expressed through its values at t_0 can be obtained *via* N^* computational steps that are similar to the technique described above but involves temporal points $t_n = nT^*$, where $T^* = (t-t_0)/N^*$, $n = 0,1,2,...,N^*$.

On the basis of initial conditions $\Phi_0^i = \zeta_i(\mathbf{r},t_0)$, $\Psi_0^i = [\zeta(\mathbf{r},t_0)]'_t$, the first step is made ($t \in [t_0,t_1]$): the current field value $\zeta_i(\mathbf{r},t) = \hat{P}[t_0,t;\Phi_0^i,\Psi_0^i]$ becomes the initial conditions $\Phi_1^i = \zeta_i(\mathbf{r},t_1)$ and $\Psi_1^i = [\zeta_i(\mathbf{r},t_1)]'_t$ for the second step. At the n-th step ($t \in [t_{n-1},t_n]$), the current field value $\zeta_i(\mathbf{r},t) = \hat{P}[t_{n-1},t;\Phi_{n-1}^i,\Psi_{n-1}^i]$ becomes the initial conditions $\Phi_{n-1}^i = \zeta_i(\mathbf{r},t_{n-1})$ and $\Psi_{n-1}^i = [\zeta(\mathbf{r},t_{n-1})]'_t$, for the n-th step. At the (N^*-1)-th step ($t \in [t_{N^*-2},t_{N^*-1}]$), the current field value $\zeta_i(\mathbf{r},t) = \hat{P}[t_{N^*-2},t;\Phi_{N^*-2}^i,\Psi_{N^*-2}^i]$ becomes the initial conditions $\Phi_{N^*-1}^i = \zeta_i(\mathbf{r},t_{N^*-1})$ and $\Psi_{N^*-1}^i = [\zeta_i(\mathbf{r},t_{N^*-1})]'_t$, for the N^*-th step. At the N^*-th step ($t \in [t_{N^*-1},t_{N^*}]$), $\zeta_i(\mathbf{r},t) = \hat{P}[t_{N^*-1},t;\Phi_{N^*-1}^i,\Psi_{N^*-1}^i]$. Now, let us take into account that, during relatively long intervals $\hat{T}_m$, waves propagate in free space and, during relatively short intervals $\hat{\hat{T}}_m$, waves are scattered by the walls of the microstructure. The effect of opaque-state interval $\hat{\hat{T}}_m$ can be approximately reduced to instantaneous correction of the initial conditions for subsequent transparent-state interval by introducing spatial operator $\hat{Z}$ (see below). Operator $\hat{Z}$ acts at the node instants $t_n = nT_B$ (that is, $T^* = nT_B$), which coincide with the instants at which the microstructure walls are switched from the transparent to opaque state.

On the basis of the initial conditions $\Phi_0^i = \zeta_i(\mathbf{r}, t_0)$, $\Psi_0^i = [\zeta(\mathbf{r}, t_0)]_t'$, the first step ($t \in [t_0, t_1]$) is made (as in [65]). At this step, the current field value $\zeta_i(\mathbf{r}, t) = \hat{P}[t_0, t; \Phi_0^i, \Psi_0^i]$ becomes the initial conditions $\Phi_1^i = \hat{Z}\zeta_i(\mathbf{r}, t_1)$ and $\Psi_1^i = \hat{Z}[\zeta_i(\mathbf{r}, t_1)]_t'$, for the second step. At the n-th step ($t \in [t_{n-1}, t_n]$), the current field value $\zeta_i(\mathbf{r}, t) = \hat{P}[t_{n-1}, t; \Phi_{n-1}^i, \Psi_{n-1}^i]$ becomes the initial conditions $\Phi_{n-1}^i = \hat{Z}\zeta_i(\mathbf{r}, t_{n-1})$ and $\Psi_{n-1}^i = \hat{Z}[\zeta(\mathbf{r}, t_{n-1})]_t'$, for the n-th step. At the (N^*-1)-th step ($t \in [t_{N^*-2}, t_{N^*-1}]$), the current field value $\zeta_i(\mathbf{r}, t) = \hat{P}[t_{N^*-2}, t; \Phi_{N^*-2}^i, \Psi_{N^*-2}^i]$ becomes the initial conditions $\Phi_{N^*-1}^i = \hat{Z}\zeta_i(\mathbf{r}, t_{N^*-1})$ and $\Psi_{N^*-1}^i = \hat{Z}[\zeta_i(\mathbf{r}, t_{N^*-1})]_t'$, for the N^*-th step. At the N^*-th step, ($t \in [t_{N^*-1}, t_{N^*}]$), $\zeta_i(\mathbf{r}, t) = \hat{P}[t_{N^*-1}, t; \Phi_{N^*-1}^i, \Psi_{N^*-1}^i]$. Operator $\hat{Z}$ is defined as follows: at the instants $t = t_n = nT_B$ ($n = 1, 2, 3, ..., N^*$), when virtual resonators are closed, potential ζ_i in the exterior of S is retained that is, and the value $\zeta_i(\mathbf{r}, t_n + 0) = \hat{Z}\zeta_i(\mathbf{r}, t_n) = [\psi(\mathbf{r})]_{x_i}'$ is immediately imparted to potential $\zeta_i(\mathbf{r}, t_n)$ in the interior of S, where $\psi(\mathbf{r})$ is a known solution to the interior Neumann problem for the Laplace equation $\psi_{x_1x_1}'' + \psi_{x_2x_2} + \psi_{x_3x_3} = 0$ satisfying the boundary condition (when $\mathbf{r} \in S$) $[\psi(\mathbf{r})]_{\mathbf{n}}' = (\mathbf{n}, \mathbf{a}_1)\zeta_1(\mathbf{r}, t^*) + (\mathbf{n}, \mathbf{a}_2)\zeta_2(\mathbf{r}, t^*) + (\mathbf{n}, \mathbf{a}_3)\zeta_3(\mathbf{r}, t^*)$. This definition of operator $\hat{Z}$ combines the continuity of the vector potential along normal $\mathbf{n}$ to S and spatially stepwise zeroing of fields $\mathbf{E}$ and $\mathbf{H}$ inside S at the instants $t_n = nT_B$ (to be more exact, during the temporal intervals $nT_B \le t < nT_B + \bar{T}$). The computational procedure described above [65] corresponds to solution of the problem of diffraction from "black body" U (**Fig. 3-18-d**). This procedure is an alternative to known versions [77-81] and does not necessitate imposing any constant impedance conditions on blackbody surface S. The action of above parametric microstructure cannot be described by some combination of electric circuits with parameters, unchanged in time. This is why our version of black body does not need to formulate special boundary conditions on the surface S^* of black body. And this surface S^* quickly fluctuates between S and $\bar{S}$ with temporal period $T_B \ll \tau$.

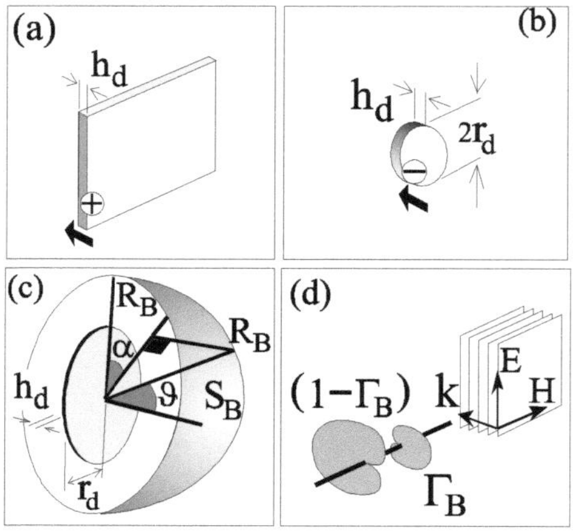

Figure 3-19: The geometry of the problem of diffraction on the black disk: Huygens plane source (a); Huygens disk source (b); coordinates of the diffraction problem (c); the directivity patterns of radiation of two equivalent Huygens disks in the presence of planar incident wave (d).

3.3.14. Scattering by a "Black" Disk

Backscattering of a plane wave from a blackbody is characterized by the coefficient Γ_{back} (see (3-1), (3-8)). Of course Γ_{back} is connected with Γ_B functionally. The 3D problem of scattering of a normally incident plane wave from a circular ideally black disk (with $|\Gamma_B| = 0$ in the 1D problem) allows an extremely simple one-step solution ($N* = 1$ see above Section 3.3.13). The disk is of radius r_d and thickness $h_d = cT_B = L << r_d, \lambda$. Due to coinciding of incident wave front with plane of disk, we can describe the CWB's action as following (**Fig. 3-14-b**):

(a) periodical impulsive regeneration of Huygens disk (**Fig. 3-19-b**, as part of Huygens plane source, **Fig. 3-19-a**) with period T_B and with amplitude directivity pattern $-(1-\Gamma_B)\mathbf{G}_{em}(\vartheta - \pi)$ of electric field (average on temporal interval $\tau >> T_B$, **Fig. 3-19-d**), which describes forward-scattering field mostly and have sign, inverted in comparison with the incident wave (**Fig. 3-19-d**);

(b) periodical impulsive regeneration of Huygens disk with period and with amplitude directivity pattern $\Gamma_B \mathbf{G}_{em}(\vartheta)$ of electric field (average on temporal interval $\tau >> T_B$, **Fig. 3-19-d**), which describes mostly back-scattering and is caused by finite reflection coefficient $\Gamma_B \neq 0$ in one-dimensional problem. Both Huygens disks have linear polarization of incident wave. The locally planar (quasi one-dimensional) character of the problem is very useful for estimations. During each pulse (pulse of initial conditions) of Huygens disks the wave field has not time to leave almost fully the disk's area during the interval $T_B = h_d / c$. The waves caused by disk's boundaries can fill very small relative volume $\sim 2h_d / r_d << 1$ during time T_B. So the next pulse of Huygens disk radiation is beginning in almost clear disk's area without fields of previous (at zero initial conditions). As another appropriate circumstance we note that all points of disk surface are touching the front of incident wave simultaneously. The analogous approach can be applied to the cases when thin black structure of arbitrary dimensions and shape coincides with the front (not planar) of incident wave. For electromagnetic waves (instead (3-8), see Section 3.1.10) we obtain

$$\Gamma_{back} \approx \int_0^{2\pi} d\alpha \int_0^{\pi/2} d\vartheta \left|\Gamma_B \mathbf{G}_{em}(\vartheta) - (1-\Gamma_B)\mathbf{G}_{em}(\vartheta - \pi)\right|^2 \sin\vartheta \quad , \tag{3–26}$$

where $\mathbf{G}_{em} = \boldsymbol{\alpha}_0 G_\alpha + \vartheta_0 G_\vartheta$, $G_\alpha(\vartheta,\alpha) = f(\vartheta)\sin(\alpha)$, $G_\vartheta(\vartheta,\alpha) = f(\vartheta)\cos(\alpha)$ (**Fig. 3-19-c**), $\boldsymbol{\alpha}_0$, ϑ_0 –mutually orthogonal vectors ($|\boldsymbol{\alpha}_0| = |\vartheta_0| = 1$) of spherical coordinates ϑ α on half-sphere S_B of radius R in far zone at $R >> r_d^2/\lambda$, $0 < \alpha \le 2\pi$ –azimuth angle, $0 < \vartheta \le \pi$ –polar angle, $f(\vartheta) \approx (2\pi)^{-1/2}\{I_1(kr_d \sin(\vartheta))/(kr_d \sin(\vartheta))\}\cos^2(\vartheta/2)$, where $k = \omega/c$, I_1 –Bessel function of 1-st order). Note that, when $kr >> 1$, the absorbing structure effectively reduces backscattering ($\Gamma_{back} << 1$) of the disk but does not reduce forward scattering (the shadow behind the disk) with the corresponding coefficient $\Gamma_{forw} = 1 - \Gamma_{back}$. Scattering of an incident wave by a black disk differs from scattering by a dielectric or metal disk of the same size. Thus, the total scattering cross section (CS) S_Σ, backscattering CS S_{back}, forward scattering CS S_{forw} of a black disk are satisfying the equation $S_\Sigma = S_{back} + S_{forw}$ always and $S_\Sigma = S_{forw} = S_g$, when $kr >> 1$, where S_g –the geometric cross section of the disk. At the same time, the total CS of a metal disk $S_\Sigma = (2\pi/3) r_d^4 \lambda^{-2}$ is much less than its geometric cross section ($S_\Sigma << S_g$) when $kr << 1$ and forward scattering and backscattering coefficients are determined by the formula $\Gamma_{back} = \Gamma_{forw} = S_\Sigma S_g^{-1}/2 = r_d^2 \lambda^{-2}/3 << 1$. In the case of ideal parametric black disk with $\Gamma_B = 0$ we obtain the coefficient $\Gamma_{back} = 1/8$ for $r_d k << 1$ and $\Gamma_{back} \approx (r_d k)^{-4}/2 << 1$ for $r_d k >> 1$, where $k = 2\pi/\lambda$.

CHAPTER 4

Approach of Transparent Body: Active Suppression of Radiation and Acoustical Scattering Fields, Produced by Some Physical Bodies in Liquids

Abstract: An algorithm for the suppression of the radiation and scattering fields created by vibration of the smooth closed surface of a body of arbitrary shape placed in a liquid is designed and analytically explored. The frequency range of the suppression allows for both large and small wave sizes on the protected surface. An active control system is designed that consists of: (a) a subsystem for fast formation of a desired distribution of normal oscillatory velocities or displacements (on the basis of pulsed Huygens' sources, Section 4.6) and (b) a subsystem for catching and targeting of incident waves on the basis of a grid (one layer) of monopole microphones, surrounding the surface to be protected (Section 4.7). The efficiency and stability of the control algorithm are considered. The algorithm forms the control signal during a time much smaller than the minimum time scale of the waves to be damped. The control algorithm includes logical and nonlinear operations, thus excluding interpretation of the control system as a traditional combination of linear electric circuits, where all parameters are constant (in time). This algorithm converts some physical bodies placed in a liquid into one that is transparent to a special class of incident waves. The active control system needs accurate information on its geometry, but does not need either prior or current information about the vibroacoustical characteristics of the protected surface, which in practical cases represent a vast amount of data. Joint suppression of radiating and scattering field by a coating of controlled thickness is considered. The problems of suppressing the sound field generated by a vibrating body in a liquid are considered in another representation too. For solving these problems, an acoustically thin active coating with a real-time thickness control is proposed (Section 4.8). The coating should be placed directly on the surface of the body to be protected. Solutions to the problems of suppressing the radiation and scattering of sound by a body are obtained in the general form based on linear operators, which characterize (i) the sound radiation by a vibrating surface, (ii) the scattering of incident waves by a fixed surface, and (iii) the vibroelastic properties of the body in an acoustic vacuum. Conditions ensuring the stability of the active system are formulated. Forming of directions of zero scattered by minimum tools is considered too.

4.1. RULES OF CHAIN SUBSTITUTION

As the basis of the suggested approach, we took the way of fast tuning of normal velocity, its matching with incident waves: if the normal velocity of the external surface of a coating is identical with the projection of particle velocities (normal to the body surface), caused by an incident wave in uniform media (*i.e.* in the absence of scattering bodies), then the scattered wave is absent. In this way, we solve the problem of radiation suppression in the absence of incident waves: normal velocity of surface must be zero (and the radiation zero). The approach, which is suggested below, can be explained by the simple example of changing of elements in oscillatory electric circuits, without changing currents (velocities) and voltages (forces) in electric or mechanical circuits. The acoustical analog of inserted (changed) elements is a transparent area in an infinite space, filled by the same homogeneous compressible liquid, but this area is planned to be filled by a protected body, which we must make transparent too, so the sources of external waves "would not" feel any difference in boundary problem. In the language of electrical oscillatory chains, this operation (called “substitution”) means such a change of some chain element, which cannot be detected by the sources, so the sources of the external waves must not "feel" any additional influences, fulfilled by change of boundary problem. Of course we are not interested in the substitution, which is trivial and copies all the characteristics of changed elements. We are interested in the cases when "invisibility" of such change may be achieved with minimum tools (**Fig. 4-1**): (a) satisfying the same (as before the substitution) electric voltage (force, pressure) in the break point of the active device; (b) satisfying the same (as before substitution) electric current (velocity) in the active device at the break point; (c) satisfying the same ratio between electric current (velocity) and electric voltage (force) (*i.e.* impedance as before substitution) (see [89]).

4.2. CHOICE OF VELOCITY OR DISPLACEMENT CONTROL (NONEQUIVALENCE OF VELOCITY AND PRESSURE CONTROL)

In the problem, which was presented above, the substitution area does not radiate or scatter waves, so for electric and mechanical chains all three ways are equivalent. Let us note also that in the absence of tangential (viscosity) stresses in the external fluid matching of both normal velocities and matching of acoustical pressure is equivalent.

Vladimir V. Arabadzhi

This is due to the inevitable mutual connection between pressure and normal velocity on the protected surface S_V. Mathematically, this means that the existence of an operator (surface integral), due to which any distribution of normal velocities (on S_V) creates only one possible distribution of acoustical pressure on this surface S_V. On the other hand, we always can find the operator (surface integral), due to which any distribution of acoustical pressure (on S_V) creates the only possible contribution of pressure on this surface S_V. Thus we cannot create pressure and normal velocity fields independently. In the opposite case, after tuning the velocity is fulfilled, we will try to tune the pressure and inevitably must break the tuned velocity, achieved above (and analogously in the inverse direction). Below we have chosen the tuning of normal velocities, because the kinematic value - normal displacement of the controlled boundary has one very important characteristic: it remains (saved) after the end of short pressure pulse (blow), which created this displacement. Now we will describe the one-dimensional boundary problem, where a short blow (uniformly distributed along the plane border) induces normal displacement of the plane border of the elastic half-space (without any dissipation in material). Normal displacement, caused by blow, will remain constant ("frozen") during an arbitrary unlimited time (**Fig. 4-2-a**). This effect can be interpreted as some plasticity of the plane boundary, separating the elastic half-space. On the other hand, a pressure pulse, which was created by the same blow, departs from the blow point with the speed c_w of sound propagation, so this boundary cannot be saved in the memory pressure pulse (caused by the same blow) after the termination of blow. The reason of this difference in behavior of wave fields (pressure and displacement) can be found on a micro level too, for example − in the acoustical monopole field: acoustical pressure $p = p_0 r^{-1} \exp(-ikr + i\omega t)$, created by a monopole is proportional only to $1/r$ (which describes radiation field), but the radial velocity field $\upsilon = (p_0 / \rho_w c_w)[r^{-1} + (ik)^{-1} r^{-2}]\exp(-ikr + i\omega t)$ of the monopole contains the term $1/r^2$ (which describes the non-radiation field) besides the radiation term $1/r$ (r - denotes the distance from monopole center to the watch point, ρ_w −mass density of media, c_w −speed of sound in compressible media, $k = 2\pi/\lambda$ −wave number, λ -wavelength). To illustrate the difference between pressure control and velocity (displacement) a simple scenario is presented in **Fig. 4-2-a**, and **Fig. 4-2-b**. There the series of fast deformations ("instant photos") of a half-infinite elastic rod with square cross-section (**Fig. 4-2-a**) and a half-infinite space filled with an elastic compressible medium (**Fig. 4-2-b**) at the moments $t = 0 < t_1 < t_2 < t_3 < ...$ after a blow by the same "hammer" (with same square shape of cross-section with dimension $D_h \times D_h$ and the same initial mechanical impulse) during the time $\tau_c << D_h / c_w$. So the hammer during the interval τ_c makes the rod shorter by the value $\tau_c << D_h / c_w$ (**Fig. 4-2-a**) and makes an imprint of cross-section $D_h \times D_h$ and depth $A_h = J\rho_w^{-1/2} E_w^{-1/2} D_h^{-2}$ in the border of half infinite elastic space (**Fig. 4-2-b**). We suppose that all the residual space in **Figure 3** (besides the elastic rod and elastic half-space) is filled with vacuum. Above we denoted: J −mechanical impulse of hammer's blow, ρ_w −mass density, E_w −Young's modulus (**Fig. 4-2-a**) of the rod material and analogously for compressibility of a half space media (**Fig. 4-2-b**), c_w −speed of longitudinal sound propagation in the rod material and the elastic half-space. Lifetime of rod's shortening is $\tau_L = \infty$ (**Fig. 4-2-a**). On the other hand the lifetime of the imprint in the plane border of compressible half-space is $\tau_L \approx D_h / c_w$. We note that:

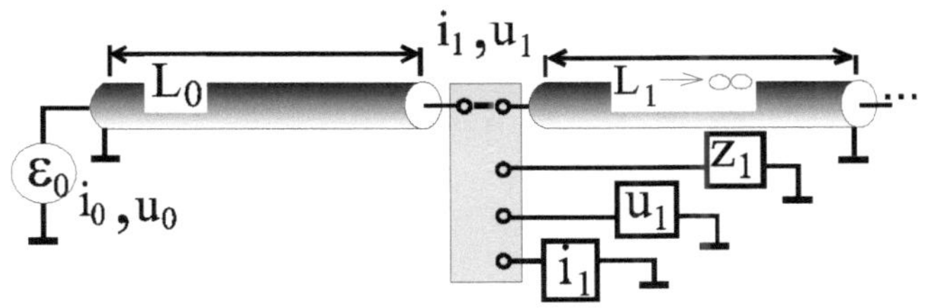

Figure 4-1: The half-infinite line of wave transmission (broken at the length L_0); i_0, u_0 -current and voltage (complex magnitudes) on the left end is induced by source ε_0 before cutting. i_1, u_1 −current and voltage (complex magnitudes) in the broken point before cutting. There are three equivalent boundary conditions (or substitutions) on the right hand end of the piece of transmission line (of length L_0), when the source ε_0 "does not feel" the difference between them and infinite ($L_1 \to \infty$) line of transmission: active source of current i_1, active source of voltage u_1, passive impedance $z_1 = u_1 / i_1$.

$$\tau_L >> \tau_c \qquad (4\text{-}1)$$

In **Fig. 4-2-c**, the evolution of the following scenario is presented: the plane boundary of the infinite compressible half space is excited by a uniform bipolar (step-like) blow of pressure with a sign change in the point $x = 0$. During pressure blow time $(0, \tau_c)$ this pulse creates a step-like bipolar deformation (imprint) $u(x, 0, \tau_c)$ of the boundary $z = 0$. After the blow the deformation becomes more smooth along with propagation (with velocity $c \leq c_w$) in all tangential directions and will have a smaller step magnitude of (up to zero at). We note that at time interval $(0, \tau_c)$ this problem (as in **Fig. 4-2-b**) is fully one-dimensional (under condition (4-1)), but at times $t > \tau_c$ the boundary problem becomes significantly two-dimensional (see Section 4.6.2.5. below).

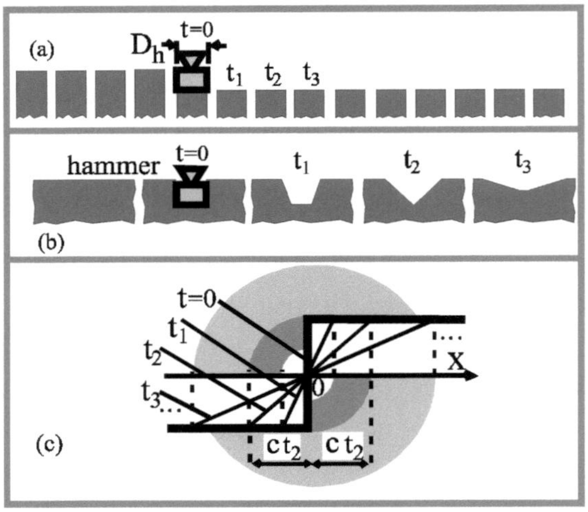

Figure 4-2: On the nonequivalence of control of velocity and control of pressure: "plasticity" in the boundary problems without dissipation.

4.3. FEATURES OF THE STATEMENT OF THE PROBLEM

Below we search for the solution in the form of an active system with the following qualitative characteristics:

a) simultaneous suppression of radiation and scattering fields of the protected body;

b) wide frequency range of suppression;

c) minimum prior and current information on the incident waves and radiation waves;

d) minimum prior and current information on the vibroacoustical characteristics of protected body;

e) neutral floatability of protected body in liquid; and

f) quick response of active system, caused by changes of the incident wave field.

The protected body is of neutral floatability and can be both much larger and much smaller than the length of wave to be suppressed. Parameters of the protected body are unknown, but its shape is smooth, closed and convex (we assume the absence of empty cavities, which are characterized by significant ringingness).

The suggested system consists of:

(I) Sub-system of rapid formation of the required distribution of normal oscillatory velocities (DNOV, or displacements, *i.e.* DNOD) on the outside surface of a two-layer piezoelectric acoustically thin coating. This coating is mounted immediately on the protected body (hull-mounted or, more correctly speaking, on the dissipative layer (see Section 4.1), supporting the absorption of high frequency sound, appearing at the active control process;

(II) Sub-system of microphones for capturing and targeting incident waves, spaced near the protected body.

(1) The main functional element of the active coating is presented by an active piston. This represents two-layer plane segments [22-27], [90] of a piezocomposite material. Longitudinal impedance and speed of propagation of longitudinal sound are the same for an active coating and the external fluid respectively [26], [27], [91], [92]. Such active pistons almost fully cover the protected surface. Pistons are spaced immediately on the protected surface, but they do not interact with it acoustically. Each active piston represents the source of unidirectional (outside) radiation or, more precisely, an acoustically transparent source of Huygens. We need the radiation to be unidirectional because the radiation helps us to exclude interactions between the protected surface and active pistons; this helps us solve together the problems of suppression scattering and radiation, due to the same directions (from outside the protected surface) of radiated and scattered waves. Such a piston creates short bipolar pulses of radiation. These pulses quickly form the desired smooth (in the sense of averaged during a period of pulses) normal displacement (velocity) of the external boundary of active pistons. The active control system creates pulses of active pistons of required magnitudes and they are intersecting each other at their borders (with duration equal to 1/3 of the pulse period in sequence), and their temporal period of pulses presents 2/3 of their duration (see Section 5.3.). By the way, the value of the central part of each pulse in the sequence remains zero and has duration equal to 1/3 of the period of pulses in sequence. The Huygens's piston almost does not interact with the protected body in the frequency range of active suppression and very quickly creates the normal displacements ("imprints") of required depth and direction (dent or knoll) in the external fluid. Due to the shortness of any local one-dimensional control action (pulses, blows, see (4-1)) we need not know a priori information about the matrix of mutual impedances of active pistons which constitute the active coating (at the synthesis of required DNOD). The pulse-like control actions (relatively rare and short pulses of force) allow us the temporal separation of the process of forming control pulses simultaneously for all the surface of the protected body. Each one of these active piston pulses has finite duration and their interactions in principle could lead to instability of the system. This effect will be estimated below in Section 5.5.

(2) The solution suggested includes a one-layer microphone grid. To target, for instance, only one incident wave from the predetermined sector of directions, requires only four microphones. This becomes possible due to the following simplifying expectations concerning the characteristics of the fields of incident waves:

(a) Incident waves present a low number of groups of plane waves with a sufficiently clear wave front. Besides this, we assume the incident waves are much more powerful than the diffusive isotropic noise sound field. In the suggested system, due to the special statement of the problem with initial conditions, the direction of incident wave is determined (measured) in real time before the first contact of the wave front with the protected body. In practice we have the small amount of plane waves, which can be targeted by the suggested system. The Malyuzhinets method needs a two-layer and dense microphone array and also requires control connections between each sensor (microphone) and each actuator (in our terminology–active piston).

(b) Using logical and threshold-like operations (*i.e.* nonlinear operations) with the microphones' signals, the active control system forms the group of bearing for each powerful incident wave plane.

Then each group of bearings becomes an independent vector-microphone. The maximum of the directivity pattern of this vector-microphone has the opposite direction with respect to the vector of propagation of its incident wave. Therefore, the parameters of the bearing groups depend on the incident waves. This operation supports the stability of the whole control system. Unlike the suggested method the Malyuzhinets's solution can be reduced to a combination of linear circuits. Parameters of these circuits do not depend of magnitude and structure of the incident waves.

4.4. THE CONTROL OF NORMAL VELOCITIES (DISPLACEMENTS) OF THE BODY'S SURFACE

Here we consider the problem of quick synthesis of a desired distribution of normal oscillatory velocities (displacements) on the outside surface of an active coating during the time, much smaller, than the characteristic temporal scale of the distribution to be formed. In the absence of incident waves the desired DNOD is zero, and in such a case the active coating solves the radiation suppression problem. In the presence of incident waves the active coating (in addition to suppression of radiation) can (see below) suppress the scattering field of the protected body. In the presence of incident waves the active coating suppresses scattered waves jointly with radiation. This is due to the formation on the external surface of the coating of the normal oscillatory velocities distribution, matched for the incident wave, on the base of signals of some group of microphones (bearing group, see Section 4.7.2). The active coating is placed immediately on the surface of the protected body. Below we point to several important characteristics and assumptions in the statement of this problem.

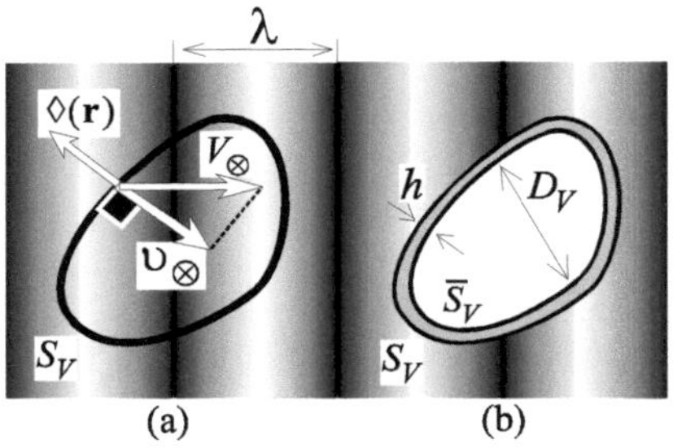

Figure 4-3: Matching of distribution of normal oscillatory velocities (DNOV) to obtain a "transparent" body (λ –wavelength).

1* Let us assume that in the absence of physical scattering surface S_V (outside surface of active coating, which protects a body with a characteristic linear dimension D_V and minimum curvature radius $R_V > 0$ (**Fig. 4-3**) the particle velocities $V_\otimes$ of a non-viscous compressible fluid have the projection $\upsilon_\otimes = (\Diamond, V_\otimes)$ on the vector $\Diamond = \Diamond(\mathbf{r})$ (which is normal to the surface S_V and with normalized length $|\Diamond| = 1$, $\mathbf{r} \in S_V$). Now, if we created (by means of the active coating) the DNOV [92]:

$$\upsilon = \upsilon_\otimes , \tag{4-2}$$

we would have suppressed the scattering field completely. By the way, the problem of radiation suppression would be solved together with the problem of scattering suppression: without incident waves the normal oscillatory velocity of external surface S_V of the coating must be zero (when nonzero velocity is possible on the internal surface of active coating). The condition (4-2) must be satisfied in the finite wide operating range of temporal ω and spatial frequencies:

$$|\omega| \in [\omega_{\min}, \omega_{\max}], \ |\chi| \in [\chi_{\min}, \chi_{\max}], \tag{4-3}$$

i.e. for a finite area of characteristic temporal and spatial scales:

$$\tau \in [\tau_{\min}, \tau_{\max}]), \ \lambda \in [\lambda_{\min}, \lambda_{\max}], \tag{4-4}$$

where $\omega_{\min} > 0$, $\chi_{\min} = \omega_{\min} / c_w$, $\chi_{\max} = \omega_{\max} / c_w$, where c_w -sound speed in fluid. Wave dimensions of the protected body with a characteristic space scale D_V can be both

$$\chi_{\min} D_V << 1 \text{ and } \chi_{\max} D_V >> 1 . \tag{4-5}$$

The minimum frequency $\omega_{\min}$ must be sufficiently large: this does not permit the displacement magnitude to overcome the border $A_{\max}$ of the linear characteristics of sensors and actuators [92], *i.e.*:

$$(\omega_{\min} T_V)^{-1} A_\otimes < A_{\max} << h , \tag{4-6}$$

where $A_{\otimes}$ –desired normal displacement of outside surface of coating, h –its thickness. At $\omega_{\min} > 0$ the relation analogous to (4-2) will be satisfied for normal displacements too.

2* The formation of the current DNOV value $\upsilon_{\otimes}(\mathbf{r},t)$ at any moment must be fulfilled by some control actions during the short previous time interval of duration $T_V \ll \tau_{\min}$, much shorter than the minimum temporal scale $\tau_{\min}$ of the wave to be suppressed.

3* The thickness h of the active coating (or distance between surfaces S_V and $\overline{S}_V$, **Fig. 4-4**) is much smaller than the length of wave, which must be suppressed, *i.e.* $h \ll \lambda$.

4* We assume that the protected body has neutral floatability in the outside liquid. The vibroacoustical characteristics of the protected body depend on the temperature and hydrostatic pressure of the outside liquid. It is clear that the control system we need should not include any acoustical interactions with the body, so we will assume below that very little information about the vibroacoustical characteristics of the protected body is available (*i.e.* it is practically unknown).

5* Literally following the condition 2*, one can get the principal contradiction: formation of a smooth trajectory (normal temporal displacement of the outside coating boundary S_V) by means very short pieces. It would seem that this condition contradicts the traditional control by slow complex magnitudes of the fields, because its spectral power is concentrated close to the current frequency (or, in other words, when both phases and magnitudes of the fields can be determined separately). We remove this contradiction by using very wide spectrum signals. In other words we will form a sequence of short pulses with a smooth average trajectory (inside the range (4-3), which we need). Further, we will control the normal displacement $u(\mathbf{r},t)$ of the coating outside boundary S_V (instead the normal velocity $\upsilon(\mathbf{r},t)$) by short periodical blows of duration τ_c and with period $T_V \gg \tau_c$ (see the description of the algorithm, see (4-26) or (5-31) below). The spectral power density of desirable displacement $u_{\otimes}(\mathbf{r},t)$ (being averaged over one or several periods T_V) is concentrated in the range (4-3), (4-4). Fast boundary displacements, induced by blow force pulses within short time intervals $t \in [nT_V, nT_V + \tau_c]$, we will alternate with relatively slow displacements during long time intervals $t \in [nT_V - \tau_c, (n+1)T_V]$, of "relaxation", *i.e.* "relaxation-blow-relaxation-blow-...", etc, ($n = 1,2,\ldots$) "mintage" like operations. Outside the frequency range (4-3) we suppose an arbitrary far field and near field of waves of a finite power admissible. As a result of the above mentioned control, the average on period (*i.e.* inside range (4-3)) normal oscillatory velocity $\overline{\upsilon}(r,t) = T_V^{-1} \int_{t-T_V}^{t} \upsilon(r,\xi)d\xi$ must be close to the desirable value $\left|\overline{\upsilon}(\mathbf{r},t) - \upsilon_{\otimes}(\mathbf{r},t)\right|_{r\in S_V} \to 0$ (or $\left|u(\mathbf{r},t) - u_{\otimes}(\mathbf{r},t)\right|_{r\in S_V} \to 0$), as described before), due to the relation:

$$\overline{\upsilon}(r,t) \cong \{u(\mathbf{r},t) - u(\mathbf{r},t-T_V)\} / T_V . \qquad \textbf{(4-7)}$$

4.5. PRIOR INFORMATION ABOUT THE CONSTRUCTION OF THE PROTECTED BODY

For the considerations that follow we, however, must make several simple general assumptions to limit the possible vibroacoustical characteristics of the protected body. Several famous essential works, for example [93], are devoted to an interaction between acoustical waves and mechanical structures. Below we will limit our consideration by the simplest rough models.

4.5.1. The Construction of the Protected Body

We assume that the structure of the protected body near its surface represents a closed elastic shell (locally this shell may be analogously the elastic uniform plate) of thickness a with internal surface $S_{\bullet}$, covered outside by polymer layer of thickness h_r (**Fig. 4-4**). This layer is a dissipative absorber of the waves with frequencies $\omega \geq \pi / T_V$. The outside surface of the dissipative layer represents the outside surface of the protected body. Many different points within the shell can be connected by various dynamical chains. Possible characteristics of these chains are limited by the following: (a) assumption about low ringingness of the shell (with chains) in liquid; (b) assumption about the spatial and temporal smoothness of the vibration fields of the protected body. Thus, the protected body presents a closed smooth elastic shell, embraced inside by discrete passive dynamic connections (chains), and covered outside by a polymer layer, which is dissipatively opaque for high frequency waves at $\omega \geq \pi / T_V$.

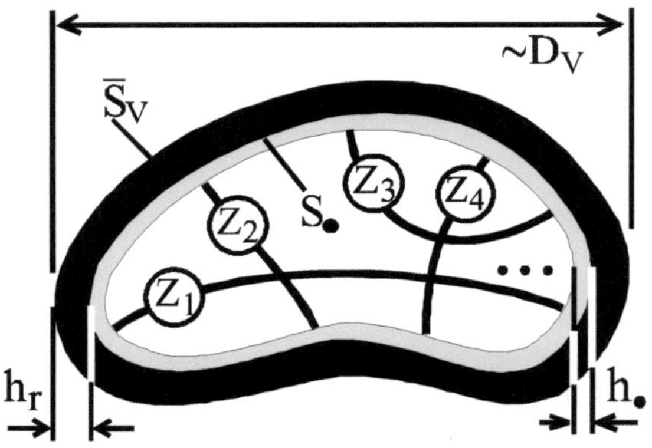

Figure 4-4: The assumed structure of protected body with internal discrete connections in the form of mechanical passive chains with impedances $Z_1, Z_2, Z_3, \ldots$

4.5.2. Ringingness of the Protected Body

Let's assume that a weightless layer (with a surface coating $\overline{S}_V$) with arbitrary distribution (independent on external pressure) of the velocity $\upsilon_0 = (d/dt)h(\mathbf{r},t)$ of change of its thickness $h(\mathbf{r},t)$ (or layer of monopole sources) with Fourier image $\tilde{\upsilon}_0(\mathbf{r},\omega)$ is limited in the module *i.e.* $|\tilde{\upsilon}_0(\mathbf{r},\omega)| \leq q_0$. This distribution $\tilde{\upsilon}_0(\mathbf{r},\omega)$ of monopole sources induces the corresponding distribution $\tilde{\upsilon}_1(\mathbf{r},\omega)$ (or $\tilde{\upsilon}_1[\tilde{\upsilon}_0]$) of normal velocities of the surface $\overline{S}_V$. After this we call the maximum ringingness Q_{LF} on low frequencies the maximum ratio $|\tilde{\upsilon}_1(\mathbf{r},\omega)| / q_0$ which was selected using all possible variations of distributions $\tilde{\upsilon}_0(\mathbf{r},\omega)$ (and $\tilde{\upsilon}_1[\tilde{\upsilon}_0]$) in all points $\mathbf{r}$ of a surface $\overline{S}_V$ and on all frequencies of range (4-3), and analogously we introduce the high frequency ringingness *i.e.*: $Q_{LF} = q_L / q_0$ (where $q_L = \max|\tilde{\upsilon}_1(\mathbf{r},\omega)|$ among any $\tilde{\upsilon}_0(\mathbf{r},\omega)$ (and $\tilde{\upsilon}_1[\tilde{\upsilon}_0]$), $\mathbf{r} \in \overline{S}_V$, $|\omega| \in [\omega_{\min}, \omega_{\max}]$) and $Q_{HF} = q_H / q_0$ (where $q_H = \max|\tilde{\upsilon}_1(\mathbf{r},\omega)|$ among any $\tilde{\upsilon}_0(\mathbf{r},\omega)$ (and $\tilde{\upsilon}_1[\tilde{\upsilon}_0]$), $\mathbf{r} \in \overline{S}_V$, $|\omega| \geq \pi / T_V$). Of course we are interested in having the values Q_{LF} and Q_{HF} as small as possible.

4.5.3. Smoothness of Vibrational Fields on the Protected Body

This assumption about smoothness is caused by the finite elasticity to bending of a body's parts (or elastic plate) with Young's modulus $E_\bullet$, Poisson's modulus $\sigma_\bullet$ and mass's density $\rho_\bullet$. Shell or elastic plates act as filter for low spatial frequencies and for suppression of high spatial frequencies. Let us consider the simplest problem about excitation of bending oscillations in an infinite plate (spatial filter) with thickness $h_\bullet$, induced by the application of normal pressure $P = P_0 \exp(i\omega t - i\chi x)$, where χ –spatial frequency, P_0 -amplitude of pressure. The ratio $\eta_\bullet = A / A_0$ we will call the transmission coefficient of plate, where A -amplitude of bending oscillations, which was caused by pressure P , and A_0 -amplitude of fluid particle displacements in a sound wave in the fluid with pressure P , mass density ρ_w and speed c_w of sound. We assume that plate and fluid give the following ratio $\eta_\bullet = A / A_0 << 1$ on the spatial frequencies $|\chi| \geq \pi / L_V$, where $|\chi| \geq \pi / L_V$ –characteristic linear scale of the active coating splitting (see Section 4.6.2 below), on all temporal frequencies ω . If the condition

$$L_V^4 \omega_{\max}^2 h_\bullet^{-2} << (2\pi)^4 12^{-1} (1 - \sigma_\bullet^2)^{-1} E_\bullet \rho_\bullet^{-1} \quad \textbf{(4-8)}$$

is satisfied, one can use the simple rough estimate:

$$\eta_\bullet \le 12(1-\sigma_\bullet^2)(2\pi)^{-4}\rho_w E_\bullet^{-1} h_\bullet^{-3} L_V^5 Q_{\max}\omega_{\max}^2 << 1, \quad \textbf{(4-9)}$$

where $Q_{\max} = \max\{Q_{LF}, Q_{HF}\}$. Inserting typical practical parameters, *i.e.* $\sigma_\bullet \approx 0.3$, $E_\bullet \approx 2\times 10^{11}$, $\rho_\bullet \approx 7\times 10^3$, $\rho_w \approx 10^3$, we obtain instead of (4-8), (4-9):

$$L_V^4 \omega_{\max}^2 h_\bullet^{-2} << 6.2\times 10^9 \text{ (in SI units)} \quad \textbf{(4-10)}$$

and

$$\eta_\bullet \le 10^{-11} \times h_\bullet^{-3} L_V^5 Q_{\max}\omega_{\max}^2 << 1 \quad \textbf{(4-11)}$$

Note, that the estimate (4-11) can only be of any practical interest at $h_\bullet << L_V, h_r$.

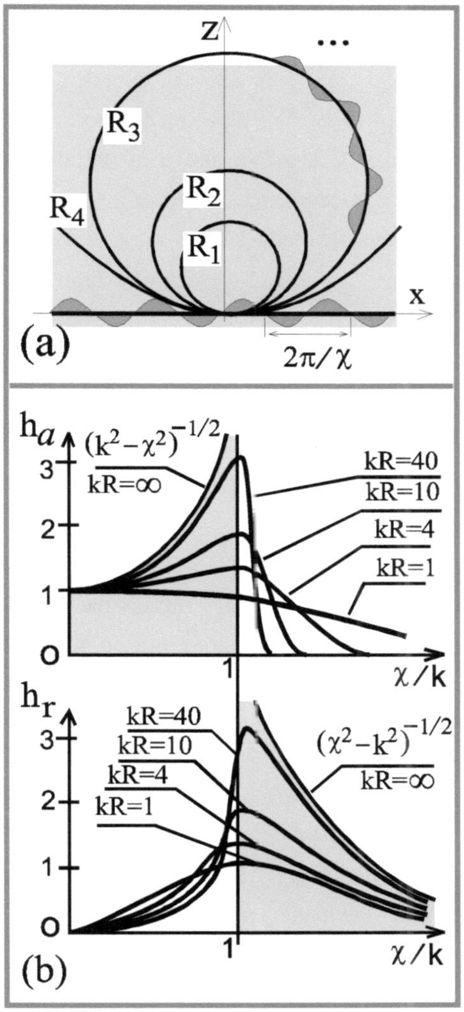

Figure 4–5: The evolution ($R = R_1 > R_2 > R_3 > R_4$) of cylindrical boundary problem to the plane boundary problem (a). The relative active h_a and reactive h_r powers of sound fields, which were created by boundary conditions on cylindrical ($kR < \infty$) and plane ($kR = \infty$, gray area) surfaces S (b).

4.5.4. Constancy of Parameters of the Protected Body

Within the bounds of the temporal interval $\bar{\tau} >> D_V / c_w$ we will assume constant parameters of the protected body. The solution suggested does not require constancy of parameters for a long time.

4.5.5. Layer of Dissipative Polymer

This polymer layer of thickness h_r and longitudinal impedance Z_r is spaced immediately on the external surface $S_\bullet$ of an elastic plate (shell). We will assume that Z_r satisfies the condition $|Z_r - Z_c| / Z_c << 1$ (where Z_c is a longitudinal impedance of the active coating (sections 4.6.2.5 and 5.3) within the frequency range (4-3). We thus assume the surface S_V transparent for the waves of the range above mentioned. Besides this, we assume that on the frequencies $\sim \pi / T_V$ the distance ℓ_r of dissipative relaxation of waves in e-times satisfies conditions $h_r << \lambda_{\min}$ and $h_r >> \ell_r$. It is clear that we can combine these conditions only if the condition $h_r >> c_r T_V$ is satisfied, where $h_r >> c_r T_V$ –speed of sound in the polymer. This polymer layer thus gives a significant reduction of high frequency ringingness $Q_{HF} < 1$ to the protected body.

4.6. ACTIVE COATING AND ITS CONTROL ALGORITHM

In this section, we consider the solution of the problem of forming of DNOV (DNOD), having minimum info about the vibroacoustical characteristics of the protected body. We place on the body external surface $S_\bullet$ immediately the high frequency dissipative polymer layer of thickness h_r and the active coating of thickness $2h_c << \lambda_{\min}$ in the form of two piezoelectric layers of thickness h_c (each layer), with identical polarization along the normal to S_V (the external surface of piezoelectric is contacting with outside compressible media). We assume that the longitudinal impedance $Z_c = \rho_c c_c$ of the piezoelectric material is equal to the impedance $Z_w = \rho_w c_w$ of the outside liquid (*i.e.* $|Z_w - Z_c| / Z_c << 1$), where $Z_c = \rho_c c_c$ –mass density of piezoelectric material and $Z_w = \rho_w c_w$ -mass density of outside liquid and correspondingly, c_c –speed of longitudinal sound in the piezoelectric material, c_w –speed of sound in the outside liquid. The total cross sectional structure of coating is presented in **Fig. 4-10-c**. For simplicity, we will assume also $c_c > c_w$. The control algorithm, stability of active system and procedure of measurement are investigated below.

4.6.1. The Simplified Models of "Near Field – Far Field" Conversion

For the further estimations of the efficiency of the DNOV synthesis (and for the more clear understanding of the general problem of Chapter 4) below we consider two simple examples of boundary problem, when spectral characteristics $\bar{\upsilon}(\boldsymbol{\chi}, \omega)$ of DNOV $\upsilon(\mathbf{r}, t)$ on the surface determines the initial radiation power W_0 (without active coating) and residual radiation power W_1 (when active coating is switched on). Here we assume, that $\mathbf{r}$ and $\boldsymbol{\chi}$ are the coordinate on S and vector $\boldsymbol{\chi}$ (scalar χ in our considerations below) of spatial frequency on S respectively.

4.6.1.1. Plane Two-Dimensional Boundary Problem

We consider the simplest two-dimensional (uniform along axis "y") boundary problem of sound radiation into half-space $z > 0$, produced by one-dimensional DNOV $\upsilon(x,t)$ on the plane S or $z = 0$. One must solve this problem for the velocity potential $\phi(x,z,t)$, which satisfies the wave equation $\phi''_{xx} + \phi''_{zz} = c^{-2}\phi''_{tt}$ (where c -sound speed in liquid half-space $z > 0$) and boundary condition $\phi'_z(x,0,t) = \upsilon(x,t)$ on the plane $z = 0$ in the absence of incident wave from $z = \infty$. Further, we obtain the spatial-temporal Fourier amplitude spectrum $\bar{\bar{\phi}}(\chi, z, \omega) = \int_{-\infty}^{+\infty} dx \int_{-\infty}^{+\infty} dt \; \exp(+i\chi x - i\omega t) \; \phi(x,z,t)$ of the wave potential $\phi(x,z,t)$:

$$\bar{\bar{\phi}}(\chi, z, \omega) = \left\{ Y(+\xi)\exp\left(-iz\sqrt{|\xi|}\right) + Y(-\xi)\exp\left(-z\sqrt{|\xi|}\right) \right\} \; \bar{\bar{\upsilon}}(\chi, \omega)$$

,

where, $\xi = k^2 - \chi^2$, $Y(\xi) = I(\xi)/\sqrt{\xi}$, $I = 1$ if $\xi > 0$, $I = 0$ if $\xi \le 0$, $k = \omega / c$, $\overline{\overline{\upsilon}}(\chi, \omega) = \int_{-\infty}^{+\infty} dx \int_{-\infty}^{+\infty} dt \, \exp(+i\chi x - i\omega t) \, \upsilon(x,t)$. Taking into account the expression $p(x,z,t) = \rho\phi'_t(x,z,t)$ (where ρ – mass density of liquid) of sound pressure p *via* wave potential ϕ, in is easy to get real h_a (active) and image h_r (reactive) parts of the energetic conversion coefficient of DNOV (on plane $z = 0$) into sound. This coefficient presents the ratio between sound power stream density (real and image) $W(\chi, \omega) = \overline{\overline{p}}^*(\chi, 0, \omega)\overline{\overline{\upsilon}}(\chi, \omega)/2$ on the plane $z = 0$, produced by DNOV $\upsilon = \exp(i\chi x - i\omega t)$, and real sound power stream density (when $\chi = 0$) $\mathrm{Re}\{W(0, \omega)\}$:

$$h_a(\chi, \omega) = \mathrm{Re}\{W(\chi, \omega)\} / \mathrm{Re}\{W(0, \omega)\} = Y(k^2 - \chi^2),$$

$$h_r(\chi, \omega) = \mathrm{Im}\{W(\chi, \omega)\} / \mathrm{Re}\{W(0, \chi)\} = Y(\chi^2 - k^2). \quad \textbf{(4–12)}$$

We see from (4–12), that $h_a > 0$, $h_r = 0$ if $k^2 - \chi^2 > 0$, and $h_a = 0$, $h_r > 0$ if $k^2 - \chi^2 < 0$. In other words the half-space $z > 0$ acts like a spatial low-pass filter for the radiation field. The value h_a can be interpreted as some coefficient of transferring in the area of spatial frequencies χ. Expressions (4-12) are valid also in general case, when we have DNOV variations along coordinate "y" also In this case we must use $\chi = \sqrt{\chi_x^2 + \chi_y^2}$, where χ_x and χ_y –spatial frequencies of DNOV along coordinates "x" and "y" respectively.

4.6.1.2. Cylindrical Boundary Problem

Now consider the boundary problem of sound radiation by azimuth DNOV $\upsilon(\vartheta, t)$ ($0 \le \vartheta \le 2\pi$) on the infinite cylindrical surface S of radius R with temporal-spatial Fourier spectrum

$$\overline{\overline{\upsilon}}(\chi_m, \omega) = (2\pi)^{-1} \int_0^{2\pi} d\vartheta \int_{-\infty}^{+\infty} dt \, \upsilon(\vartheta, t) \, \exp(-i\omega t - im\vartheta).$$

Unlike the plane case described above, spatial Fourier spectrum of sound field is concentrated on discrete spatial frequencies $\chi = \chi_m = m / R$ ($m = 0, \pm 1, \pm 2, \ldots$). For quantities $h_a(\chi, \omega) = \mathrm{Re}\{W(\chi, \omega)\} / \mathrm{Re}\{W(0, \omega)\}$, $h_r(\chi, \omega) = \mathrm{Im}\{W(\chi, \omega)\} / \mathrm{Re}\{W(0, \chi)\} = Y(\chi^2 - k^2)$ (analogs of (4-12)) and DNOV of the form $\upsilon(\vartheta, t) = \exp(i\omega t - im\vartheta)$ we obtain the following expressions

$$h_a(\chi_m, \omega) = \mathrm{Re}\{g_m(kR)\} / \mathrm{Re}\{g_0(kR)\}, \; h_r(\chi_m, \omega) = \mathrm{Im}\{g_m(kR)\} / \mathrm{Re}\{g_0(kR)\}, \quad \textbf{(4-13)}$$

where $g_m(kR) = H_m^{(2)}(\zeta) / [H_m^{(2)}(\zeta)]'_{\zeta = kR}$; $H_m^{(2)}(kR)$ –Hankel function of second type, m –th order and argument kR; $[H_m^{(2)}(\zeta)]'_{\zeta = kR}$ –derivative of this Hankel function at $\zeta = kR$. **Fig. 4–5–b** gives us plots h_a and h_r as functions of χ / k for several values of parameter $kR = 1, 2, 10, 40$ [79]. Plots h_a and h_r are smooth, but, of course, only points $\chi = \chi_m$ have a physical sense. At $kR \to \infty$ functions h_a and h_r are closing to the analogous functions of plane case. Using **Fig. 4-5-b**, we can state: if the surface S has big wave dimensions, we should not split it (and coating) into a lot of pistons of scale $L_V \sim 2\pi c / \omega_{\max}$ (see below), in order to exclude the growth of radiation of surface S in this case.

4.6.1.3. The Algorithm of the Sound Pressure Compensation

The algorithm of sound pressure compensation on pistons reveals itself in the iterative tuning of the vibrational velocity complex amplitudes of each piston on the basis of the measured complex amplitude of the sound pressure either on the piston of directly near it and "knows nothing" about the form of the radiating surface S. The pistons are connected with each other not organizationally (*via* the control circuits) but only *via* the sound field (typical case of the individual local strategy). The algorithm is local spatially but nonlocal temporally.

4.6.1.3.1. Analytical Description

We consider a closed surface S with the rigidly assigned (on some angle frequency ω) normal oscillating velocities distribution of complex amplitudes $V_S(\mathbf{r})$ on which there are pistons with controlled velocities $\{V_j\}$ in the points $\{\mathbf{r}_j\}$ ($j=1,2,...,N$) equipped with pressure transducers with the complex amplitudes $p(\mathbf{r}_j)$. The total DNOV on the surface S is,

$$V(\mathbf{r}) = V_S(\mathbf{r}) + \upsilon(\mathbf{r}). \tag{4-14}$$

Here, $\upsilon(\mathbf{r}) = \sum_{j=1}^{N} \Pi(\mathbf{r}-\mathbf{r}_j)V_j$ is presented in the form of the sum of the surface S normal velocity $V_S(\mathbf{r})$ (we assume this velocity distribution independent of the velocities V_j of pistons) and the active piston addition $V_j\Pi(\mathbf{r}-\mathbf{r}_j)$, where V_j is piston complex amplitude of normal velocity (velocity of piston thickness change for instance) and $\Pi(\mathbf{r}-\mathbf{r}_j)=1$ at $\left|\mathbf{r}-\mathbf{r}_j\right| \le \varphi_j > 0$, $\Pi(\mathbf{r}-\mathbf{r}_j)=0$ at $\left|\mathbf{r}-\mathbf{r}_j\right| > \varphi_j$ (φ_j –radius j-th damping piston). Total power W radiated by the surface S is presented in the form of integral:

$$W = (1/2)\,\mathrm{Re}\left[\oiint_{S} dS(\mathbf{r}_0) \oiint_{\overline{S}} dS(\mathbf{r})\{V(\mathbf{r})G(\mathbf{r},\mathbf{r}_0)V^*(\mathbf{r})\}\right], \tag{4-15}$$

where, $G(\mathbf{r},\mathbf{r}_0)$ is Greens function of the surface S, $\overline{S}$ is the surface embracing closely the protected surface S together with damping pistons. We assume that the radiated power is a certain functional of the pistons velocities $W=\hat{\Omega}[V]$. The change of piston velocities $\Delta\upsilon(\mathbf{r}) = \sum_{j=1}^{N} \Pi(\mathbf{r}-\mathbf{r}_j)\Delta V_j$, where ΔV_j is the increase of j-th damping piston velocity complex amplitude, causes the corresponding increase ΔW of the radiated power $\Delta W = \hat{\Omega}[V+\Delta\upsilon] - \hat{\Omega}[V]$. We consider that the piston radiuses φ_j ($j=1,2,...,N$) are so small that the sound pressure change on the distance φ_j ($\varphi_j < \lambda_{\min}$) is negligible. For the small increase $\Delta\upsilon$ we obtain ΔW in the following form $\Delta W = (1/2)\,\mathrm{Re}\oiint_{\overline{S}}\left[(\hat{G}(\Delta\upsilon))^*\upsilon + (\hat{G}(\upsilon))^*\Delta\upsilon\right]d\overline{S}$, where $\hat{G}$ is the Green's operator (see 4.8.4.). The algorithm of the sound pressure compensation on pistons $p(\mathbf{r})$:

$$\Delta V_j = \gamma_j p(\mathbf{r}_j) \quad (\gamma_j = const < 0) \tag{4-16}$$

Substituting (4-16) into (4-15) we get the expression:

$$\Delta W = \sum_{j=1}^{N} \gamma_j \left|p(\mathbf{r}_j)\right|^2 < 0, \tag{4-17}$$

from which, taking a certain equal for all j sign γ_j, we can conclude, that any small increase $\Delta\upsilon$ of the piston velocities in the form (4–16) leads to the negative increase ΔW e of radiating power W. When such an operation is repeated many times, the radiating power will decrease monotonously, at each subsequent stage by less than at the previous stage reaching a certain stationary level. One may easily make sure that W is quadratic form of the piston velocities V_j. Therefore the radiating power has the only one minimum at some $\upsilon(\mathbf{r})$ and this is achieved within the limit by the operations (4-15), when the sound pressure on pistons is a set of zero.

The discrete operation of tuning the piston complex amplitudes V_j takes place in time and has the physical sense if it occupies many periods of the sound field. Such a tuning makes the ASPC nonlocal in time. On the other hand, for many $j=1,2,...,N$ ($j=1,2,...,N$) (or for all $j=1,2,...,N$, excerpt $j=1,2,...,N$) the coefficient γ_j can be equal to zero, not changing the tendency to decrease the radiating power, as we see from (4-17).

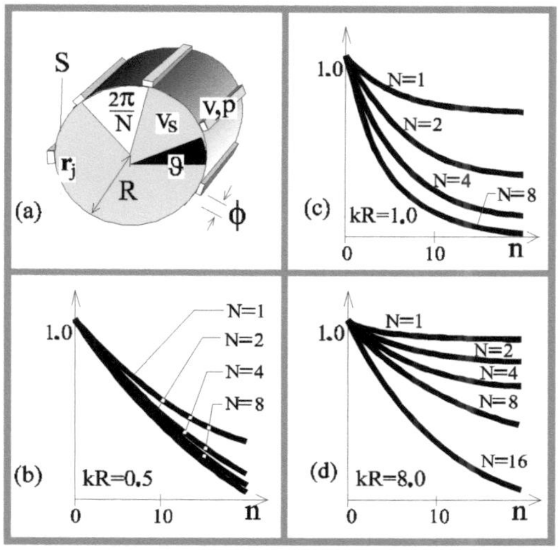

Figure 4-6: Relative decreasing of the radiating power (by cylinder with surface S of radius R, equipped with $N = 1, 2, 4, 8, 16$ equidistant pistons, **Fig. 4-6-a**) after the algorithm of sound pressure compensation switching on (with the tuning coefficient $\gamma = 0.1/(\rho c)$). The cylinder wave dimensions and numbers of initial DNOV modes of $V_S(\vartheta)$ were the following: $kR = 0.5$, $\overline{m} = 0$ (**Fig. 4-6-b**), $kR = 1.0$, $\overline{m} = 0$ (**Fig. 4–6–c**), $kR = 8.0$, $\overline{m} = 6$ (**Fig. 4-6-d**). n is the ordinal number of tuning steps.

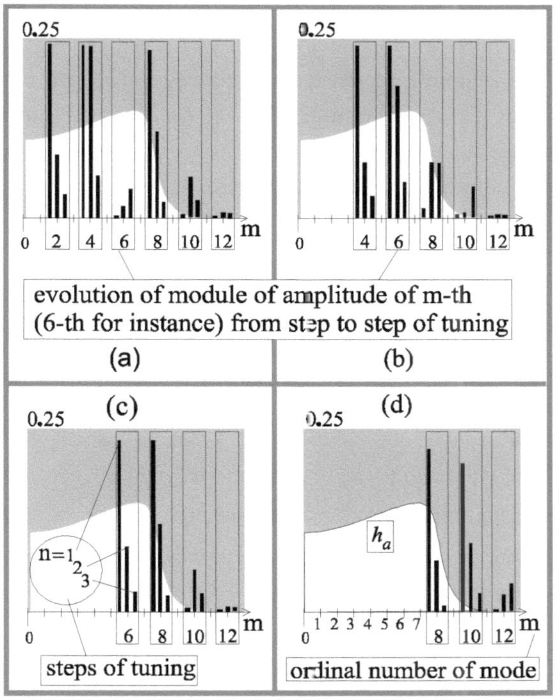

Figure 4-7: Azimuth DNOV spectrum evolution on a cylinder (of wave dimension $kR = 8.0$) after the algorithm of sound pressure compensation (with $N = 16$ damping pistons) switching on. Relative level $|A_m(n)|/|A_{\overline{m}}|$ of azimuth modes $m = 2, 4, 6, 8, 10, 12$ and ordinal numbers of iterations $n = 0, 1, 2, 3$ with tuning coefficient $\gamma = 0.1/(\rho c)$. Initial (*i.e.* at $n = 0$) DNOV azimuth spectrum $A_m(0) = A_m \delta(m - \overline{m})$ $A_m(0) = A_{\overline{m}}\delta(m - \overline{m})$ is presented by "pure" azimuth mode of number $\overline{m} = 2$ (**Fig. 4-7-a**), $\overline{m} = 4$ (**Fig. 4-7-b**), $\overline{m} = 6$ (**Fig. 4-7-c**), $\overline{m} = 8$ (**Fig. 4-7-d**). For all uneven numbers $m = 1, 3, 5, 7, 9, 11, \ldots$ we have $A_m(n) = 0$ (due to equidistant spacing of pistons of even number $N = 16$ damping pistons and even numbers $m = 2, 4, 6, 8,$ of initial modes). h_a is the active part of the transmission coefficient of conversion of vibration field into a sound field (see section 4.1).

Therefore, the piston velocities tuning can be performed both by all pistons simultaneously and by each piston independently. Taking into account the fact that each piston is tuned on the basis of measured values of the vibration velocity and the sound pressure directly on the piston and "knows nothing" about the form of the surface S, we can come to the conclusion that algorithm (4-16) is spatially local. The iterative operation of the piston velocities tuning by (4-16) has the following form:

$$[V_j]_{n+1} = [V_j]_n + \gamma_j [p(\mathbf{r}_j)]_n \ (\gamma_j < 0), \quad \textbf{(4-18)}$$

where, j is the piston number, n is an ordinal number of iteration step. Algorithm (4-17) provides the monotonous convergence to the minimum of radiating power (to zero radiation if we have sufficient number of pistons controlled). Now we want to know, what factors determine one of two possible results of algorithm's (4-18) action: (a) convergence to the zero total radiating power; (b) convergence to the partial decreasing of total radiating power. To get answer we will use numerical simulation below. The above algorithm can be classified as {LS/NT/CA/F} (see **Fig. 1-3**).

4.6.1.3.2. Numerical Check

To verify the algorithm (4−18) we shall confine ourselves to the simplest case of equidistant pistons (strips of width φ and of controlled thickness) location in points $\mathbf{r}_j$ ($j = 1,2,3,..,N$) on the cylindrical surface S of radius R (**Fig. 4-6-a**), *i.e.* the two-dimensional case with DNOV $V_S(\vartheta)$. We shall judge about the result of algorithm (4-18) action by a relative (with respect to the initial one) drop of the radiating power (per unit length of the cylinder). The size of pistons is $k\varphi = 0.01$, the wave size of the cylinder takes values $kR = 0.5$, $kR = 1.0$, $kR = 8.0$ and the number of pistons $N = 1,2,4,8,16$.

Figs. 4-6-b-d give the plots of radiating power decrease as the function of iteration number n at the tuning parameter $\gamma = 0.1/(\rho c)$, zero initial velocities of pistons. The initial DNOV $V_S(\vartheta)$ is presented by azimuth modes of numbers: $\overline{m} = 0$ (**Fig. 4-6-b**); $\overline{m} = 0$ (**Fig. 4-6-c**); $\overline{m} = 6$ with (**Fig. 4-6-d**). In particular, it is seen from **Fig. 4-6-b**, **Fig. 4-6-c**, **Fig. 4-6-d**), that for the efficient suppression of the radiation power the number of damping pistons must satisfy the conditions

$$N > kR \ \textbf{and} \ N > 2\overline{m}. \quad \textbf{(4-19)}$$

Figs. 4-7-a-d give the plots of up conversion of azimuth spectrum of DNOV, caused by the steps of algorithm (4-18). There $V_S(\vartheta)$ is presented by one pure azimuth mode $\overline{m} = 2$ (**Fig. 4-7-a**), $\overline{m} = 4$ (**Fig. 4-7-b**), $\overline{m} = 6$ (**Fig. 4-7-c**), $\overline{m} = 8$ (**Fig. 4-7-d**). The sequence of three iteration steps ($n = 1,2,3$) is presented in frame and describes the evolution of the relative value of m -th mode amplitude ($m = 2,4,6,8,10,12$, **Fig. 4-7-a**; $m = 4,6,8,10,12$, **Fig. 4-7-b**; $m = 6,8,10,12$, **Fig. 4-7-c**; $m = 8,10,12$, **Fig. 4-7-d**). There we assume $\delta(\xi) = 0$ at $\xi \neq 0$, $\delta(\xi) = 1$ at $\xi = 0$. So we see in **Fig. 4-7**, that step by step low modes $|m| < kR$ (with radiation) become smaller and high modes $|m| \sim kR$ (without radiation) appeared to become bigger. For all uneven numbers $m = 1,3,5,7,9,11,...$ we have $A_m(n) = 0$ (due to equidistant spacing of pistons of even number $N = 16$ damping pistons and even numbers $m = 2,4,6,8,$ of initial modes).

4.6.1.4. Algorithm of the Volume Velocity Compensation

Here we will use the above two-dimensional cylindrical model of boundary problem (as in the item 4.6.1.) with strip-like pistons (uniform along " y "). We consider the simplest algorithm of control of DNOV for suppression of radiation of cylindrical surface S (**Fig. 4-8-a**). We assume that surface S has one-dimensional DNOV along " ϑ " (cylindrical boundary value problem, see **Fig. 4-5-a**). Below we will assume also, that initial DNOV $\upsilon(R,\vartheta,t)$ (**Fig. 4-8-d**) on cylindrical surface S does not depend one of any external sound pressure in outside waveguiding media (for example: metal surface vibrating in outside air). Like in **Fig. 4-6-a** on the cylindrical surface S of radius R, we place $N \gg 1$ strip-like pistons (with centers in angular points $\vartheta_j = j\Lambda$, $j = 1,2,...,N$, **Fig. 4-8-c**) of controlled thickness $h_j(t)$ periodically (with angular period $\Lambda = 2\pi / N \ll 1$). Linear width of piston is φ (and angular width is

$\varepsilon = \varphi / R \le (2\pi / N)$). Now we place on S closely to each active piston the sensor of normal oscillatory velocity with output signal $u_j(t) = \upsilon(R, \vartheta_j, t)$ in the points (R, ϑ_j) of the surface S. So instead smooth DNOV $\upsilon(R,\vartheta,t)$ with minimum angular scale $\Delta_{\min}$ we obtain (by adding of DNOV of active pistons) the DNOV of the form $\upsilon_S(R,\vartheta,t) = \upsilon(R,\vartheta,t) + \sum_{j=1}^{N} \upsilon_j(R,\vartheta,t)$, where $\upsilon_j(R,\vartheta,t) = \Pi_j(\vartheta)(d/dt)h_j = \Pi_j(\vartheta)V_j$, $\Pi_j(\vartheta) = 1$ if $j\Lambda - (\varepsilon/2) < \vartheta < j\Lambda + (\varepsilon/2)$, $\Pi_j(\vartheta) = 0$ if $j\Lambda - (\varepsilon/2) > \vartheta > j\Lambda + (\varepsilon/2)$. Signal $u_j(t)$ excites the corresponding active piston and produces the velocity of change of its thickness of the form $(d/dt)h_j(t) = \nu\, u_j(t)$, or

$$V_j(t) = -\nu\, u_j(t) \qquad \textbf{(4-20)}$$

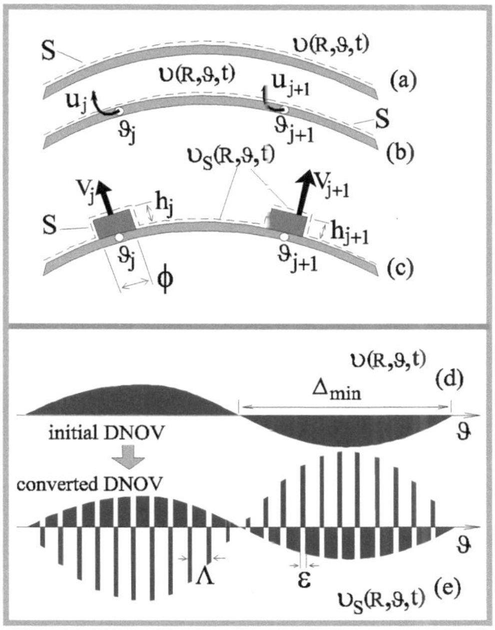

Figure 4-8: An example of conversion of DNOV by the piston system in accordance with algorithm (4-20): Radiating cylindrical surface S with DNOV $\upsilon(R,\vartheta,t)$, which does not depend on sound pressure in outside liquid (a); equidistant spacing in points $\vartheta_j = j\Lambda$ ($j = 1,2,...,N$) of sensors of normal oscillatory velocity with output signals $u_j(t) = \upsilon(R,\vartheta_j,t)$ (b); spacing of active strip-like (of width φ) pistons (actuators) of controlled thickness $h_j(t)$ in accordance with signals $u_j(t)$, centered in points ϑ_j, surface S now is overlapping actuators; initial (d) DNOV υ on S and converted (e) DNOV υ_S on S.

where $\nu = \Lambda / \varepsilon$ [94]. We can call (4-20) an algorithm of volume velocity compensation (by active pistons) with classification {LS/LT/CA/F} (see **Fig. 1-3**). Due to (4-20) we obtain the zero spatial average volume velocity on the surface S (**Fig. 4-8-e**). Taking into account **Fig. 4-5-b**, we choose $\Lambda \le \lambda / 2R$, $\Delta_{\min} > \Lambda$ and $kR >> 1$ and obtain zero radiation from surface S (see **Fig. 4-9**). This effect (zero radiation) can be achieved at very low ratio $\varepsilon / \Lambda << 1$, because $\nu = \Lambda / \varepsilon >> 1$. Note that we assumed that active pistons are supported by vibrating surface, but do change its initial DNOV $\upsilon(R,\vartheta,t)$. It is easy to see from **Fig. 4-8-e,** that it resembles **Fig. 1-3-c**, which can be interpreted as spatial and temporal analogs respectively.

The algorithm (4-20) can be interpreted in spectral terms also (using section 4.6.1.2.):

1) Equidistant active piston damping system converts spatial spectrum $\bar{\bar{\upsilon}}(\chi,\omega)$ of initial DNOV $\upsilon(R,\vartheta_j,t)$ to the area of high $|\chi| > k$ spatial frequencies χ (χ –"linear spatial frequency" along the

surface S) with zero radiation. This means up-conversion of DNOV spectrum in the area of spatial frequencies (**Fig. 4-9-a**);

2) Damping system converts spectral coefficient $h_a(\chi,\omega)$ of radiation transmission (**Fig. 4-9-b**).

4.6.2. Discrete Coating

All active coatings (based on scheme "sensor-computer-actuator") can be presented only in spatially discrete form. Let's consider the case of sound radiation by some vibrating closed smooth surface S_V and divide this surface S_V into a set of $N_V >> 1$ monopole-like radiating pistons $\overline{S}_i$ ("equivalent pistons")

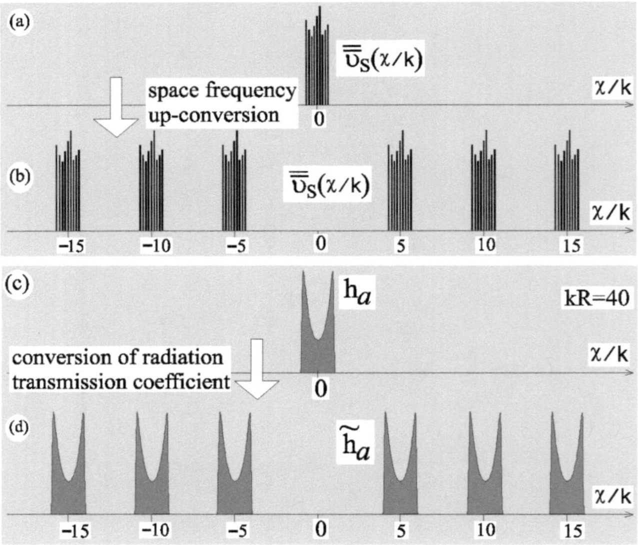

Figure 4-9: An examples of conversion of spatial NDOV spectrum $\overline{\overline{\upsilon}}(\chi,\omega)$ (a, b) by the piston system; conversion of spectral coefficient $h_a(\chi,\omega)$ of radiation transmission (c, d) by the piston system, when $kR = 40$. Note that physical sense have only discrete values $\chi = n/R$ ($n = 0, \pm 1, \pm 2, \ldots$).

$$S_V = \bigcup_{i=1}^{N_V} \overline{S}_i, \; \overline{S}_i \bigcap_{i \neq j} \overline{S}_j = 0 \tag{4-21}$$

($i, j = 1, 2, \ldots, N_V$, **Fig. 4-10-a**). Each equivalent piston $\overline{S}_i$ is represented by the prominent polygon with square $\overline{\sigma}_i$ [92] and characteristic linear scale $\sim L_V$, satisfying the following conditions $L_V < \lambda_{\min}/2$ and $L_V >> h_r + 2h_c$. Each prominent polygon must have multipolarity order not larger than monopole. In particular, this means, that at any moment of time power stream density has the constant sign within the frequency range (4-3). Within each equivalent piston $\overline{S}_i$ (*i.e.* for $\mathbf{r} \in \overline{S}_i$) we suppose also DNOV $\upsilon(\mathbf{r},t) = \upsilon_i(t)$ is constant and also another constant DNOV within other pistons. Normal velocity of external surface of each i-th piston (spatial constant on $\overline{S}_i$) is equal to:

$$\upsilon_i(t) = (\overline{\sigma}_i)^{-1} \iint_{\overline{S}_i} \upsilon(\mathbf{r},t) dS, \tag{4-22}$$

where $\upsilon(\mathbf{r},t)$ is the initial velocity, forming the smooth DNOV on S_V (*i.e.* before the beginning of action of control system), $\overline{\sigma}_i$ -square of equivalent piston. This velocity $\upsilon_i(t)$ corresponds to the piston's "mass center"

$$\overline{\boldsymbol{\rho}}_i = (\overline{\sigma}_i)^{-1} \iint_{\overline{S}_i} \mathbf{r} dS. \tag{4-23}$$

Outside Green operator $\hat{G}_e$ (4.8.4, see below) acts like a low pass filter of spatial frequencies (section (above sections 4.6.1.1, 4.6.1.2). So at the distance $\overline{r} > L_V^2 \chi_{\max}$ from S_V. So at the distance $\overline{r} > L_V^2 \chi_{\max}$ from S_V the sound, produced by DNOV $\upsilon(\mathbf{r},t)$, practically does not differ from the sound, produced by corresponding system of N_V "equivalent pistons".

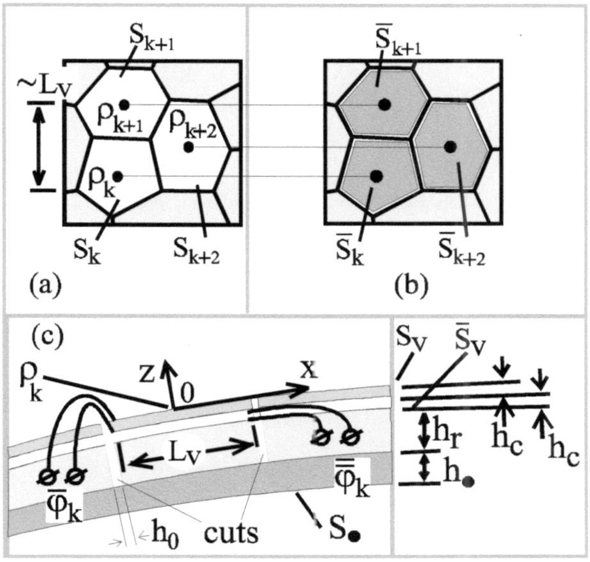

Figure 4-10: Discrete structure of active coating (a, b), cross-sectional structure of active coating (c).

Now we place "active pistons" S_i (or actuators), identically covering the "equivalent pistons" $\overline{S}_i$, described above. Each S_i is of the same shape, square $\sigma_i = \overline{\sigma}_i$ and "mass center" $\boldsymbol{\rho}_i = \overline{\boldsymbol{\rho}}_i$ like $\overline{S}_i$, *i.e.* $S_V = \bigcup_{i=1}^{N_V} S_i$, $S_i \bigcap_{i \neq j} S_j = 0$ $(i, j = 1, 2, ..., N_V$, **Fig. 4-10-b**). The characteristic linear dimension $\sim L_V$ of each piston is much smaller than the value $2\pi / \omega_Q$ (i. e. $L_V << 2\pi / \omega_Q$), where the frequency ω_Q will be determined in Sections 6.2., 6.3. below. Each active piston S_i represents a so-called "Huygens's piston". Such a piston produces waves of radiation only on one side (external). All together $N_V >> 1$ Huygens' pistons form the active coating of the protected body. This coating is placed between the closed surfaces $\overline{S}_V$ (internal) and surface S_V (external). The problem of control of radiation and scattering means the formation of the desired (DNOV) distribution of normal oscillatory velocities (displacements) on the surface S_V.

As we are interested in distributions of normal oscillatory velocities (DNOV) we make periodical cuts (filled with outside liquid) *via* active pistons (depth $z \in [0, -2h_c]$ from outside liquid) and *via* polymer layer (depth $z \in [-2h_c, -h_r - 2h_c]$) up to the steel plate (depth $z = -h_r - 2h_c$). This allows us to consider the waves in depths $z \in [0, -h_r - 2h_c]$ as the waves in compressible media (much more simple).

4.6.2.1. The Cross-Section Structure of the Protected Body

Each Huygens' piston (or active piston) S_i represents two layers (two piezoelectric layers, which were polarized identically along the axis "z" **Fig. 4-10-c**) in cross-section In top view Huygens' piston presents a planar convex polygon. In other words the active coating is divided into a lot ($N_V >> 1$) of Huygens' pistons (**Fig. 4-10-b**) analogous to the pistons $\overline{S}_i$, called above equivalent radiating pistons. Equivalent pistons $\overline{S}_i$ and active pistons S_i (Huygens's) have the same "mass's center" $\boldsymbol{\rho}_i$, the same square σ_i and the same characteristic linear spatial scale $\sim L_V$. Surfaces of each piezoelectric layer are covered by thin metal weightless films and equipped by electric contacts (**Fig. 4-10-c**). Between contacts of each i-th Huygens' piston the differences $\overline{\phi}_i(t)$ and $\overline{\overline{\phi}}_i(t)$ of the electric potentials are supported by active control system. Active pistons are separated by narrow cuts (slots of width h_0), which are filled with outside liquid (**Fig. 4-10-c**).

4.6.2.2. The Control Algorithm for the Synthesis of DNOV

The electric voltages, which are applied to each layer of the m-th Huygens' piston ($m = 1, 2, ..., N_V$), have a view:

$$\overline{\phi}_m(t) = \sum_{n=0}^{[t/T_V]} B_{m,n}\overline{\phi}_B(t - nT_V), \ \overline{\overline{\phi}}_m(t) = \sum_{n=0}^{[t/T_V]} B_{m,n}\overline{\overline{\phi}}_B(t - nT_V), \quad \textbf{(4-24)}$$

and can be presented by the temporal sequence of pulses (**Fig. 4-11-a, Fig. 4-11–b**):

$$\overline{\phi}_B(t) = \{I(t) - I(t - \tau_c)\}\phi_0 \text{ and } \overline{\overline{\phi}}_B(t) = -\overline{\phi}_B(t - \tau_c) \quad \textbf{(4-25)}$$

with period T_V and duration $3\tau_V$ (or $3T_V / 2$), where we introduced the time τ_V during which wave overcomes the distance between the boundaries of one piezoelectric layer. There we supposed $I(\xi) = 1$ at $\xi > 0$, $I(\xi) = 0$ at $\xi > 0$, $\phi_0 = const$. The magnitude $B_{m,n}$ of these pulses is determined by the [90]:

$$B_{m,n} = (\psi_0 T_V / 2)^{-1} \int_{nT_V - T}^{nT_V} \overline{F}(Y_m - Y_{\otimes m})dt, \quad \textbf{(4-26)}$$

where $Y_m(t) = \sigma_m^{-1} \iint_{S_m} u(\mathbf{r}, t)dS_m$, $Y_{\otimes m}(t) = \sigma_m^{-1} \iint_{S_m} u_{\otimes}(\mathbf{r}, t)dt$, T –temporal interval of averaging, $\psi_0 = const$ –amplitude of basic bipolar pulse of displacement, $[\xi]$ means the whole part of number ξ, $u_{\otimes}(\mathbf{r}, t)$ and $u(\mathbf{r}, t)$ –perfect (desired) and real (measured) normal displacements of surface $\psi_0 = const$. The operator $\overline{F}[q] = \int_0^t F(t - \xi)q(x, y, \xi)d\xi$ represents a linear filter with frequency transmission coefficient $\tilde{F}(\omega) = \int_0^t F(t)\exp(-i\omega t)dt$. The module of the last near zero frequency has the form $|\tilde{F}(\omega)| \sim |\omega^{2\nu}|$ (ν –whole positive number). On the frequencies of range (4-3) (and up to the frequencies $\sim 1/T_V$) its module is $|\tilde{F}(\omega)| = 1$. This algorithm (4-26) can be classified as {LS/LT/CA/F} (see **Fig. 1-3**). The relation between the mechanical ψ_0 and electric ϕ_0 magnitudes is determined by the formula $\psi_0 / \phi_0 = (\tau_c / h_c)(\varepsilon_0 \varepsilon_c \eta_c / E_c)^{1/2}$, where ε_0 –dielectric constant of vacuum, ε_c –dielectric constant of piezoelectric, E_c –Young's modulus of piezoelectric material, $\psi_0 / \psi_0 = (\tau_c / h_c)(\varepsilon_0 \varepsilon_c \eta_c / E_c)^{1/2}$ –coefficient of electro-mechanical coupling in piezoelectric ($\eta_c = 0.2 \div 0.8$). In analog representation the filter $\overline{F}$ can be described as electric differential RC-circuit with time scale $\tau_{RC} = \tau_{\max}$. In accordance with (4-26) the electric pulses $\overline{\phi}_B(t - nT_V)$ and $\overline{\overline{\phi}}_B(t - nT_V)$ are acting correspondingly in the temporal intervals:

$$nT_V \le t < nT_V + \tau_c \textbf{ and } nT_V + \tau_V \le t < nT_V + \tau_V + \tau_c \ (n = 1, 2, ...). \quad \textbf{(4-27)}$$

$\psi_0 / \psi_0 = (\tau_c / h_c)(\varepsilon_0 \varepsilon_c \eta_c / E_c)^{1/2}$ is the error signal, which one had averaged spatially (instantly in time) along the surface of piston S_m and also averaged in time [90] on the previous temporal interval $nT_V - T \le t < nT_V$, where $T = j(3\tau_V + \tau_c)$ ($j = 1, 2, ...$). Time interval of averaging satisfies the condition $\int_0^T \psi_B(t)dt$. Voltages $\overline{\phi}_m(t)$ and $\overline{\overline{\phi}}_m(t)$ form the field of normal displacements (DNOD)

$\psi(\mathbf{r},t) = \sum_{n=0}^{[t/T_V]} B_{n,m}\psi_B(\mathbf{r},\boldsymbol{\rho}_m;t-nT_V)$ (4-28) of normal displacements of the surface $\psi_0 = const$, induced by m-th piston. The function $B_{n,m}\psi_B(\mathbf{r},\boldsymbol{\rho}_m;t-nT_V)$ represents the normal displacement in the point $\mathbf{r} \in S_m$ (**Fig. 4-11-c**), created by pulses $B_{m,n}\overline{\phi}_B(t-nT_V)$, $B_{m,n}\overline{\overline{\phi}}_B(t-nT_V)$ of electric voltage of duration $\tau_c \ll \tau_V$, respectively, on the external and internal layers of the active piston S_m (with center in the point $\boldsymbol{\rho}_m$) at the moments $t = nT_V$ and $t = nT_V + \tau_V$. It was shown [90], that in a one-dimensional problem (at $L_V \to \infty$) such a double layer of piezoelectric material (or piston) with the above mentioned magnitude-temporal relation (**Fig. 4-11-c**) between inducing electric pulses (**Figs. 4-11-a**, **4-11-b**) creates a Huygens' source (or Huygens' piston).

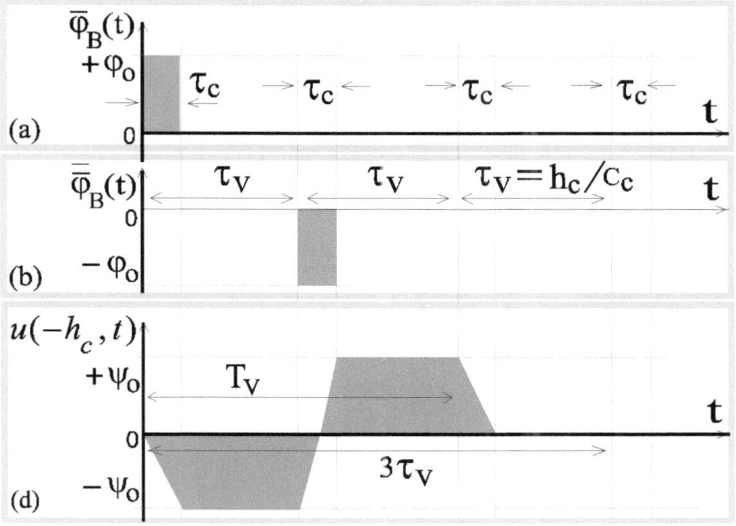

Figure 4-11: The shape of pulse of electric voltage $\overline{\phi}_B(t)$, which is applied to external layer of $z \in [-h_c, 0]$ active piston (a); shape of pulses of electric voltage $\overline{\overline{\phi}}_B(t)$, which is applied to internal layer $z \in [-2h_c, -h_c]$, of active piston (b); shape of "basic pulses" $B_{n,m}\psi_B(\mathbf{r},\boldsymbol{\rho}_m;t)$ of normal displacement of outside surface S_m of piston, which was created by electric pulses above mentioned on a distance $|\mathbf{r}-\boldsymbol{\rho}_m| < c_w t, L_V$ from the border of piston S_m (or in the locally one-dimensional problem) (c).

This double layer provides a bipolar pulse of media particles displacement of duration $T_B = 3\tau_V + \tau_c$. These pulses are moving through space (like wavelets [36], **Fig. 4-12**) with speed c_w. A wavelet (we will also call it a "basic pulse") is only radiated forwards (outside the surface $\psi_0 = const$) and must always contain an interval of zero magnitude and of duration τ_V, between two pulses of mutually opposite polarity and same duration τ_V. At the back ($z < -2h_c$) the pulses, induced by both piezoelectric layers, move and add to mutually opposite phases giving a zero sum in the backspace (inside $\psi_0 = const$, **Fig. 4-12**).

The main difference between the suggested approach and traditional approaches is concerned with the following. Usually the desired trajectory (temporal) of the normal displacement of $\psi_0 = const$ can be achieved by the temporal sequence of pulses without their intersection and with nonzero mean value of each pulse (for instance - delta pulse). In the algorithm of the synthesis of desired trajectory, described below, we use bipolar pulses mutually intersecting with: zero mean value, duration $T_B = 3\tau_V + \tau_c$ (where $\tau_c \ll \tau_V$), temporal period $T_V = 2\tau_V$ of sequence. Each main pulse consists of three consequent identical intervals τ_V. The first (left part) interval of the next basic pulse is covered by the third (right part) interval of the previous basic pulse. The absence of the source (one-dimensional [90]) and wave field in back-space means that we can generate forward radiation waves without any mechanical support of both piezoelectric layers, due to the special construction of the Huygens' source. Besides one-side radiation this Huygens' source is an acoustically transparent source. The wave, which was radiated by a Huygens' source, does not depend on impedance in back-space and can be simply added to any wave passing without any interaction with the source. In Section 5.2.9 we can see the same effect of the one side radiation without mechanical support for the monochromatic excitation (on some frequency ω) of two piezoelectric layers (with the phase difference $\Delta\phi = \omega h_c / c_c$ of excitation). A very small phase difference (necessary for one side radiation) is defined by stationary device tuning. This phase difference does not depend on acoustical field, measured by sensors. So the measurement errors $\sim \Delta\psi$ (which determines phase and magnitude of one-side radiation) of the acoustical field is included coherently (with mutually opposite signs) into both voltages, exciting both piezoelectric layers, and cannot cause the growth of the unbalance of waves in $\sim \Delta\psi / \Delta\phi >> 1$ times. Note that in the absence of incident waves the algorithm (4-26) can be used for radiation suppression, when $Y_{\otimes m}(t) = 0$.

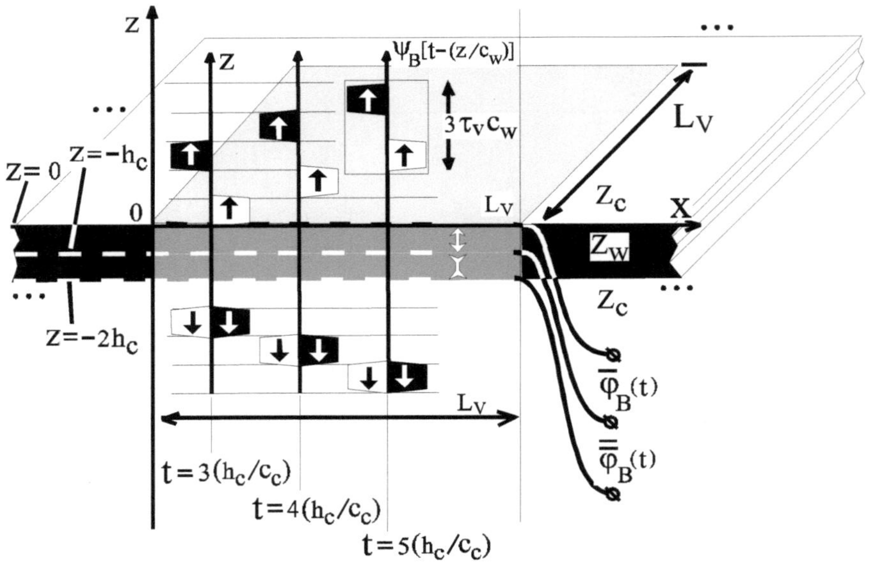

Figure 4-12: "Huygens' piston" of finite dimensions $L_V \times L_V$ in the homogeneous free compressible space and the instant spatial distributions of media particles displacements are represented by a running "basic pulses" $B_{n,m}\psi_B(z,t)$ (at $|\mathbf{r} - \boldsymbol{\rho}_m| < c_w t, L_V$, on external side of an active piston), induced by piezoelectric layers (and zero pulses at internal side of active piston). Darker pulses (waves) are produced by the electric voltage $\bar{\phi}_B(t)$ layer of active piston. Less dark pulses (waves) are induced by the electric voltage $\bar{\bar{\phi}}_B(t)$.

4.6.2.3. Some Notes About the Stability of the Synthesis of DNOV (DNOD)

The interaction between elements of an active damping system always presents possible channels of instability, so it follows that the absence of interaction would also mean the absence of instability. Now, we will consider very briefly the main types of interactions in the system of DNOV synthesis. We will consider below the two planes $z = 0$ (analog of surface S_V) and $z = -2h_c$ (analog of surface $\bar{S}_V$) filled with Huygens' pistons (two layers of piezoelectric material). This pair of planes is in contact outside with a half-infinite compressible homogeneous space. In addition we assume that the compressibility and speed of propagation of acoustical waves are same in the piezoelectric material and the outside compressible media. This becomes possible due to the use of piezoelectric materials, containing polymers [26], [27]. Now we can achieve the continuum spectrum of electromechanical characteristics of such piezocomposite materials with ceramics and polymers [26]. Below we try to explain the smallness of the interaction effects when using the algorithm (4-26).

4.6.2.4. Interaction Between Adjacent Pistons

The control algorithm (4-26) is a three-dimensional one (discrete spatially and temporally) and presents the generalization of the one-dimensional solution [90] without diffraction waves. In general the dynamics of the surface S_V can be described as a sequence of: (a) fast (during a time $\tau_c << T_V$) one-side blows, spatially uniformly distributed along a Huygens' piston, when the error $(Y_m - Y_{\otimes m}) \neq 0$ of the m -th piston jumps to zero (during the time τ_c), forming the "imprint"; (b) slow linear growth of $|Y_m - Y_{\otimes m}|$ in a time not quicker, where $\overline{T}_V = L_V / c_w$ is a life-time of "imprint" (time of its diffraction relaxation), $A_{\otimes}$ –characteristic magnitude of the desired displacement $u_{\otimes}(t)$ of the surface S_V, Q_{LF} –body's ringingness at low frequencies (Sections 4.1.1. and 4.1.2.). The condition $\overline{T}_V = L_V / c_w >> T_V$ is the principal one for our active system, due to which we can neglect the interaction between adjacent pistons and the value of the corresponding positive feedback $\gamma_1 = (T_V / \overline{T}_V)^2 << 1$.

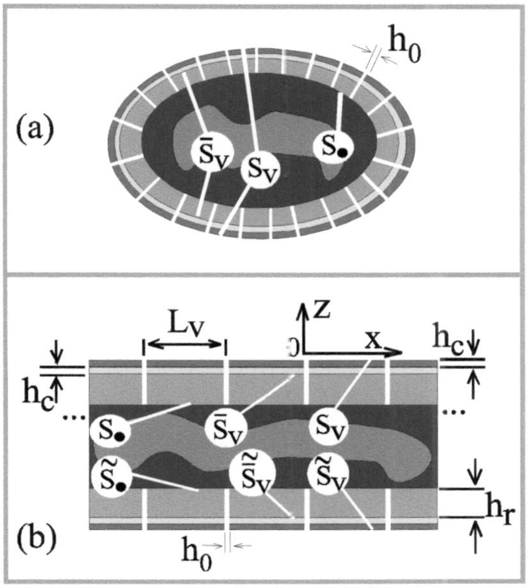

Figure 4-13: The reduction of boundary value problem from three-dimensional version (a) to the two-dimensional version (b): the synthesis of DNOVs $u_{\otimes}(x,t)$ and $\tilde{u}_{\otimes}(x,t)$ uniform along axis "y" on the upper S_V and lower $\tilde{S}_V$ parts of external infinite plane surface of coated body respectively.

4.6.2.5. Synthesis of One-Dimensional DNOD in a Two-Dimensional Boundary Value Problem

The synthesis of one-dimensional boundary value problem This case is thoroughly described in [90] and Section 5.3, and we present here only result of the spatially uniform along the plane) synthesis shown in **Fig. 5-9**. There the desired continuous trajectory $u_{\otimes}(t)$ we approximate by the sequence (with temporal period $T_V = 2\tau_V$) of "rectangular" (at time) displacement wavelets of duration $3\tau_V$, formed from two pulses (of duration τ_V) of mutually opposite polarity τ_V with pause of duration τ_V between them (Section 5.3.6., **Fig. 5-9-b**). Spatially one-dimensional case is appropriate for description of three-dimensional case, when $0 < t << L_V / c_w$ (see **Fig. 4-12**).

Below we will describe the active synthesis of distribution of normal oscillatory displacements (DNOD) on three-dimensional body (**Fig. 4-13-a**) by the following simplified (**Fig. 4-13-b**) boundary value problem. The elastic shell, satisfying the conditions (4-10), (4-11), with external surface $S_{\bullet}$ (see **Fig. 4-4**) and arbitrary filling, presents the body to be protected. Space between surfaces $S_{\bullet}$ and $\overline{S}_V$ is filled with polymer. This polymer (see section 4.5.5.) is characterized by sound speed $c_r = c_w$ and mass density $\rho_r = \rho_w$.

The distance $\ell_r(\omega)$ of dissipative relaxation satisfies simultaneously three conditions:

$$\ell_r(\omega) >> h_r \qquad \textbf{(4-29)}$$

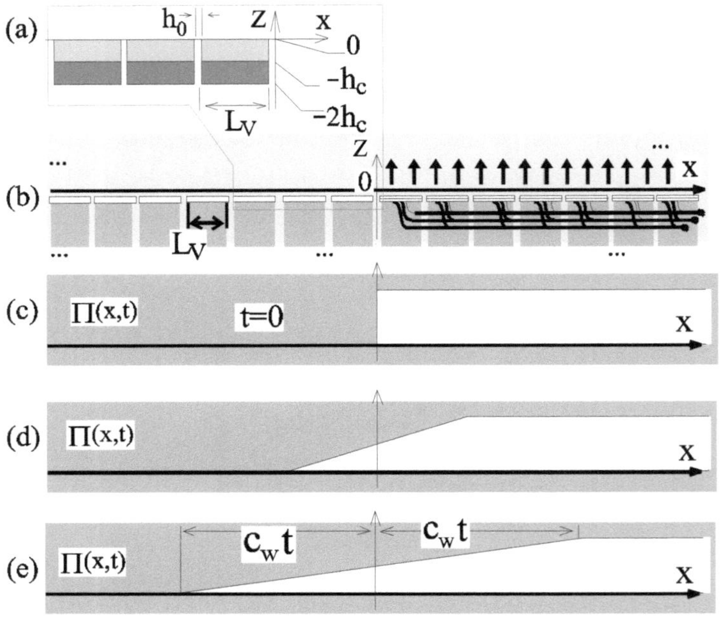

Figure 4-14: The structure of piezoelectric layers of coating (a). Supply of electric voltage (b) to the elements of piezoelectric array producing the step like spatial distribution of one (first electric pulse in **Fig. 4-11-a**, beginning at the moment $t = 0$) normal blow (black arrows). The evolution (c–e) in time of spatial DNOD $\Pi(x,t)$ from initial step like $\Pi(x,\tau_V) = I(x)$ ($I = 1$, when $x > 0$ $I = 0$) on the surface S_V (**Fig. 4-13-b**), caused by one electric pulse with $\tau_V << L_V / c_w$.

for the frequencies of the range (4-3),

$$\ell_r(\omega) << h_r \tag{4-30}$$

for the frequencies $\omega \geq 2\pi / T_V$ and

$$\ell_r(2\pi / T_V) >> T_V c_w \tag{4-31}$$

The space between surfaces $\overline{S}_V$ and S_V is filled by two identical layers (both of the thickness h_c) of piezoelectric with normal polarization. The last is characterized by longitudinal sound speed $c_c = c_w$ and mass density $\rho_c = \rho_w$. The surface S_V is in contact with outside liquid characterized by sound speed c_w and mass density ρ_w. To simplify the consideration we reduce three-dimensional boundary problem to the two-dimensional problem. So instead one three-dimensional closed single surfaces S_V, $\overline{S}_V$, $S_\bullet$ (**Fig. 4-13-a**), we obtain pairs of infinite parallel planes (**Fig. 4-13-b**): S_V and $\tilde{S}_V$, $\overline{S}_V$ and $\tilde{\overline{S}}_V$, $S_\bullet$ and $\tilde{S}_\bullet$, respectively. We assume all parameters of boundary value problem uniform along axis " y ".

Both external and internal surfaces of piezoelectric layers are covered by very thin and light metal film (**Fig. 4-14**). All content between surfaces S_V and $S_\bullet$ is divided into a lot of identical (and equidistantly spaced with spatial period $L_V + h_0$) active Huygens pistons (see section 4.6.2.) of the same characteristic linear scale L_V and the gaps of width $h_0 << L_V$ between them. These gaps are filled with outside liquid. Due to gaps (cuts), we can approximate the waves in the polymer layers by the longitudinal waves in compressible media (to simplify the problem, in comparison with elastic waves). As a final result, we must approximate the assigned arbitrary DNODs $u_\otimes(x,t)$ and $\tilde{u}_\otimes(x,t)$ on the external surfaces S_V and $\tilde{S}_V$, respectively, by discrete infinite plane array of Huygens strip-like pistons. Below, for simplicity, we will be limited by synthesis of arbitrary $u_{\otimes}(x,t)$ on one plane S_V (when $\tilde{u}_{\otimes}(x,t)=0$).

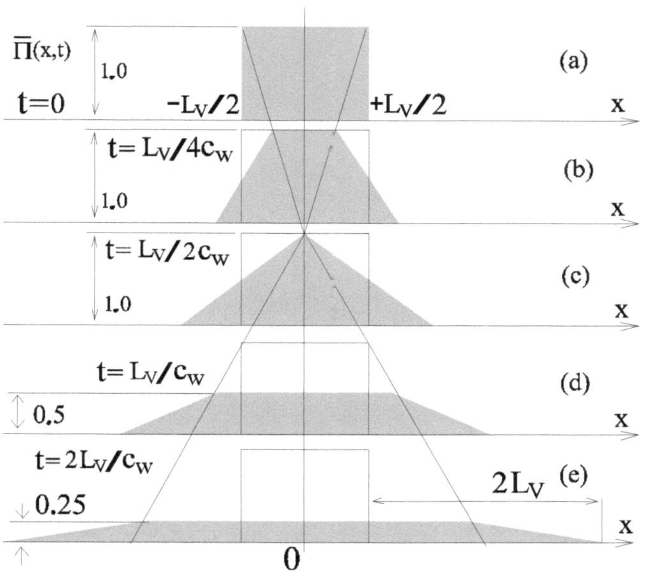

Figure 4–15: The temporal evolution (a–e) of initial spatial DNOD $\overline{\Pi}(x,\tau_V)=I[x+(L_V/2)]-I[x-(L_V/2)]$ (a) for the case, when the piezoelectric blow initially is concentrated uniformly inside the area $-(L_V/2)<x<+(L_V/2)$.

We must note: the synthesis (discrete approximation) of arbitrary assigned DNOD (during the time τ_V) is possible, if we can form the *one piston* DNOD (*delta-like* DNOD). So we need to prove the possibility of synthesis of delta-like DNOD by the sequence (5-30) of bipolar base pulses (see **Fig. 5-9-a**) in accordance with algorithm (4-26).

All processes in **Fig. 4-13**, **Fig. 4-14**, **Fig. 4-15**, **Fig. 4-16**, of duration $\sim\tau_c$ we will mean instant processes.

Now we consider some simple examples of excitation of the surface S_V by very short electric pulses to understand and show the specific role of diffraction effects in the evolution of DNOD fields in the boundary problems with initial conditions.

Fig. 4-14-b, **Fig. 4-14-c**, **Fig. 4-14-d**, **Fig. 4-14-e** shows the evolution in time (for $t\geq\tau_V$) of the initial (imprint in **Fig. 4-14-b** immediately after the blow, *i.e.* for $t=\tau_V$) spatial DNOD $\Pi(x,\tau_V)=I(x)$ ($I=1$, when $x>0$ and $I=0$, when $x<0$), produced by piezoelectric blow by one electric pulse (see Section 5.3.3.) of duration $\tau_c<<L_V/c_w$ and applied to the half-plane $x>0$. This blow excludes any back space radiation due to the unidirectional characteristics of Hyugens piston. The another half-plane $x<0$ remains free from blow before and after the electric pulse.

By the combination

$$\overline{\Pi}(x,t)=\Pi[x+(L_V/2)]-\Pi[x-(L_V/2)] \tag{4-33}$$

of above functions $\Pi(x,t)$ we solve the boundary problem, where area $|x|>L_V/2$ is free from blow, but the excitation (blow) is uniformly concentrated in the area $|x|<L_V/2$ (see **Fig. 4-15**). Initial imprint $\overline{\Pi}(x,\tau_V)=I[x+(L_V/2)]-I[x-(L_V/2)]$ (**Fig. 4-15-a**) sprawls in space and decreases (**Fig. 4-15-b**, **Fig. 4-15-c**, **Fig. 4-15-d**, **Fig. 4-15-e**).

Now we can consider the DNOD (**Fig. 4-16**), caused by two electric pulses (of the same duration τ_V) of the same amplitude module and mutually opposite polarity with temporal separation by the interval $2\tau_V$ and applied to the spatial area $|x| < L_V / 2$. So this means the response of boundary S_V to the basic pulse (**Fig. 5-8-b**) of electric voltage or its blow. We use for this purpose the combination

$$\bar{\bar{\Pi}}(x,t) = \bar{\Pi}(x,t) - \bar{\Pi}(x,t-2\tau_V)I(t-2\tau_V) \qquad \textbf{(4-4)}$$

of above functions $\bar{\Pi}(x,t)$. We see from **Fig. 4-16**, that the base pulse (**Fig. 4-11**) produces nonzero spatially integrated pulse $\int_{-\infty}^{+\infty} \bar{\bar{\Pi}}(x,t)dx > 0$ of DNOD ($\bar{\bar{\Pi}}(x,t) > 0$

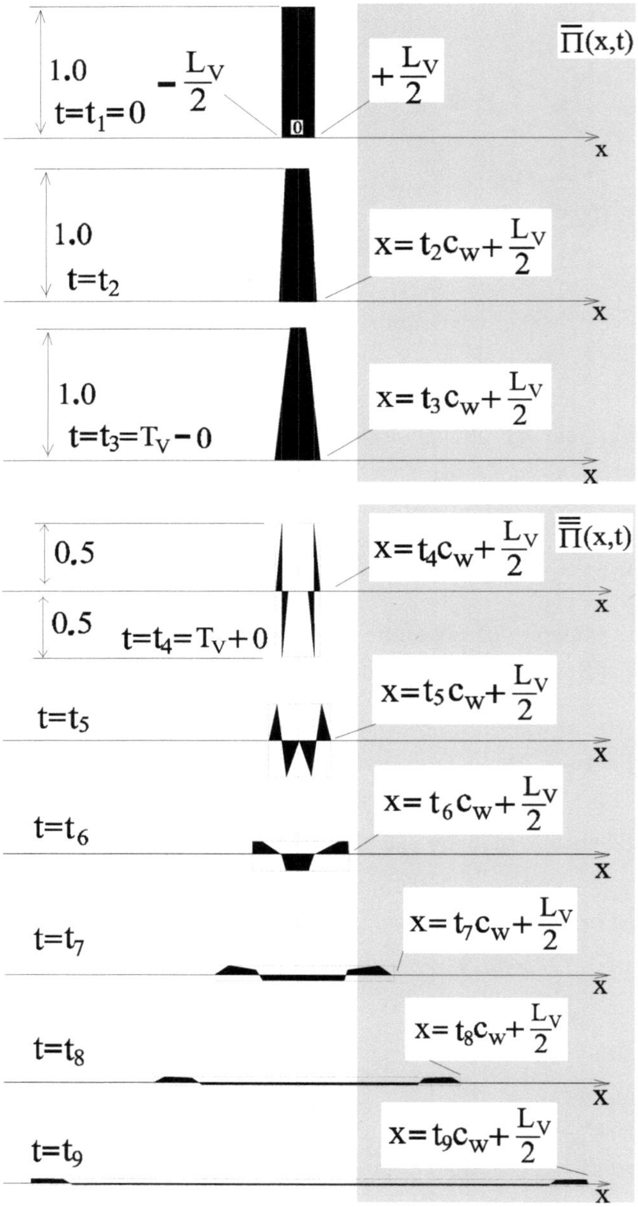

Figure 4-16: The temporal evolution of DNOD $\bar{\bar{\Pi}}(x,t)$ for $0 \le t < 7L_V / c_w$ and $\tau_V = 2^{-2} L_V / c_w$.

inside the area $|x| < (L_V / 2) + c_w \tau_V$ of) in the temporal frame $0 < t \le \tau_V$. On the other hand, for $t > \tau_V$ (and at $t \le 0$ of course) we have $\int_{-\infty}^{+\infty} \overline{\overline{\Pi}}(x,t)dx = 0$.

Now let us assume an infinite number of strip-like pistons on the plane S_V (**Fig. 4-13-b**). And all pistons are excited by the temporal sequences of base pulses in accordance with algorithm (4-26) and create the DNOD

$$u(x,t) = \sum_{n=0}^{n=\mathrm{ent}[t/T_V]} \tilde{B}_{nm} \overline{\overline{\Pi}}(x - mL_V, t - nT_V), \quad \textbf{(4-35)}$$

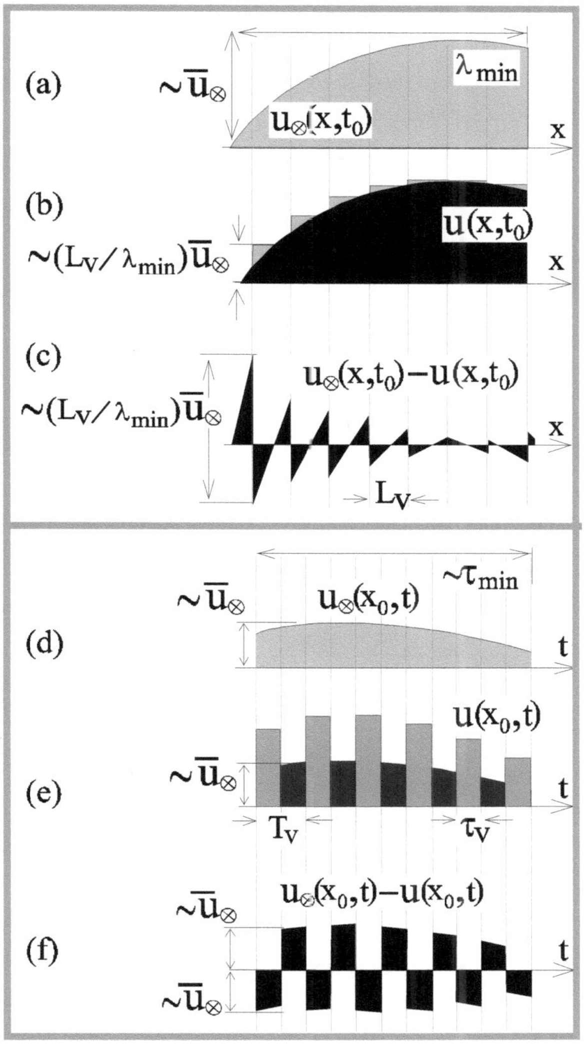

Figure 4-17: Spatial (a–c, first blow) and temporal (d–f, many blows) synthesis of desired DNOD $u_\otimes(x,t)$ (painted gray in (a) and (d), painted black in (b) and (e)) by the array of strip-like Hyugens pistons ($\overline{u}_\otimes$ denotes the characteristic amplitude of $u_\otimes(x,t)$). Discrete approximation $u(x,t)$ of desired DNOD is painted gray in (b), (e). The error $\zeta = u(x,t) - u_\otimes(x,t)$ of synthesis is painted black in (f) and $\zeta = u(x,t) - u_\otimes(x,t)$ for the first blow in (c). $\tau_{\min}$ and $\lambda_{\min}$ are taken from the range (4-4).

where coefficients $\tilde{B}_{nm}$ are defined by algorithm analogous to (4-26). The last provides zeroing $\zeta \to 0$ of error $\zeta = u(x,t) - u_\otimes(x,t)$ (where $u_\otimes(x,t)$ is the desired or assigned DNOD) on the each n-th temporal cycle within each

m -th piston. Strictly blows do not render no influences on rear half-space. Only process of evolution (diffraction) of DNOD after (in a time τ_c) blow generates radiation back ($z < -2h_c$, see **Fig. 4-14-b**). However, the power of this radiation is concentrated on very high temporal frequencies $\omega \sim \pi / \tau_V >> \omega_{\max}$ and very high spatial frequencies $\chi \sim \pi / L_V >> \chi_{\max}$ (see **Fig. 4-16**, **Fig. 4-17**). On these frequencies the conditions (4-29)-(4-31) are satisfied simultaneously, and the array of Hyugens pistons on S_V (or $z = 0$) does not act on rear half-space ($z < -2h_c - h_r$, *i.e.* does not act on the surface $S_\bullet$ of shell) due to viscous relaxation in the polymer layer. The amplitude of this high frequency to the back is about double characteristic amplitude

$$\sim 2\overline{u}_\otimes \tag{4-36}$$

(on the depth $z = -2h_c$) of desired DNOD $u_\otimes(x,t)$. On the other hand we have the same amplitude (4-36) of the forward ($z > 0$) high frequency radiation. We described the one step (two blows, **Fig. 4-17-a**, **Fig. 4-17-b**, **Fig. 4-17-c**) in the process of delta-like DNOD (spatial-temporal) formation. It is easily to see that next steps will not allow the DNOV to overcome the spatial frequency $|\chi| > 2\pi / L_V$. On these spatial frequencies waves are relaxed due to high frequency dissipation in the polymer layer and can not achieve the surface $S_\bullet$. Therefore we described the procedure of real-time synthesis of arbitrary predetermined DNOV. Note, that above described synthesis of desired DNOD is local in time and space unlike the model of G.D.Maljuzhinets [15-19], where all sensors are connected electrically with all actuators. This algorithm can be classified as {LS/LT/CA/F} (see **Fig. 1-3**).

4.6.2.6. About Ringingness of a Body

Instability can be caused by unlimited growth of the previous blow contributions of all pistons. Such a scenario can be possible in the following cases: (a) if $Q_{LF} >> 1$, when an active coating is placed on the surface of resonator mirrors (for instance, as an analog of the protected body), and the desired DNOV would "intersect" with some own resonator's mode of infinite ringingness; (b) when the high frequency ringingness is too large (see Sections 4.1.1, 4.1.2), *i.e.* $Q_{HF} >> 1$. Thus, the body's ringingness Q_{LF} in a passive system can be interpreted as a natural (passive) positive feedback with coefficient $\gamma_+ \approx Q_{LF} / (1 + Q_{LF}) < 1$. In a passive system always $\gamma_+ < 1$, but in active systems the feedback coefficient $\gamma_+ > 1$ is possible, and may cause instability.

4.6.2.7. Hierarchy of Scales in the Active Coating

Besides the algorithm (4-26), to ensure the effective formation of the desired DNOV (DNOD) on S_V we need to use the following hierarchy of system parameters:

$$A_\otimes / c_w << \tau_c << T_V << L_V / c_w << h_r / c_r << R_V / c_w, \tau_{\min}, \tag{4-37}$$

where τ_c –duration of electric voltage pulse on piezoelectric layers of the Huygens' piston (or duration of a blow on S_V), T_V –temporal period of control actions (blows), $A_\otimes / c_w << \tau_c << T_V << L_V / c_w << h_r / c_r << R_V / c_w, \tau_{\min}$ –thickness of polymer layer of high frequency dissipation, c_w –speed of propagation of longitudinal sound in polymer, L_V – characteristic tangential linear dimension of a piston, c_w –sound speed in the outside compressible media (liquid), $A_\otimes / c_w << \tau_c << T_V << L_V / c_w << h_r / c_r << R_V / c_w, \tau_{\min}$ –minimum period of the waves damped (4-4), L_V –minimum curvature radius of the surface S_V, $A_\otimes$ –characteristic magnitude of a desired normal displacement $u(\mathbf{r},t)$.

4.6.2.8. The Measuring Section of the Active Control Damping System

The control system tries to minimize the difference between three, generally speaking, physically non-identical values: (1) desired (in accordance with control algorithm (4-26) normal displacement of surface S_V; (2) real normal displacement of surface S_V; (3) measured normal displacement of surface S_V. In this section we will formulate the way to measure the value $Y_k(t)$ (see (4-26). Let's direct the axis "z" to the external media (**Fig. 4-10-c**) along the normal of the $k=1,2,...,N_V$-th ($k=1,2,...,N_V$) active piston S_k with point $z=0$ (which is identical with the piston's center $\boldsymbol{\rho}_k$) and with square σ_k. Now we cover the boundaries $z=0$ (surface S_V), $z=-h_c$, $z=-2h_c$ (surface $\overline{S}_V$), $z=-2h_c-h_r$ (surface $S_\bullet$) by thin weightless metal film between the outside media, the piezoelectric layers $-h_c \le z \le 0$ and $-2h_c \le z \le -h_c$ (with relaxed thickness $\overline{h}_c(\mathbf{r})$ and $\overline{\overline{h}}_c(\mathbf{r})$ respectively), and polymer layer $-2h_c-h_r \le z \le -2h_c$ (with relaxed thickness $\overline{\overline{\overline{h}}}_r(\mathbf{r})$), and outside surface $S_\bullet$ of the shell. So we obtain, respectively, three plane electric capacities:

$$\overline{C}_k(t)=\varepsilon_0\varepsilon_c\sigma_k\iint_{S_k}[\overline{h}_c(\mathbf{r})+\overline{h}(\mathbf{r},t)]^{-1}dS_k=\overline{J}_k(t)/[d\overline{\phi}_k(t)/dt], \quad \textbf{(4-38)}$$

$$\overline{\overline{C}}_k(t)=\varepsilon_0\varepsilon_c\sigma_k\iint_{S_k}[\overline{\overline{h}}_c(\mathbf{r})+\overline{\overline{h}}(\mathbf{r},t)]^{-1}dS_k=\overline{\overline{J}}_k(t)/[d\overline{\overline{\phi}}_k(t)/dt],$$

$$\overline{\overline{\overline{C}}}_k(t)=\varepsilon_0\varepsilon_r\sigma_k\iint_{S_k}[\overline{\overline{\overline{h}}}_r(\mathbf{r})+\overline{\overline{\overline{h}}}(\mathbf{r},t)]^{-1}dS_k=\overline{\overline{\overline{J}}}_k(t)/[d\overline{\overline{\overline{\varphi}}}_k(t)/dt],$$

where ε_0, ε_c, –relative dielectric constants of vacuum, piezoelectric and polymer, respectively; $\overline{h}(\mathbf{r})$, $\overline{h}(\mathbf{r})$, $\overline{\overline{\overline{h}}}(\mathbf{r})$ are the instant fields of thickness changes of layers $-h_c \le z \le 0$, $-2h_c \le z \le -h_c$, $-2h_c-h_r \le z \le -2h_c$ of piezoelectric material and polymer, respectively; $\overline{\phi}_k(t)$, $\overline{\overline{\phi}}_k(t)$, $\overline{\phi}_k(t)$ –the measured voltages on the corresponding capacities; $\overline{J}_k(t)$, $\overline{\overline{\overline{J}}}_k(t)$, $\overline{\overline{J}}_k(t)$–the measured charging currents. It is easy to see that at $\overline{h}(\mathbf{r},t) << \overline{h}_c(\mathbf{r})$, $\overline{\overline{h}}(\mathbf{r},t) << \overline{\overline{h}}_c(\mathbf{r})$, $\overline{\overline{\overline{h}}}(\mathbf{r},t) << \overline{\overline{\overline{h}}}_r(\mathbf{r})$, the spatially averaged value (along S_k) $H_k(t)=\sigma_k^{-1}\iint_{S_k}[\overline{h}(\mathbf{r},t)+\overline{\overline{h}}(\mathbf{r},t)+\overline{\overline{\overline{h}}}(\mathbf{r},t)]dS_k$ of change of total thickness of two piezoelectric layers and one polymer layer is expressed *via* measured capacities $\overline{C}_k(t)$, $\overline{\overline{C}}_k(t)$, $\overline{\overline{C}}_k(t)$ as the following:

$$H_k(t)=\overline{h}_c^0(\boldsymbol{\rho}_k)\left\{\left[\overline{C}_k(t)/\overline{C}_k^0\right]-1\right\}+\overline{\overline{h}}_c^0(\boldsymbol{\rho}_k)\left\{\left[\overline{\overline{C}}_k(t)/\overline{\overline{C}}_k^0\right]-1\right\}+\overline{\overline{\overline{h}}}_r^0(\boldsymbol{\rho}_k)\left\{\left[\overline{\overline{\overline{C}}}_k(t)/\overline{\overline{\overline{C}}}_k^0\right]-1\right\},$$

where $\overline{C}_k^0$, $\overline{\overline{C}}_k^0$, $\overline{\overline{\overline{C}}}_k^0$-relaxed values of capacities $\overline{C}_k(t)$, $\overline{\overline{C}}_k(t)$, $\overline{\overline{C}}_k(t)$, $\overline{h}_c^0(\boldsymbol{\rho}_k)=\sigma_k^{-1}\iint_{S_k}\overline{h}_c(\mathbf{r})dS_k$, $\overline{\overline{h}}_c^0(\boldsymbol{\rho}_k)=\sigma_k^{-1}\iint_{S_k}\overline{\overline{h}}_c(\mathbf{r})dS_k$, $\overline{\overline{\overline{h}}}_r^0(\boldsymbol{\rho}_k)=\sigma_k^{-1}\iint_{S_k}\overline{\overline{\overline{h}}}_r(\mathbf{r})dS_k$. Knowing the value $H_k(t)$ and the normal displacement $U_k(t)$ of the boundary $z=-2h_c-h_r$, averaged along the area S_V, we would obtain the value $Y_k(t)=H_k(t)+U_k(t)$ (used in algorithm (4-26). However, the statement of the problem in this work (see Introduction) excludes the usage not only of mechanical support(s) for active pistons (dynamical vibrostat), but also excludes the usage of any inertial coordinate system (cinematic vibrostat), relatively with which we would like to measure normal displacements of controlled surface. We can measure the displacement $G(\boldsymbol{\rho}_k,t)$ of boundary $z=-2h_c-h_r$ in the center $\boldsymbol{\rho}_k$ with an inertial accelerometer, if this displacement is temporally smooth (*i.e.*, the lengths of the corresponding acoustical waves in piezoelectric material and in inertial body of accelerometer are much greater than the accelerometer's dimensions). We need the spatially averaged value $U_k(t)=\sigma_k^{-1}\iint_{S_k}G(\mathbf{r},t)dS_k$. We will obtain it as a signal $U_k(t)\approx G(\boldsymbol{\rho}_k,t)$ under the assumptions about the wave field $G(\mathbf{r},t)$: spatial and temporal smoothness. Smoothness of $G(\mathbf{r},t)$ in space is given by finite bending hardness of the elastic plate (local model of the shell). Temporal smoothness of $G(\mathbf{r},t)$ is given by the absence of high frequency sources (within $S_\bullet$) and by the dissipative opacity of the polymer layer $-2h_c-h_r \le z \le -2h_c$ on the technological frequencies $\omega_n \ge 2\pi n/T_V$ of the active system ($n=1,2,...$). Now we determine $Y_k(t)$ (for algorithm (4-26) as a sum $Y_k(t) = H_k(t) + G(\mathbf{\rho}_k, t)$ of signals $H_k(t)$ of a spatially distributed capacity frequency wide band sensors of thickness (formed by metallic planes $z = -2h_c - h_r$, $z = -2h_c$, $z = -h_c$, $z = 0$) and signal $G(\mathbf{\rho}_k, t)$ of a local low frequency inertial accelerometer, which is spaced on the outside surface $z = -2h_c - h_r$ (or $S_\bullet$) of elastic plate (shell) in the piston center $\mathbf{\rho}_k$. Here $G(\mathbf{\rho}_k, t) = \int_0^t d\varsigma \int_0^\varsigma \overline{\Theta}[q(\mathbf{\rho}_k, \xi)]d\xi$, $q(\mathbf{\rho}_k, \xi)$ –signal (acceleration of the point $\mathbf{\rho}_k$ on $S_\bullet$, ξ -time) of accelerometer, operator $\overline{\Theta}[q] = \int_0^t \Theta(t-\xi)q(\mathbf{\rho}_k, \xi)d\xi$ represents a special linear low pass filter (analogous to the one used in algorithm (4-26). The module $|\tilde{\Theta}(\omega)| = 1$ of its frequency transmission coefficient $\Theta(\omega) = \int_{-\infty}^{+\infty} \Theta(t)\exp(i\omega t)dt$ near zero frequency has the form $|\tilde{\Theta}(\omega)| \sim |\omega|^{2\alpha}$ (α –whole positive number) and is close to $|\tilde{\Theta}(\omega)| = 1$ in the frequencies of range (4-3) and up to the frequencies $\sim 2\pi / T_V$. This filter excludes the possibility of components $G = const$ and $G \sim t$ in the function $G(\mathbf{\rho}_k, t)$. Note that the relative error of the value $Y_k(t)$ does not depend on the smallness of the relative thickness $h_c / \lambda_{\min} << 1$ of the piezoelectric layer.

In this section, we consider the solution of the problem of forming of DNOV (DNOD), having minimum info about the vibroacoustical characteristics of the protected body. We place on the body external surface $S_\bullet$ immediately the high frequency dissipative polymer layer of thickness h_r and the active coating of thickness $2h_c << \lambda_{\min}$ in the form of two piezoelectric layers of thickness h_c (each layer), with identical polarization along the normal to S_V (the external surface of piezoelectric is contacting with outside compressible media). We assume that the longitudinal impedance $Z_c = \rho_c c_c$ of the piezoelectric material is equal to the impedance $Z_w = \rho_w c_w$ of the outside liquid (*i.e.* $|Z_w - Z_c| / Z_c << 1$), where $Z_c = \rho_c c_c$ -mass density of piezoelectric material and $Z_w = \rho_w c_w$ -mass density of outside liquid and correspondingly, c_c -speed of longitudinal sound in the piezoelectric material, c_w –speed of sound in the outside liquid. The total cross sectional structure of coating is presented in **Fig. 4–10–c**. For simplicity, we will assume also $c_c > c_w$. The control algorithm, stability of active system and procedure of measurement are investigated below.

4.7. SUPPRESSION OF THE SCATTERING FIELD

Below we suggest an approach, which requires a much smaller number of microphones and allows placement of active sources immediately on the body's surface, unlike those described in [18], [19], [95], [96]. This active system consists of: (a) a subsystem of fast formation of a desired distribution of normal oscillatory velocities (DNOV) on the external surface of an active coating; (b) a subsystem of detection and targeting of incident waves. The active coating, described above, is spaced immediately on the surface of protected body. This coating creates the desired normal velocity (algorithm (4-26)) on external surface S_V.

4.7.1. Incident Wave Field

We will use the following assumptions for the incident wave field. We consider the incident waves appearing, for example, as a result of ray-like (**Fig. 4-18**) far propagation in the undersea sound channel [97]. We assume that the protected body (area) of characteristic size is fully covered by the cross section of tube of rays (of diameter $\sim D_R$), *i.e.* $D_R >> D_V$. The wave within the tube of rays represents locally a running plane wave. The protected body can be spaced in the intersection of two (or more) tubes of rays and therefore the scattering problem can include two (or more) incident plane waves. We expect wave dispersion effects are very small, and speed of sound propagation is constant within the area occupied by the protected body.

The source of incident waves can radiate both video (without filling by sinusoidal shapes) pulses and radio-like sound pulses (with filling by sinusoidal shapes). In any case any video pulse will be converted *via* far propagation into a group of radio pulses, running along ray tubes. In addition we assume that each plane wave has finite duration. In other words each wave has a front border of plane shape moving with the speed of sound (of smooth or step-like shape), behind which the sound field is zero and also nonzero behind the border. A step-like front border is more preferable for our approach, but it is impossible due to wave dispersion in long range propagation. The magnitudes of incident waves in the active acoustical detection problem (and in the absence of anti-sonar activity) must ensure the registration of scattered waves by the receiver system spaced on the same large distance between the scatterer (body) and the source of incident waves. Therefore if the distance between the source of incident waves and the protected body increases, the incident wave needs to shake the scatterer surface more powerfully for registration by the receiving microphones, without any increase of their sensitivity. This ensures the necessary accuracy of bearing and targeting (see below) of incident waves. Now we'll assume that the incident waves represent a small number $N_w > 1$ of powerful, smooth, plane wave groups of finite duration and with amplitudes much greater than natural noise background σ_P of sound pressure. In the absence of a scattering body the sound pressure field of incident waves has the form:

$$p(\mathbf{r},t) = \sum_{i=1}^{N_W} \Xi_i[t-(\mathbf{r},\mathbf{w}_i)c_w^{-1}], \quad \textbf{(4-39)}$$

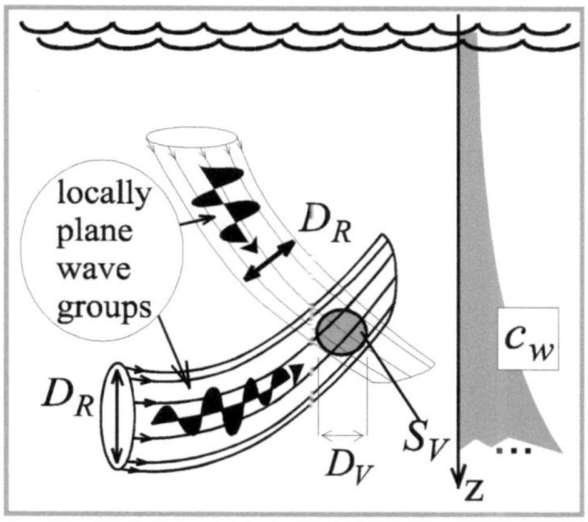

Figure 4-18: The protected body in the intersection of two tubes of rays with two incident wave groups (locally plane). This is in the undersea acoustical channel with non-uniform distribution $c_w(z)$ (along the depth " z ") of speed of propagation of acoustical waves.

where $i = 1,2,...,N_w$, (**Figs. 4-18, 4-19-a**) $\mathbf{w}_i$ −the normalized vectors of incident waves ($|\mathbf{w}_i| = 1$), c_w −speed of sound propagation in the fluid, $(\mathbf{r},\mathbf{w}_i)$ −scalar product of vectors [98].

4.7.2. The Scattering Suppression System and Algorithm

Let's assume that a closed salient surface S_P (**Fig. 4-19-a**) surrounds the closed surface (outside surface of active coating of protected body). N_P small ("nonscattering") microphones with output signals $p_m(t)$ of acoustical pressure are placed in the *known* points $(\mathbf{r},\mathbf{w}_i)$ ($i = 1,2,...,N_w$) of the surface S_P. At the output of the linear low pass filter without distortions and with frequency bandwidth ω_Q and with group delay τ_Q (the role of this filter is described in Section 6.3. below) we obtain $P_m(t) = \overline{Q}[p_m(t)]$.

4.7.2.1. Targeting of One Incident Wave

The first incident wave (with number $i = 1$) forms its bearing group, which consists of the four microphones (**Fig. 4-19-a**). These microphones differ from each other by their participation in the following events:

1. First contact between the front of an incident wave with any microphone on S_P means, that at this moment $t = \tilde{t}(1,0)$) (first for $\forall t \leq \tilde{t}(1,0)$) on the microphone ★ with number $m = \tilde{m}(1,0)$ at the point the modulus $|\overline{Q}\Xi_1[t-(\mathbf{r},\mathbf{w}_1)c_w^{-1}|$ of the acoustical pressure $P_m(t) = \overline{Q}[p_m(t)]$ of a first incident wave had crossed (from the bottom to the top) the threshold level $P_\bullet$ of fixing. There $P_\bullet$, temporal error $\tau_\bullet$, noise mean-square value of acoustic noise pressure σ_P, all satisfy the conditions:

$$\sigma_P << P_\bullet << \left|\Xi_i\right|_{\max}, \; \tau_\bullet \le \tau_{\max} P_\bullet \left|\Xi_i\right|_{\max}^{-1} << \tau_{\min} . \quad \textbf{(4-40)}$$

2. 2-nd contact between front of an incident wave with any (another) microphone on S_P means, that at this moment $t = \tilde{t}(1,1)$ (first for $\forall t \le \tilde{t}(1,1)$) on the microphone ☆ with number $m = \tilde{m}(1,1)$ at the point $\mathbf{r} = \mathbf{R}_{\tilde{m}(1,1)}$ the modulus $\left|\bar{Q}\Xi_1[t - (\mathbf{r},\mathbf{w}_1)c_w^{-1}\right|$ 1the acoustical pressure $P_{\tilde{m}(1,1)}(t)$ of a first incident wave had crossed (from the bottom to the top) the threshold level $P_\bullet$ of fixing.

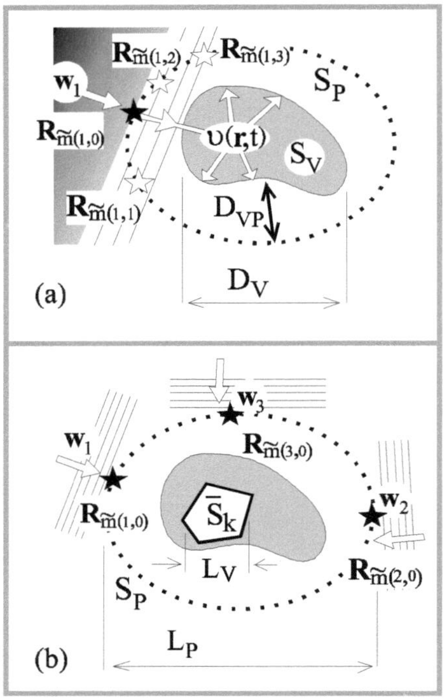

Figure 4-19: Targeting of one (a) and three (b) incident waves. Incident wave (a) forms the bearing group of microphones (one title microphone and three benchmark ones) on the surface S_P to match with the surface S_V of a protected body, covered by an active coating with pistons S_k ($k = 1, 2, ..., N_V$). Three incident waves (b), arriving from different directions $\mathbf{w}_1$, $\mathbf{w}_2$, $\mathbf{w}_3$ create three nonintersecting bearing groups.

3. 3-rd contact between front of an incident wave with any (another) microphone on S_P means, that at this moment $t = \tilde{t}(1,2)$ (firstly for $\forall t \le \tilde{t}(1,2)$) on the microphone ☆ with number $m = \tilde{m}(1,2)$ at the point $\mathbf{r} = \mathbf{R}_{\tilde{m}(1,2)}$ the modulus $\left|\bar{Q}\Xi_1[t - (\mathbf{r},\mathbf{w}_1)c_w^{-1}\right|$ of acoustical pressure $P_{\tilde{m}(1,2)}(t)$ of a first incident wave had crossed (from the bottom to the top) with the threshold level $P_\bullet$ of fixing.

4. Fourth contact between front of incident wave with any (another) microphone on S_P means, that at this moment (firstly for $\forall t \le \tilde{t}(1,2)$) on the microphone ☆ with number $m = \tilde{m}(1,3)$ in the point $\mathbf{r} = \mathbf{R}_{\tilde{m}(1,3)}$ the modulus $\left|\bar{Q}\Xi_1[t - (\mathbf{r},\mathbf{w}_1)c_w^{-1}\right|$ of acoustical pressure $P_{\tilde{m}(1,3)}(t)$ of a first incident wave had crossed (from the bottom to the top) with the threshold level $P_\bullet$ of fixing.

The microphone ★ in the point $\mathbf{r} = \mathbf{R}_{\tilde{m}(1,0)}$ we'll call the *title* microphone with pressure signal $P_m(t) = \overline{Q}[p_m(t)]$ and other microphones ☆ in the points $\mathbf{R}_{\tilde{m}(1,1)}$, $\mathbf{R}_{\tilde{m}(1,2)}$, $\mathbf{R}_{\tilde{m}(1,3)}$ we'll call *benchmark* microphones with pressure signals $P_{\tilde{m}(1,1)}(t)$, $P_{\tilde{m}(1,2)}(t)$, $P_{\tilde{m}(1,3)}(t)$ respectively. Taking into account the definitions 1–4, we obtain vector $\mathbf{w}_1$ as a solution of the system of equations:

$$\{\mathbf{w}_1, [\mathbf{R}_{\tilde{m}(1,k)} - \mathbf{R}_{\tilde{m}(1,0)}]\} = [\tilde{t}(1,k) - \tilde{t}(1,0)],\ (k = 1,2,...,N_V). \tag{4-43}$$

We assume, that all the microphones are very small (like points), therefore we can assume that the moments $t = \tilde{t}(1,1)$, $t = \tilde{t}(1,2)$, $t = \tilde{t}(1,3)$ and the points $\mathbf{R}_{\tilde{m}(1,1)}$, $\mathbf{R}_{\tilde{m}(1,2)}$, $\mathbf{R}_{\tilde{m}(1,3)}$ must be mutually connected with each other, respectively, due to condition (4-40). In the absence of a protected body (or at condition (4-2)), the title microphone ★ "knows" the incident wave field earlier than other microphones on the surface S_P. Knowing the wave pressure $P_m(t) = \overline{Q}[p_m(t)]$ in the point $\mathbf{r} = \mathbf{R}_{\tilde{m}(1,0)}$ and knowing the vector $\mathbf{w}_1$, one can determine the incident wave field at the moment $P_m(t) = \overline{Q}[p_m(t)]$ in the point , which satisfies the condition $\{\mathbf{w}_1, [\mathbf{r} - \mathbf{R}_{\tilde{m}(1,0)}]\} > 0$, if we use the formula (which can be interpreted as the "diffraction" on the transparent body):

$$\Xi_1(\mathbf{r},t) = \Xi_1\{\mathbf{R}_{\tilde{m}(1,0)}, t - (\mathbf{w}_1, [\mathbf{r} - \mathbf{R}_{\tilde{m}(1,0)}]c_w^{-1})\}. \tag{4-44}$$

4.7.2.2. Vector Microphone

To ensure the stability of active system, we form a vector microphone (with selective directional pattern) [99] with its direction of maximum sensitivity coinciding with the vector $\mathbf{w}_1$ of the first incident wave, using the microphones of the bearing group, so we need two microphones, spaced on the axis $\mathbf{w}_1$ at some distance h_1 from each other (**Fig. 4-20**). At first we take the title microphone ★ with output signal $P_{\tilde{m}(1,0)}(t)$. The signal $\overline{P}_{\tilde{m}(1,0)}(t)$ of the 2-nd microphone, which is spaced in some point $\mathbf{R}_{\tilde{m}(1,0)} + h_1\mathbf{w}_1$ (so far we only know that $h_1 > 0$) will form a combination $\overline{P}_{\tilde{m}(1,0)}(t) = \mu(1,1)P_{\tilde{m}(1,1)}(t) + \mu(1,2)P_{\tilde{m}(1,0)}(t) + \mu(1,3)P_{\tilde{m}(1,3)}(t)$ of signals of benchmark microphones ☆, where:

$$\mu(1,1) + \mu(1,2) + \mu(1,3) = 1,\ \mu(1,1), \mu(1,2), \mu(1,3) > 0. \tag{4-45}$$

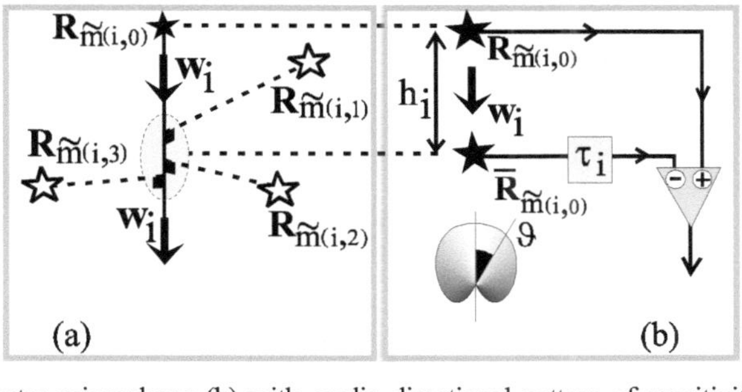

Figure 4-20: Forming of virtual vector microphone (b) with cardic directional pattern of sensitivity by means the combination (a) of signals of microphones group of bearing of the i-th incident wave. The mass center of points $\mathbf{R}_{\tilde{m}(1,1)}$, $\mathbf{R}_{\tilde{m}(1,2)}$, $\mathbf{R}_{\tilde{m}(1,3)}$ (with weights $\mu(1,1)$, $\mu(1,2)$, $\mu(1,1) + \mu(1,2) + \mu(1,3) = 1$ must be laying on the axis, coming from the point $\mathbf{R}_{\tilde{m}(1,0)}$ in direction of the vector $\mathbf{w}_i$.

Ensuring the equation:

$\mu(1,1)\mathbf{R}_{\tilde{m}(1,1)} + \mu(1,2)\mathbf{R}_{\tilde{m}(1,2)} + \mu(1,3)\mathbf{R}_{\tilde{m}(1,3)} = \mathbf{R}_{\tilde{m}(1,0)} + h_1\mathbf{w}_1$ (4-46) (*i.e. center of masses*) $\mu(1,1)$, $\mu(1,2)$, $\mu(1,1) + \mu(1,2) + \mu(1,3) = 1$, spaced at points $\mathbf{R}_{\tilde{m}(1,1)}$, $\mathbf{R}_{\tilde{m}(,2)}$, $\mathbf{R}_{\tilde{m}(1,3)}$ coincides with the point $\mathbf{R}_{\tilde{m}(1,0)} + h_1\mathbf{w}_1$ we obtain concrete weight coefficients $\mu(1,1)$, $\mu(1,2)$, $\mu(1,1)+\mu(1,2)+\mu(1,3)=1$ and distance h_1. Taking into account the smoothness of the incident wave we form the signal $\overline{P}_{\tilde{m}(1,0)}(t)$ of the virtual microphone, which is spaced at the point $\overline{\mathbf{R}}_{\tilde{m}(1,0)} = \mathbf{R}_{\tilde{m}(1,0)} + h_1\mathbf{w}_1$. Now we obtain the output signal $U(t) = P_{\tilde{m}(1,0)}(t) - \overline{P}_{\tilde{m}(1,0)}(t-\tau_1)$ of the vector microphone. For the wave running in direction $\mathbf{w}_1$ (or $\vartheta = 0$, *i.e.* the first incident wave) the signal $U(t)$ represents time derivative of sound pressure $P_{\tilde{m}(1,0)}(t)$ of incident wave in the point $\mathbf{R}_{\tilde{m}(1,0)}$ with delay $\tau_1 = h_1 / c_w$: $U(t) \cong 2\tau_1 (d/dt) P_{\tilde{m}(1,0)}(t-\tau_1)$. The incident wave pressure $\Psi_{\tilde{m}(1,0)}(t-\tau_1)$ in the point $\mathbf{R}_{\tilde{m}(1,0)}$ we obtain as integral $\Psi_{\tilde{m}(1,0)}(t-\tau_1) = (1/2\tau_1) \int_{t'=\tilde{t}(1,3)}^{t'=t} U(t')dt'$.

Below we will use the following symbols:

(a) $\upsilon_i(\boldsymbol{\rho}_k, t)$ -the desired normal velocity of the k -th active piston (temporally averaged on the interval T_V and spatially averaged along the external surface of piston S_k), used in algorithm (4–26), when targeted to i -th incident wave (simultaneously at $i > 1$) and

$$u_i(\boldsymbol{\rho}_k, t) - u_i(\boldsymbol{\rho}_k, t - T_V) = T_V \upsilon_i(\boldsymbol{\rho}_k, t) \textbf{ at } \tilde{t}(i,3) < t < \tilde{t}[(i+1),0] \,;$$

(b) $\upsilon_0(\boldsymbol{\rho}_k, t)$ -the desired normal velocity of k -th active piston *before* the arrival (or touch to microphone) of the first incident wave (averaged temporally on the interval T_V and spatially along the external surface of piston S_k). In other words we have $\upsilon_0(\boldsymbol{\rho}_k, t) = 0$, $u_0(\boldsymbol{\rho}_k, t) = 0$ in the problem of suppression of radiation by surface S_k in the absence of incident waves *i.e.* if $t < \tilde{t}(1,0)$.

4.7.2.3. Renovation of DNOV on S_k for the First Incident Wave

The desired velocity of k -th piston is (in accordance with (4-2)):

$$\upsilon_1(\boldsymbol{\rho}_k, t) = \upsilon_0(\boldsymbol{\rho}_k, t) + (\rho_w c_w \sigma_k)^{-1} \iint_{S_k} dS_k(\mathbf{w}_1, \Diamond) \Psi_{\tilde{m}(1,0)}(t - \overline{\tau}_1) \,, \tag{4-47}$$

where $\overline{\tau}_1 = c_w^{-1}(\mathbf{w}_1, \boldsymbol{\rho}_k - \mathbf{R}_{\tilde{m}(1,0)}) - \tau_1 - \tau_Q > 0$, $\overline{\tau}_1 = c_w^{-1}(\mathbf{w}_1, \boldsymbol{\rho}_k - \mathbf{R}_{\tilde{m}(1,0)}) - \tau_1 - \tau_Q > 0$ −scalar products of vectors. Simultaneously with the control processes, described above, the surface of the protected body (or internal surface S_k of active coating) oscillates with the same velocity, which must be without an active coating. DNOV is induced by one incident wave and internal sources of body vibrations. If we know exactly the microphones' coordinates and exactly know the shape of the surface, then we can estimate the relative error $(\Delta\upsilon / \upsilon)$ of DNOV as the following expression:

$$(\Delta\upsilon / \upsilon) \approx (\Delta P / P)^2 (\lambda_{\max} D_V / \ell_P^2) \,, \tag{4-48}$$

where ℓ_P –characteristic distance between adjacent microphones in group of bearing, $\lambda_{\max}$ –maximum length of the waves to be damped, $(\Delta P / P)$ –relative error of the measured acoustical pressure. The formula (4-48) can be used as the rough simple estimate of the relative value of the residual acoustical scattered field. The magnitude $B_{n,k}$, of signal of excitation of k -th piston of active coating at the moment $t = nT_V$ is in accordance with algorithm (4-26), determined by the formula $B_{n,k} = (\psi_0 T_V / 2)^{-1} \int_{nT_V - T}^{nT_V} \overline{F}[Y_k(t) - \upsilon_1(\boldsymbol{\rho}_k, nT_V)]dt$, where $Y_k(t)$ -current value (measured) spatially averaged along the k -th active piston, $\upsilon_1(\boldsymbol{\rho}_k, nT_V)$ -the desired velocity of the k -th piston.

For the special case of acoustically rigid protected surface $S_\bullet$ can be used the simple one layer active coating of controlled thickness with more simple control algorithm [92].

4.7.2.4. Microphone Signal Renovation for Catching the Second Incident Wave

To prepare other microphones with numbers $m \neq \tilde{m}(1,j)$, ($j = 0,1,2,3$) for registration and targeting of next waves (beginning from the moment $t \neq \tilde{t}(1,0)$) we exclude acoustical pressure of the first incident wave, *i.e.*

$$P_m(t) = P_m(t) - \Psi_{\tilde{m}(1,0)}(t - \overline{\overline{\tau}}_1), \qquad \textbf{(4-49)}$$

where $\overline{\overline{\tau}}_1 = c_w^{-1}(\mathbf{w}_1, \mathbf{R}_m - \mathbf{R}_{\tilde{m}(1,0)}) - \tau_1 - \tau_Q$.

So the first incident wave is described by the following group of numbers and vectors (**Fig. 4-19-b**): $\{\tilde{t}(1,0)$, $\tilde{m}(1,0)$, $\mathbf{R}_{\tilde{m}(1,0)}\}$, $\{\tilde{t}(1,1)$, $\tilde{m}(1,1)$, $\mathbf{R}_{\tilde{m}(1,1)}\}$, $\{\tilde{t}(1,2)$, $\tilde{m}(1,2)$, $\mathbf{R}_{\tilde{m}(1,2)}\}$, $\{\tilde{t}(1,3)$, $\tilde{m}(1,2)$, $\mathbf{R}_{\tilde{m}(1,2)}\}$, $\overline{\overline{\tau}}_1 = c_w^{-1}(\mathbf{w}_1, \mathbf{R}_m - \mathbf{R}_{\tilde{m}(1,0)}) - \tau_1 - \tau_Q$, τ_1, h_1, $\mu(1,1)$, $\mu(1,2)$, $\mu(1,1) + \mu(1,2) + \mu(1,3) = 1$. Here $\tilde{t}(i,j)$, $\tilde{m}(i,j)$, $\mu(i,j)$ −functions of discrete arguments $i = 1,2,...,N_W$, $j = 0,1,2,3$; $\tilde{t}(i,j)$, $\mu(i,j)$ −real numbers, $\tilde{m}(i,j)$ −whole number.

Now we remind the sense of the used symbols: $k = 1,2,..., N_V$ −ordinal numbers of damping pistons; $n = 1,2,3,...$ −ordinal number of basic temporal cycle (of duration T_V) of active system; $i = 1,2,..., N_w$ −ordinal numbers of incident waves in sequence of their first contacts (touching) with microphones system; $\tilde{m}(i,j)$ -ordinal numbers of microphones in the group of bearing of the i-th incident wave in dependence of concrete scenario ($j = 0,1,2,3$); $m = 1,2,..., N_P$ -ordinal numbers of "physical" microphones independently of any scenario. Let's assume that system of incident waves (4-39) and microphones coordinates $\mathbf{R}_m$ ensure the following conditions: $\tilde{t}(i,0) \le \tilde{t}(i,1) \le \tilde{t}(i,2) \le \tilde{t}(i,3)$ for $\forall i$; $\tilde{m}(i,j) \neq \tilde{m}(k,\ell)$ at $\forall i \neq k$ and $\forall j,\ell = 0,1,2,3$; $T_V << |\tilde{t}(i,1) - \tilde{t}(i,0)|, |\tilde{t}(i,2) - \tilde{t}(i,0)|, |\tilde{t}(i,3) - \tilde{t}(i,0)| < D_{VP} / c_w$. In other words we assume that the microphone system and system of incident waves always permit us to form bearing groups without intersection of microphones for targeting each incident wave.

4.7.2.5. Two Incident Wave Targeting

At the moment $t = \tilde{t}(2,0)$ the second incident wave arrives (**Fig. 4-19-b**). Among a lot of microphones with numbers $m \neq \tilde{m}(1,j)$ (where $j = 0,1,2,3$) we elect one title microphone ★ and three bench-mark microphones ☆, forming the group of bearing in accordance with the rule (above described) of first crossing (from the bottom to the top) of threshold level $P_\bullet$ (**Fig. 4-21**) by the modulus $|\overline{Q}\Xi_2[t - (\mathbf{r}, \mathbf{w}_2)c_w^{-1}|$ of acoustical pressure of 2-nd incident wave on the moments $t = \tilde{t}(2,0)$, $t = \tilde{t}(2,1)$, $t = \tilde{t}(2,2)$, $t = \tilde{t}(2,3)$. We assume, that these moments are connected with front border of second incident wave in the spatial points $\mathbf{R}_{\tilde{m}(2,0)}$, $\mathbf{R}_{\tilde{m}(2,1)}$, $\mathbf{R}_{\tilde{m}(2,2)}$, $\mathbf{R}_{\tilde{m}(2,3)}$ respectively. From the system of equations:

$$\{\mathbf{w}_2, [\mathbf{R}_{\tilde{m}(2,k)} - \mathbf{R}_{\tilde{m}(2,0)}]\} = [\tilde{t}(2,k) - \tilde{t}(2,0)], \ (k = 1,2,..., N_V) \qquad \textbf{(4-50)}$$

we obtain the vector $\mathbf{w}_2$ of 2-nd incident wave. Knowing the wave pressure in the point $\mathbf{R}_{\tilde{m}(2,0)}$ and knowing the vector $\mathbf{w}_2$, one can determine the incident wave field at the moment $t = \tilde{t}(2,3)$ in the point $\mathbf{r}$, which satisfy the condition $(\mathbf{w}_2, (\mathbf{r} - \mathbf{R}_{\tilde{m}(2,0)})) > 0$. We obtain this from the formula

$$\Xi_2(\mathbf{r}, t) = \Xi_2\{\mathbf{R}_{\tilde{m}(2,0)}, t - (\mathbf{w}_2, [\mathbf{r} - \mathbf{R}_{\tilde{m}(2,0)}]c_w^{-1})\}. \qquad \textbf{(4-51)}$$

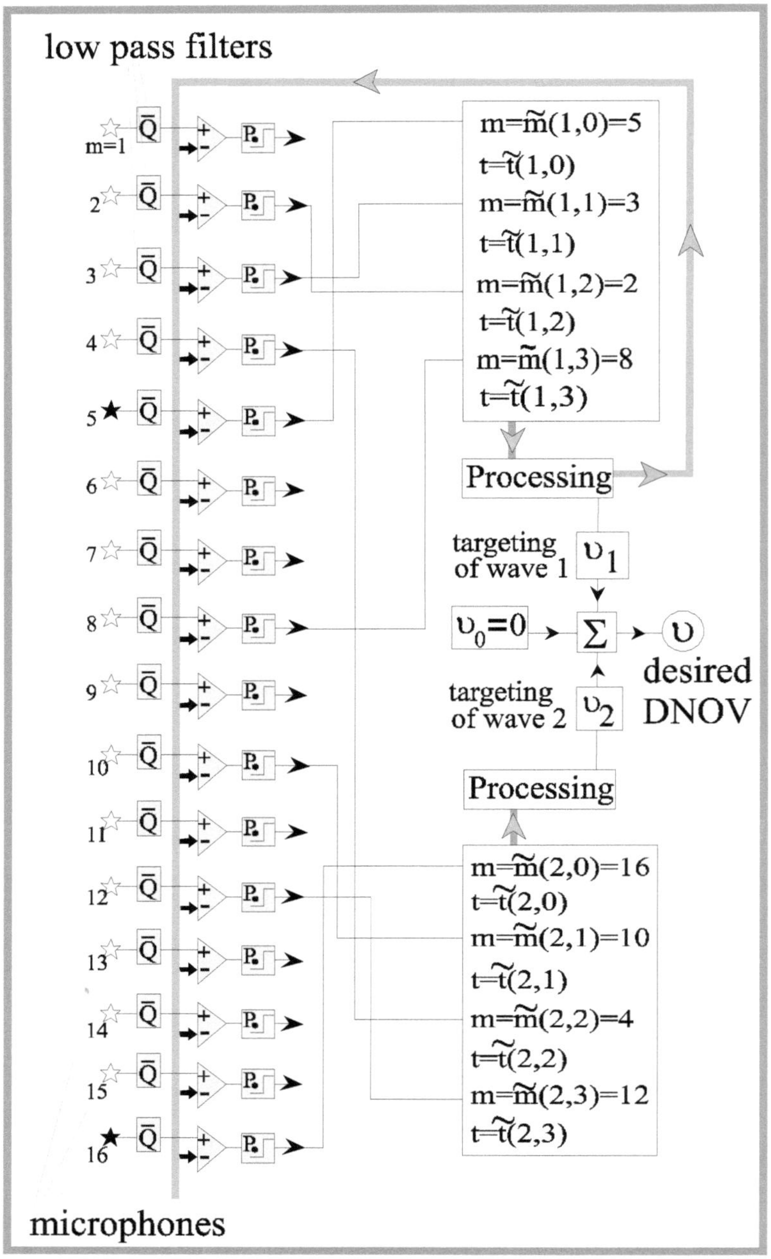

Figure 4-21: The example of forming groups of bearing for renovation of signals of microphones for $N_P = 16$ (with targeting of two incident waves).

As in the procedure of the vector microphone, described above, we form the signal $P_{\tilde{m}(2,0)}(t-\tau_2)$ and then form signal $\overline{P}_{\tilde{m}(2,0)}(t-\tau_2)$ of the view

$$\overline{P}_{\tilde{m}(2,0)}(t) = \mu(2,1)P_{\tilde{m}(2,1)}(t) + \mu(2,2)P_{\tilde{m}(2,0)}(t) + \mu(2,3)P_{\tilde{m}(2,3)}(t)\,. \qquad \textbf{(4-52)}$$

The parameters $\mu(2,1)$, $\mu(2,2)$, $\mu(2,3)$ ($\mu(2,1), \mu(2,2), \mu(2,3) > 0$), h_2, we obtain from the system of equations:

$$\mu(2,1) + \mu(2,2) + \mu(2,3) = 1\,, \qquad \textbf{(4-53)}$$

$$\mu(2,1)\mathbf{R}_{\tilde{m}(2,1)} + \mu(2,2)\mathbf{R}_{\tilde{m}(2,2)} + \mu(2,3)\mathbf{R}_{\tilde{m}(2,3)} = \mathbf{R}_{\tilde{m}(2,0)} + h_2\mathbf{w}_2 \,. \quad \textbf{(4-54)}$$

The output signal of vector microphone has the form $\Psi_{\tilde{m}(2,0)}(t-\tau_2) = (1/2\tau_2)\int_{t'=\tilde{t}(2,3)}^{t'=t} U(t')dt'$, where $U(t) = P_{\tilde{m}(2,0)}(t) - \overline{P}_{\tilde{m}(2,0)}(t-\tau_2)$, $\tau_1 = h_1 / c_w$.

4.7.2.6. Renovation of DNOV on T_V for Two Incident Waves

The desired velocity of the k-th piston is (in accordance with (4-2)):

$$\upsilon_2(\boldsymbol{\rho}_k, t) = \upsilon_1(\boldsymbol{\rho}_k, t) + (\rho_w c_w \sigma_k)^{-1} \iint_{S_k} dS_k(\mathbf{w}_2, \Diamond)\Psi_{\tilde{m}(2,0)}(t-\overline{\tau}_2), \quad \textbf{(4-55)}$$

$\overline{\tau}_2 = c_w^{-1}(\mathbf{w}_2, \boldsymbol{\rho}_k - \mathbf{R}_{\tilde{m}(2,0)}) - \tau_2 - \tau_Q > 0$, $\overline{\tau}_2 = c_w^{-1}(\mathbf{w}_2, \boldsymbol{\rho}_k - \mathbf{R}_{\tilde{m}(2,0)}) - \tau_2 - \tau_Q > 0$ –scalar products of vectors. Simultaneously with the control processes, described above, the surface of the protected body (or internal surface S_V of an active coating) oscillates with the same velocity, which must be without active coating. DNOV is induced by two incident waves and internal sources of vibrations of body. The magnitude $B_{n,k}$ of signal of excitation of k -th piston of active coating at the moment $\tau_1 = h_1 / c_w$ is in accordance with algorithm (4-26) is determined by the formula $B_{n,k} = (\psi_0 T_V / 2)^{-1} \int_{nT_V - T}^{nT_V} \overline{F}[Y_k(t) - \upsilon_2(\boldsymbol{\rho}_k, nT_V)]dt$, where $Y_k(t)$ -current value (measured) spatially averaged along the k -th active piston, $\upsilon_2(\boldsymbol{\rho}_k, nT_V)$ –the desired velocity of k -th piston.

4.7.2.7. Microphone Signal Renovation for Catching the Third Incident Wave

From the time $t = \tilde{t}(2,0)$ we cancel acoustical pressure of the 2-nd incident wave in all microphones, if they are not included in the groups of bearing, to target on both 1-st and 2-nd incident waves, *i.e.*:

$P_m(t) = P_m(t) - P_{\tilde{m}(2,0)}(t - \overline{\overline{\tau}}_2)$, (4-56) where $m \neq \tilde{m}(1,j), \tilde{m}(2,j)$ ($k = 1,2,..., N_V$), $\overline{\overline{\tau}}_2 = (\mathbf{w}_2, \mathbf{R}_m - \mathbf{R}_{\tilde{m}(2,0)})c_w^{-1}$, $(\mathbf{w}_2, \mathbf{R}_m - \mathbf{R}_{\tilde{m}(2,0)})$. Due to operation (4-56) other microphones (they are not involved in any group of bearing formed already) can register the arrival of next incident waves. In the absence of scattering body (or at condition (4-2)) the title microphone "knows" incident wave field before other microphones on surface S_P. So we prepare other microphones with numbers $m \neq \tilde{m}(1,j), \tilde{m}(2,j)$ ($k = 1,2,..., N_V$) for registration and targeting of third incident wave.

4.7.2.8. Calibration

To ensure the accuracy of our measurement of acoustical pressure $p(\mathbf{r},t)$, we need to have identical calibration within each bearing group. We assume that the output of each microphone depends linearly on acoustical pressure: $P_m(t) = a_m p_m(t) + b_m$ ($m = 1,2,..., N_P$). Coefficients a_m, b_m have slow temporal drift and need fast identification of them. The relation (like (4-39), (4-44), (4-51)) between time delay and media particle displacement is constant along the propagation of front plane border of wave. These relations give us parameters a_m, b_m ..., if the coordinates of microphones are known exactly and natural acoustical noise is very weak in comparison with the amplitude of incident waves.

4.7.2.9. Reset of Active System

If at some moment we notice, that the "silence" (*i.e.* $|P_m(t)| < P_\bullet$) has been continued simultaneously on each microphone of some group of bearing during the time $\sim L_P / c_w$ (without exclusions), so the microphones of this group of bearing should be returned to their initial status (*i.e.* waiting) and do not control active coating. In other words every moment $t = nT_V$ the active system is checking the output of the counter

$$I_i(t_n) = N_0^{-1} \sum_{\ell=n-N_0}^{\ell=n} \prod_{m=\tilde{m}(i,k),k=0,1,2,3} I\left(\left|P_{\tilde{m}(i,k)}(t_\ell)\right| - P_\bullet\right) \quad \textbf{(4-57)}$$

for each i-th group of bearing (where $N_0 = L_P / c_w$, $I(\xi) = 0$ at $\xi < 0$. $I(\xi) = 1$ at $\xi \geq 0$). The signal of i-th group of bearing resets the waiting status, when gets jump from position $I_i = 1$ into position– $I_i = 0$. So the algorithm, described in section 4.7.2, can be classified as {NS/LT/CA/F} (see **Fig. 1-3**).

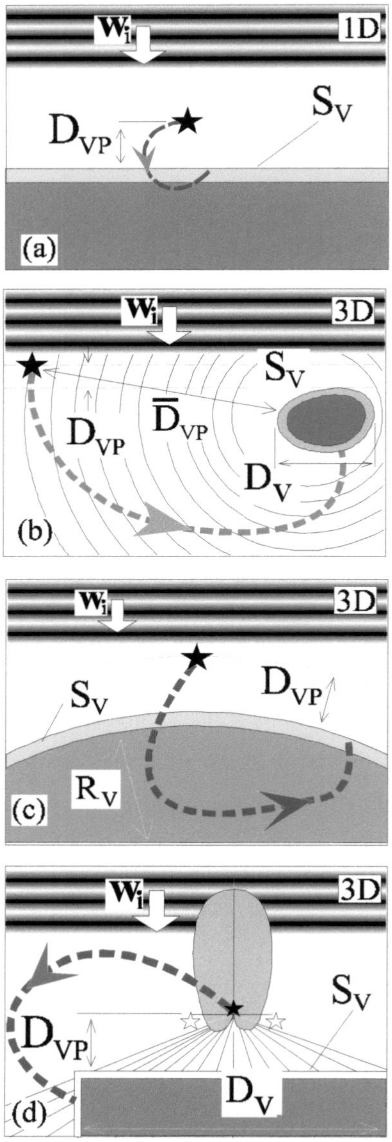

Figure 4-22: Various scenarios of the wave feedback in active (star means the microphone (a-c) or bearing group of microphones (d), big grey arrow denotes the electronic channel of control the DNOV by the microphone (feedback).

4.7.2.10. Notes about the Stability of an Active System with Microphones

The description above was made under the assumption that active coating forms desired DNOV on the surface S_V sufficiently quickly, precisely and safely. Now we consider possible instability channels, created by microphones. Together these channels must cover all frequencies. Only in this case their blocking could guarantee system's stability. There the three channels can generate instability *via* microphones.

(i) Active coating generates the acoustical field in the range (4-3) and this field penetrates *via* microphones. This channel of instability is blocked by the vector-microphone, which ensures the feedback coefficient has a value of less than 1.

(ii) Active coating generates the sound field outside of the range (4-3) too. On these frequencies the directional pattern of vector-microphone has too many of narrow petals, but does not ensure the desired integral (in body angle– $\pm 2\pi$) and anisotropy with respect to the vector $\mathbf{w}_i$ of incident wave, if $h_i\omega / c_w >> 1$, for instance. Besides this, it is impossible to form the desired DNOV by pistons of finite dimension– L_V. To block this channel of instability we use low pass filter with time constant τ_Q (under the condition– $D_V << c_w \tau_Q << 2\pi / \chi_{\max}$) on the output of each microphone without inertial distortions of desired signals and inserting of only delay– τ_Q. Control system takes into account this delay without problems with targeting of incident wave with condition $\tau_Q < D_{VP} / c_w$.

(iii) Reactive acoustical field (near field), created by active coating in the range (4-3), penetrates the microphones too. This field can't be reduced by a vector microphone. However this field is exponentially decreasing as $\gamma = \exp\left[-D_{VP}\sqrt{(2\pi / L_V)^2 - \chi_{\max}^2}\right]$ at the distance D_{VP} from coating. So at not very large distance D_{VP} we can get $\gamma << 1$ by decreasing L_V.

The items (i), (ii) are illustrated in **Fig. 4-22** by the various versions of boundary problem are presented. We assume that one can quickly create any predetermined DNOV on S_V by means of active coating and actuators. We consider the boundary problem with initial conditions with incident waves of the form (4-39).

Version in **Fig. 4-22-a** (spatially 1D case): At first moment $t = 0$ incident wave pressure and signal of microphone ★ is zero. Surface S_V is at rest. In the absence of noise and errors in measuring and control the signal supports the continuous DNOV (without discrete pistons) $\upsilon_0(t)$ on S_V which satisfy the condition (4-2), and scattering field is not produced by anything. Now let's suppose some small error in measuring or control $\upsilon_0(t) + \upsilon_1(t)$ instead– $\upsilon_0(t)$. This error $\upsilon_1(t)$ will produce the scattering sound pressure field $p_1(t)$ of the 1-st order on the microphone. Passing *via* microphone this field will produce the velocity error and scattering field of 2-nd order, i. e. $\upsilon_2(t)$ and $p_2(t)$ respectively,…,*etc.* So we obtain the row $\sum_{n=1} \upsilon_n(t)$ of velocity errors (or scattering fields $\sum_{n=1} p_n\left[\sum_{m=1} \upsilon_n\right]$) on the frequencies– $\omega_n = \pi nc / D_{VP}$, where $n = 1, 2, 3, \ldots$ (frequencies of instability or self-excitation), D_{VP} –minimum distance between microphone and outside surface of active coating. This row gives the unlimited value of summarized error. So the version in **Fig. 4-22-a** is unstable.

Version in **Fig. 4-22-b** (spatially 3D case): microphone ★ is spaced at the great distance $\bar{D}_{VP} >> D_V$ from the protected body and its surface– S_V. So, if $D_{VP} << D_V << \bar{D}_{VP}$, we will obtain the maximum relative error of each n-th order $\sim (D_V / \bar{D}_{VP})^n << 1$ (where $n \sim c_w t / (2D_{VP})$) is concentrated on the very low frequencies $\bar{\omega}_n = \pi nc / \bar{D}_{VP}$. So this scheme is stable and gives suitable resulting accuracy. Although this case is not of practical interest of course? Due its very big dimensions.

Version in **Fig. 4-22-c** (spatially 3D case, spherical version of **Fig. 4-22-b**): microphone ★ is spaced at the great distance D_{VP} from the sphere S_V of radius R_V with active coating with corresponding ideal DNOV $\upsilon_0(\mathbf{r}, t)$. So, if $D_{VP} << R_V$ we will obtain large maximum relative error of n-th order $\sim [R_V / (D_{VP} + R_V)]^n < 1$ (where $n \sim t / (2D_{VP})$).

Version in **Fig. 4-22-d**: vector microphone is formed by the bearing group of microphones. Let's assume that maximum suppression coefficient of vector microphone (on the base of bearing group) is $0 < \gamma < 1$. So we obtain the following estimate of scattering sound pressure field– $\left|\sum_{n=1}^{\infty} p_n\right| \approx |p_0|\gamma^2(1-\gamma)^2 << |p_0|$, where $\gamma = \gamma(D_V, D_{VP}, \omega)$, or shorter $\left|\sum_{n=1}^{\infty} p_n\right| << |p_0|$.

Version in **Fig. 4-22-b** (spatially 3D case): microphone ★ is spaced at the great distance $\bar{D}_{VP} >> D_V$ from the protected body and its surface S_V. So, if $D_{VP} << D_V << \bar{D}_{VP}$, we will obtain the maximum relative error of each n-th order $\sim (D_V / \bar{D}_{VP})^n << 1$ (where $n \sim t/(2D_{VP})$) is concentrated on the very low frequencies $\bar{\omega}_n = \pi nc / \bar{D}_{VP}$. So this scheme is stable and gives suitable resulting accuracy. Although this case is not of practical interest of course? Due its very big dimensions.

Version in **Fig. 4-22-c** (spatially 3D case, spherical version of **Fig. 4-22-b**): microphone ★ is spaced at the great distance D_{VP} from the sphere S_V of radius R_V with active coating with corresponding ideal DNOV $\upsilon_0(\mathbf{r},t)$. So, if $D_{VP} << R_V$ we will obtain large maximum relative error of n-th order $\sim [R_V / (D_{VP} + R_V)]^n < 1$ (where $n \sim t/(2D_{VP})$).

Version in **Fig. 4-22-d**: vector microphone is formed by the bearing group of microphones. Let's assume that maximum suppression coefficient of vector microphone (on the base of bearing group) is– $0 < \gamma < 1$. So we obtain the following estimate of scattering sound pressure field $\left|\sum_{n=1}^{\infty} p_n\right| \approx |p_0|\gamma^2(1-\gamma)^2 << |p_0|$ or shorter $\left|\sum_{n=1}^{\infty} p_n\right| << |p_0|$.

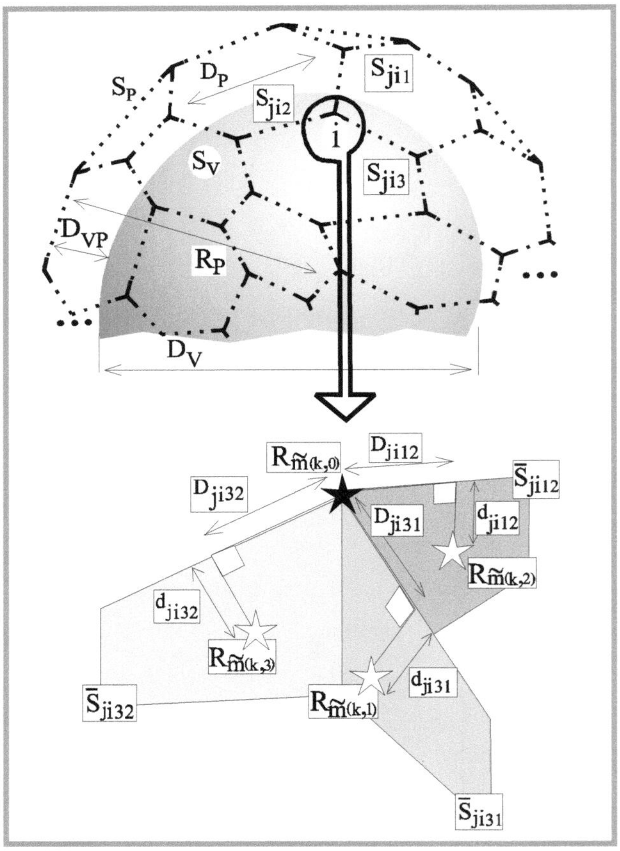

Figure 4-23: Placement of microphones on the surface of prominent hexagonal polyhedron S_P near its apexes. S_P, S_{ji2}, S_{ji3} –the sides of S_P closing to its j-th apex with i-th group of bearing with sector S_P for catching and target of i-th incident

wave. Planes $\overline{S}_{ji12}$, $\overline{S}_{ji31}$, $\overline{S}_{ji32}$ –bisectors of angles between sides S_P and S_{ji2}, S_{ji3} and S_P, S_{ji3} and S_{ji2}. Title microphone is spaced in i-th apex of S_P, benchmark microphones are spaced on planes $\overline{S}_{ji12}$, $\overline{S}_{ji31}$, $\overline{S}_{ji32}$. Here we assumed j–ordinal number of apex, i–ordinal number of incident wave (in order of their arrival).

4.7.2.11. Hierarchy of Scales in the System of Scattering Suppression

Summarizing the above considered, we can state, that, to ensure stability and efficiency of active system with microphones (if the stability and efficiency of system of synthesis of DNOV on S_V was earlier ensured (4-37), we must satisfy the following conditions:

$$D_{VP} / c_w > \tau_Q >> T_V \text{ , } \gamma = \exp\left[-D_{VP}\sqrt{(2\pi / L_V)^2 - \chi_{\max}^2} \right] << 1 \text{ ,} \quad \textbf{(4-58)}$$

$$L_V \chi_{\max} << 1 \text{ , } L_{pel} \chi_{\max} << 1 \text{ .}$$

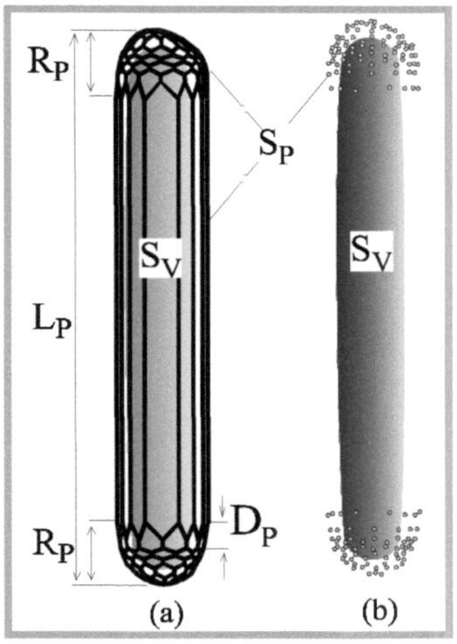

Figure 4-24: An example of prominent hexagonal polyhedron (a) surface S_P embracing the protected surface S_V of sufficiently stretched shape. Microphones are concentrated only near apexes of polyhedron S_P (number of bearing groups (b) is equal to the number of apexes of S_P) in the end areas of S_P. Microphones are absent in the lateral area of S_P, where its curvature (along the longitudinal coordinate) is zero.

4.7.2.12. Placement of Microphones

Now we will consider questions of rational placement of microphones, which allows us to target the required number N_w of incident waves with a minimum number of microphones. Each i-th bearing group is targeting only one incident wave. We characterize this group by its targeting and catching sector Ω_j. The last one represents multitude Ω_j of directions $\mathbf{w}_i$ of incident waves. The i-th bearing group can target only one incident wave (with $\mathbf{w}_i \in \Omega_j$) till first contact of this wave with the protected body. In other words it presents a lot of directions for which one bearing group could form a vector-microphone or solve the system of equations (4-45), (4-46), (4-52), (4-53), (4-54) with sufficient accuracy.

4.7.2.13. Geometry of the Microphone Grid

Numerous microphone placement variants are possible. Below we'll describe only one concrete version. In this version the surface S_P is represented by a prominent hexagonal polyhedron with plane six cornered sides (**Fig. 4-23**) of characteristic linear scale S_P and apexes with ordinal number– $j = 1, 2, .., N_w$. Each group of bearing is spaced near polyhedron's apexes. Title microphone of i-th group of bearing is spaced in j-th apex which unites three sides S_P, S_{ji2}, S_{ji3}. Planes $\overline{S}_{ji12}$, $\overline{S}_{ji31}$, $\overline{S}_{ji32}$ are bisectors of angles between sides S_P and S_{ji2}, S_{ji3} and S_P, S_{ji3} and S_{ji2} respectively. Benchmark microphones are spaced on planes– $\overline{S}_{ji12}$, $\overline{S}_{ji31}$, $\overline{S}_{ji32}$. Distances D_{ji12}, $\overline{S}_{ji31}$, D_{ji32} do not exceed quarter of corresponding S_P 's edges. Lines d_{ji12}, d_{ji31}, d_{ji32} are normal with corresponding edges of S_P and spaced in planes $\overline{S}_{ji12}$, $\overline{S}_{ji31}$, $\overline{S}_{ji32}$ correspondingly. Small (but finite) values– $d_{ji12} \sim D_{ji12}^2 / R_P$, $d_{ji12} \sim D_{ji12}^2 / R_P$, $d_{ji12} \sim D_{ji12}^2 / R_P$ are necessary to ensure targeting of incident wave, when its direction is normal to some side S_P, S_{ji2}, S_{ji3} of S_P. Besides this, the finite distances $d_{ji12}, d_{ji23}, d_{ji31} > 0$ are necessary to ensure sufficient accuracy of the solution of equations (4-45), (4-46), (4-52), (4-53), (4-54). R_P denotes the averaged curvature radius of S_P (*i.e.* curvature radius of some smooth surface embracing S_P closely). We choose concrete values $D_{ji12}, D_{ji23}, D_{ji31}, d_{ji12}, d_{ji23}, d_{ji31}$ to ensure: (a) covering of the full area of directions $\mathbf{w}_i$ of possible arrival of incident waves; (b) simultaneous targeting of several incident waves by several groups of bearing without any conflicts between each other. Total number N_P of microphones, number N_{SP} of apexes of polyhedron– S_P, total number N_{PG} of groups of bearing, total number N_w (expected) of incident waves simultaneously targeted are connected by relation $N_P / 4 = N_{PG} = N_{SP} = N_w$. The distance D_P between adjacent title microphones in any area of surface S_P with curvature radius R_P one can estimate approximately as $D_P \le 2R_P\theta_w$, where R_P refers to some smooth surface embracing all apexes of S_P. There θ_w denotes the expected minimum difference between directions of incident waves, *i.e.* $\theta_w = \min[\arccos(\mathbf{w}_n, \mathbf{w}_m)]_{n \ne m}$ $(m, n = 1, 2, ..., N_w)$.

4.7.2.14. Total Number of Microphones

One can see, that in the areas of S_P with smaller curvature radius R_P the density of microphones is larger, than in the areas of S_P with larger curvature radius– S_P. It is the most obviously in the case of surface S_P (and– S_V) of stretched shape (**Fig. 4-24**) with length S_P and curvature radius $R_P << L_P$ at its ends. The microphones are absent on the side areas of surface S_P due to our assumption about plane incident waves. In other words, our active system protects *directions* in space, unlike protection of *areas* of space in the traditional systems. When our active system targets $N_w \approx 4\pi / \theta_w^2 >> 1$ incident waves, using our approach, the active system needs $N_w \approx 4\pi / \theta_w^2 >> 1$ microphones. This amount is smaller by $N_P / N_{P0} \approx \lambda_{\min}^2 / \theta_w^2 L_P R_P$ times than when using the approach of G.D. Malyuzhinets.

4.7.2.15. Microphone Noise

By the reduction of the number of microphones included in the control of the active coating, we reduce the total power of control noise $N_P / N_{P0} >> 1$ times. This noise can be significantly less than in the case, where we are targeting all $N_w \approx 4\pi / \theta_w^2$ expected incident waves simultaneously (*i.e.* all microphones are involved in control). Really the number $\overline{N}_w << N_w$ of incident waves can be much less (for example $\overline{N}_w = 1$ or $\overline{N}_w = 2$). In this real case the system suggested can give the reducing of total power of control noise in $(N_{P0} / N_P)(N_w / \overline{N}_w) >> 1$ times, at comparing with the method of G.D. Malyuzhinets. Of course we assume that microphone noises are statistically independent with zero mean value, and power of sum of noises is equal to sum of noise powers of each microphone. And this sum of noise powers does not depend of signs, with which these noises are included in sum with signals.

4.7.2.16 Noise of a Vector Microphone

The spatial base max $h_i << \lambda_{\max}$ (**Fig. 4-20**) of vector microphones is of the order $h_i \sim \ell_P^2 / R_P$. The output signal of the vector microphone represents the differential between two signals and has characteristic magnitude $\Psi \sim P_\Xi h_i / \lambda_{\max}$, where P_Ξ –characteristic magnitude of incident wave pressure. We need a small relative control error σ_P / Ψ, where σ_P –square noise value or square root of mean power of noise. We use the assumption about large magnitude of incident wave to have $\sigma_P / \Psi << 1$. So we need ensure the condition $\sigma_P / P_\Xi << h_i / \lambda_{\max}$, which is analogous to (4-40).

4.7.2.17 Conclusions

Above we considered the solution in the form of active system. Below we shall remind the differences (advantages and drawbacks) between Malyuzhinets' method [15-19] and the approach described above. Let's assume that incident field presents the plane wave with direction $\mathbf{w}_i$ inside the sector Ω_i (**Fig. 4-20-b**, of catching and targeting). The approach suggested requires only four microphones, unlike Malyuzhinets' method, where– $(N_{P0} / N_P)(N_w / \bar{N}_w) >> 1$. In both cases the bodies have vibrations (caused by internal sources and incident waves) as in the absence of an active system. Active control by four microphones (bearing group, **Fig. 4-20-b**) gives much smaller control noises than in Maliuzhinets' method. In addition, the means of the fastening of the four microphones are greatly more transparent (transparency is necessary), than the microphones array in Malyuzhinets' method. In both approaches the total amount of actuators (active pistons) is very high– $N_V >> 1$. However, Malyuzhinets' method does not permit placement of actuators immediately on the protected body (to keep their unidirectional characteristics), and this leads to the great technical problem of *transparent fastening* thereof.

4.8. SUPPRESSION OF THE SOUND FIELD OF A VIBRATING BODY: A ONE-LAYER ACTIVE COATING

The problems of suppressing the sound field generated by a vibrating body in a liquid are considered. For solving these problems, an acoustically thin active coating with a real-time thickness control is proposed [66]. The coating should be placed directly on the surface of the body to be protected. Solutions to the problems of suppressing the radiation and scattering of sound by a body are obtained in the general form on the basis of linear operators, which characterize (i) the sound radiation by a vibrating surface, (ii) the scattering of incident waves by a fixed surface, and (iii) the vibroelastic properties of the body in an acoustic vacuum. Conditions ensuring the stability of the active system are formulated.

4.8.1. The Principle of Solving the Problem

In this section, we try to formulate a rather complicated problem of vibroacoustic interaction of the incident wave, the body submerged in the liquid, and the elements of active control of the boundary-value problem. Let us assume that, in a boundless liquid, a certain wave with the pressure $P_e(\mathbf{r},t)$ and particle velocity $\mathbf{V}_e(\mathbf{r},t)$ is incident on a homogeneous space region bounded by a certain closed film with the surface and filled with the same liquid.

Assuming that the film is thin, flexible, light, and tensile, we neglect the elastic stresses that are introduced by it on the liquid and are normal to the surface S as compared to the corresponding dynamic stresses in the incident wave. The stresses tangential to the surface S and arising from low viscosity of the liquid are also neglected. This means that the lines of power flux of the incident wave (*i.e.*, the wave generated by sources outside S) penetrate through the surface S an even number of times without producing any additional fields, including near fields, and the normal particle velocity $\upsilon(\mathbf{r},t)$ of the liquid at any point $\mathbf{r} \in S$ of the film coincides with the projection $(\Diamond, \mathbf{V}_e)$ of the particle velocities in the incident wave onto the unit normal $\Diamond = \Diamond(\mathbf{r})$ ($\mathbf{r} \in S$, $|\Diamond| = 1$ to the surface S (see (4-2) and **Fig. 4-3** in Section 4.4.):

$$\upsilon = V_e = (\Diamond, \mathbf{V}_e)_{\mathbf{r} \in S}. \quad \textbf{(4-59)}$$

As will be shown below, the condition of equality of the pressure $p(\mathbf{r},t)$ at the surface S to the pressure P_e in the undistorted incident wave

$$p = (P_e)_{\mathbf{r} \in S} \quad \textbf{(4-60)}$$

is equivalent to (4-59). If we create distribution (4-59) of normal particle velocities (DNPV) or pressure at the surface S of a certain protected body with arbitrary elastic properties (for this purpose, the active forces will be needed), the scattered field will not arise. Thus, for combined suppression of radiation and scattering, the following are necessary:

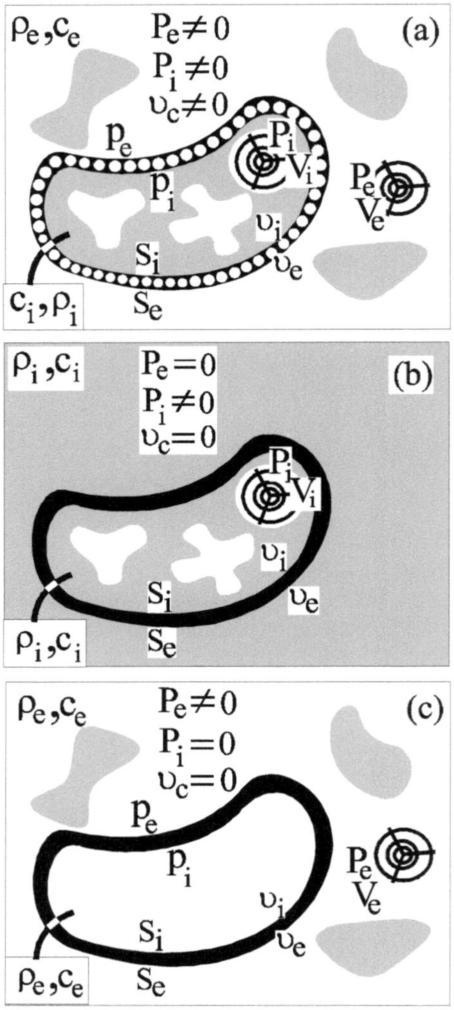

Figure 4-25: Geometry of the initial boundary-value problem: all the wave sources in the regions "i" and "e" with inhomogeneities and the active coating are switched on (a); passive matching of the region with the region "i", *i.e.*, $p_i = P_i$ and $\upsilon_i = V_i$ (b); passive matching of the region "i" with the region "e", *i.e.* $p_e = P_e$ and $\upsilon_e = V_e$ (c).

(i) to know distribution (4-59) (or the corresponding pressure distribution (4-60)) and

(ii) to be able to produce it at the body surface S. The mechanical structure inside S is assumed to be arbitrary.

Note that condition (4-59) (or (4-60)) should be satisfied only in certain limited operating range of time and space frequencies (4-2), (4-3), (4-4), (4-5).

4.8.2. Formulation of the Problem

We consider the following boundary-value problem. A simply connected space region "i" (**Fig. 4-25-a**) bounded by the surface S_i is filled with a homogeneous liquid medium with density ρ_i and sound velocity c_i, which contains discrete passive elastic inhomogeneities. A simply connected space region "e" bounded by the surface S_e is filled with a homogeneous liquid medium with density ρ_e and sound velocity c_e which contains discrete passive elastic inhomogeneities. The inhomogeneities in "i" and "e" have no contact with the surfaces S_i and S_e respectively. In the homogeneous parts of the regions "i" and "e", wave sources are present.

The surfaces S_i and S_e are the inner and outer surfaces of a certain acoustically thin, light, flexible, and tensile membrane separating the regions "i" and "e". The acoustic thinness means the smallness of the membrane thickness compared to the wavelengths of waves to be suppressed; *i.e.*, $L_c \chi_{\max} << 1$, where $L_c \chi_{\max} << 1$ is the membrane thickness. The membrane weightlessness means the smallness of the inertial forces of membrane particles compared to the pressure in the waves to be suppressed; *i.e.*,

$$\omega_{\max} L_c \rho_c < (\rho_e c_e), (\rho_i c_i) , \quad \textbf{(4-61)}$$

where ρ_c is the membrane density. The flexibility and tensility of the membrane mean that the pressure caused by its surface tension and bent is negligible compared to the pressure in the sound waves. The shape of the surfaces S_i and S_e is determined by the hydrostatic equilibrium of all the components of the regions "i" and "e" for specified flexural elasticity and tensility of the membrane. The natural vibration frequencies of the membrane due to its tension and flexural elasticity are much lower than $\omega_{\min}$. The aforementioned properties of the membrane ensure its acoustic transparency in the frequency range (4-3).

Since the surface S_e (and S_i) is closed, we consider the region "i" together with the membrane as a model of the body to be protected. The membrane will also serve as an active coating of the protected body.

Let us suppose that the difference

$$\upsilon_e - \upsilon_i = \upsilon_c \quad \textbf{(4-65)}$$

between the normal velocities υ_i and υ_e (projections onto the normal directed from "i" to "e") of the surfaces S_i and S_e or the velocity $\upsilon_c(\mathbf{r},t) = \{L_c(\mathbf{r},t)\}'_t$ describing the variation of thickness $L_c(\mathbf{r},t)$ of the coating (the membrane) is set from the outside and does not depend on the sound pressure at the surfaces S_i and S_e. In other words, the active coating represents a monopole layer (**Fig. 4-25-a**) with a certain prescribed volume velocity. The active coating switched off means $\upsilon_c = 0$ or the constancy of its thickness in time.

Below, we consider only the fields of pressure p_i, p_e and normal velocity υ_i, υ_e (projections onto the normal $\Diamond(\mathbf{r})$ from "i" to "e") at the surfaces S_i, S_e respectively. The wave sources in the regions "i" and "e" will not be defined, but we will formulate the definitions of the fields generated by them, which may be useful in the following consideration.

The fields $p_i = P_i$ and $\upsilon_i = V_i$ of sound pressure and normal velocity at the surface S_i define the pressure and velocity (**Fig. 4-25-b**) produced by the sources located in the region "i" for:

(a) sources switched off in the region "e",

(b) active coating switched off, and

(c) homogeneous filling of the region "e" with the liquid of density ρ_i and sound velocity c_i and conservation of the previous inhomogeneities in "i". Such a formulation of the problem implies a passive matching of the region "e" with the region "i".

The fields $p_e = P_e$ and $\upsilon_e = V_e$ of sound pressure and normal velocity at the surface S_e define the pressure and velocity (**Fig. 4-25-c**) produced by the sources located in the region "e" for:

(a) sources switched off in the region "i",

(b) active coating switched off, and

(c) homogeneous filling of the region "e" with the liquid of density ρ_e and sound velocity c_e and conservation of the previous inhomogenieties in "e". This problem formulation means a passive matching of the region "i" with the region "e". Note that the passive matching of "e" to "i" is not identical to the passive matching of "i" to "e".

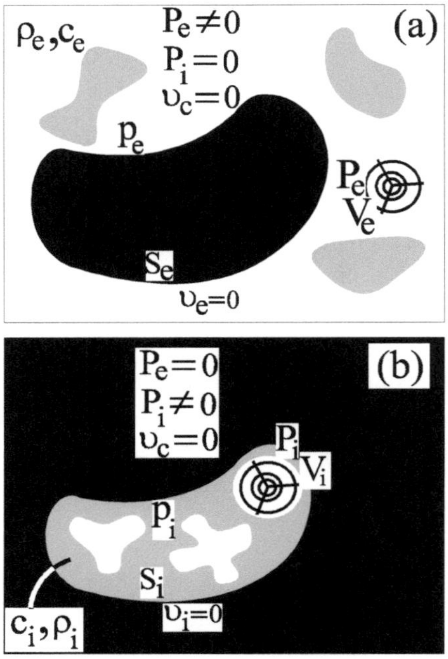

Figure 4-26: Determination of the external scattering operator $\hat{H}_e$ (a), determination of the internal scattering operator $\hat{H}_i$ (b).

Now, in the most general description of the initial boundary-value problem (**Fig. 4-25-a**), the fields p_i, p_e, υ_i, υ_e can be represented by certain linear functionals of the distributions P_i, P_e, υ_c at the surfaces S_e and S_i respectively:

$$p_e = p_e[P_e, P_i, \upsilon_c],\ \upsilon_e = \upsilon_e[P_e, P_i, \upsilon_c],\ p_i = p_i[P_e, P_i, \upsilon_c],\ \upsilon_i = \upsilon_i[P_e, P_i, \upsilon_c]. \quad \textbf{(4-66)}$$

Obtaining these expressions in an explicit form will make it possible to form the radiation and scattering fields by controlling the quantity υ_c on the basis of the measured values of p_i, p_e, υ_i, υ_e and υ_c. Using the superposition principle, we divide the initial complicated problem into several simpler problems (functional blocks). Below, from these functional blocks, we find the solution to the initial boundary-value problem.

4.8.3. Scattering Operators

Let the total pressure field p_e formed at the scattering of wave P_e (**Fig. 4-26-a**) by a fixed closed surface S_e (as distinct from the surface of an ideal rigid body, which could vibrate without deformation) represent the linear functional

$$p_e = (\hat{1} + \hat{H}_e)P_e \quad \textbf{(4-67)}$$

of the field P_e, where $\hat{1}$ is the unit operator and $\hat{H}_e$ is the linear operator of external scattering in the form

$$\hat{H}_e p_e(\mathbf{r},t) = \int_{-\infty}^{t} dt_0 \oiint_{S_e} dS_e H_e(\mathbf{r},\mathbf{r}_0;t-t_0)P_e(\mathbf{r}_0,t_0)$$

with kernel $H_e(\mathbf{r},\mathbf{r}_0;t-t_0)$.

The scattering of wave P_i by the fixed closed surface S_i (**Fig. 4-26-b**) produces the total pressure field

$$p_i = (\hat{1} + \hat{H}_i)P_i \qquad \textbf{(4-68)}$$

where the operator $\hat{H}_i$ of internal scattering acts on S_i in a way similar to that of $\hat{H}_e$. The operators $\hat{H}_e^{-1}$ and $\hat{H}_i^{-1}$ with kernels $H_e^{-1}(\mathbf{r},\mathbf{r}_0;t-t_0)$ and $H_i^{-1}(\mathbf{r},\mathbf{r}_0;t-t_0)$ acting on the surfaces S_e and S_i determine the quantities P_e and P_i, respectively, in terms of $p_e - P_e$ and $p_i - P_i$:

$$P_e = \hat{H}_e^{-1}(p_e - P_e),\ P_i = \hat{H}_i^{-1}(p_i - P_i) \qquad \textbf{(4-69)}$$

The operators $\hat{H}_e^{-1}$ and $\hat{H}_i^{-1}$ are inverse of the operators $\hat{H}_e$ and $\hat{H}_i$, *i.e.*,

$$\hat{H}_e^{-1}\hat{H}_e = \hat{1},\ \hat{H}_i^{-1}\hat{H}_i = \hat{1}. \qquad \textbf{(4-70)}$$

4.8.4. Radiation Operators

The total pressure p_e of the waves radiated by the surface S_e into the region "e" with the distribution of normal velocities υ_e specified at S_e (with switched-off sources in the region "e") is represented by the external impedance Green operator $\hat{G}_e$ of the form

$$p_e = +\hat{G}_e\upsilon_e \qquad \textbf{(4-71)}$$

$$\text{or } \hat{G}_e\upsilon_e = \int_{-\infty}^{t} dt \oint\!\!\oint_{S_e} dS_e G_e(\mathbf{r},\mathbf{r}_0;t-t_0)\upsilon_e(\mathbf{r}_0,t_0) \qquad \textbf{(4–72)}$$

with the kernel $G_e(\mathbf{r},\mathbf{r}_0;t-t_0)$. The total pressure p_i of the waves radiated by the surface S_i into the region "i" with the distribution of normal velocities υ_i specified at S_e (with switched-off sources in the region "i") is represented by the internal impedance Green operator $\hat{G}_e$ of the form

$$p_i = -\hat{G}_i\upsilon_i \qquad \textbf{(4-73)}$$

analogous to the operator (4-72), but with the kernel $G_i(\mathbf{r},\mathbf{r}_0;t-t_0)$ and acting at the surface S_i.

In the alternative formulation of the problem, the fields of velocities υ_e and υ_i are generated by the preset pressure fields p_e and p_i, respectively, at the surfaces S_e and S_i and are described by the operators

$$\upsilon_e = \hat{G}_e^{-1}p_e \text{ and } \upsilon_i = -\hat{G}_i^{-1}p_i \qquad \textbf{(4-74)}$$

with the kernels $G_e^{-1}(\mathbf{r},\mathbf{r}_0;t-t_0)$ (it should be pointed out that $G_e^{-1}(\mathbf{r},\mathbf{r}_0;t-t_0) \neq 1/G_e(\mathbf{r},\mathbf{r}_0;t-t_0)$ and $G_i^{-1}(\mathbf{r},\mathbf{r}_0;t-t_0) \neq 1/G_i(\mathbf{r},\mathbf{r}_0;t-t_0)$) acting similarly to operator (4–72). The above pressure fields P_e and P_i undistorted by the regions "i" and "e", respectively, correspond to the following undistorted fields V_e and V_i of projections of the particle velocities in the wave fields P_e and P_i onto the normal to the surfaces S_e and S_i:

$$V_e = \hat{G}_e^{-1}P_e,\ V_i = -\hat{G}_i^{-1}P_i \qquad \textbf{(4-75)}$$

Let us assume that a certain linear operator of the form similar to (4-68) and (4-72) exists if it uniquely determines the output function from the input function. In this sense, the operators $\hat{G}_i$, $\hat{G}_e$ and $\hat{G}_e^{-1}$, $\hat{G}_i^{-1}$ exist in both time and monochromatic representations. They are connected by obvious relationships

$$\hat{G}_e^{-1}G_e = \hat{1}, \; \hat{G}_i^{-1}G_i = \hat{1}. \qquad \textbf{(4-76)}$$

The operators $\hat{H}_e$ and $\hat{H}_i$ introduced above can be determined numerically if the surfaces S_e and S_i and the inhomogenieties of the medium are preset. The process of inversion of the operators $\hat{H}_e$ and $\hat{H}_i$ becomes sensitive to the calculation errors when, at the frequencies corresponding to large wave sizes of the surfaces S_e and S_i, a considerable area of geometrical shadow is formed. However, time representation (4–68) of the operators $\hat{H}_e$ and $\hat{H}_i$ covers all the frequencies and is not limited by the suppression range (4–2). Therefore, the operators $\hat{H}_e$, $\hat{H}_i$ also exist and their analytic operations are correct.

4.8.5. Dynamic Equations of the Boundary-Value Problem

The linear operators $\hat{H}_e$, and $\hat{G}_e$, introduced above completely characterize the effect produced by the distributions of parameters of the medium in the regions "e" and "i" on the vibrations of the boundaries S_e and S_i and allow us to write the following four equations determining the dynamics of the initial boundary-value problem (**Fig. 4-25-a**):

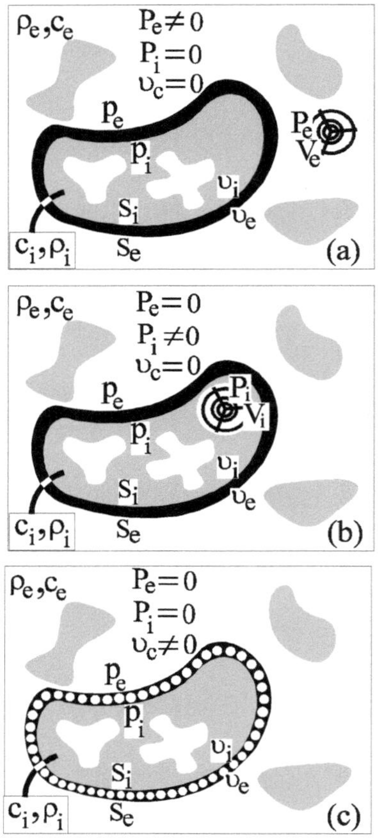

Figure 4–27: Sound field excitation in the boundary-value problem by the sources in the region "e" (a), the sources in the region "i" (b), and the active coating (c).

(a) the dynamics of the boundary S_e of the active coating is determined by the equation

$$p_e = (\hat{1} + \hat{H}_e)P_e + \hat{G}_e \upsilon_e; \quad \textbf{(4-77)}$$

(b) the absence of tangential stresses and the weightlessness (4-61) of the active layer determine the equality of pressures at S_i and S_e, *i.e.*,

$$p_i = p_e; \quad \textbf{(4-78)}$$

(c) the υ_e and υ_i velocities at S_e and S_i are connected by the equation

$$\upsilon_e = \upsilon_i + \upsilon_c; \quad \textbf{(4-79)}$$

(d) the dynamics of the boundary S_i of the active coating is determined by the equation

$$p_i = (\hat{1} + \hat{H}_i)P_i - \hat{G}_i \upsilon_i. \quad \textbf{(4-80)}$$

4.8.6. Partial Solutions of the Boundary-Value Problem

From the set of equations (4-77)-(4-80), we obtain the partial expressions for the quantities p_e, p_i, υ_e, υ_i in terms of the fields P_e, P_i, υ_c and operators $\hat{H}_e$, $\hat{G}_e$, $\hat{H}_i$, $\hat{G}_i$ (see **Fig. 4–27**):

(i) the fields of the sources located in the region "e" (**Fig. 4–28–a**)

$$p_i[P_e,0,0] = p_e[P_e,0,0] = (\hat{1} + \hat{H}_e)(\hat{G}_e + \hat{G}_i)^{-1}\hat{G}_i P_e, \quad \textbf{(4-81)}$$

$$\upsilon_i[P_e,0,0] = \upsilon_e[P_e,0,0] = -(\hat{1} + \hat{H}_e)(\hat{G}_e + \hat{G}_i)^{-1} P_e. \quad \textbf{(4-82)}$$

where $(\hat{G}_e + \hat{G}_i)^{-1}$ designates the operator inverse to $(\hat{G}_e + \hat{G}_i)$;

(ii) the fields of the sources located in the region "i" (**Fig. 4–28–b**)

$$p_i[0,P_i,0] = p_e[0,P_i,0] = (\hat{1} + \hat{H}_i)(\hat{G}_e + \hat{G}_i)^{-1}\hat{G}_e P_i, \quad \textbf{(4-83)}$$

$$\upsilon_i[0,P_i,0] = \upsilon_e[0,P_i,0] = (\hat{1} + \hat{H}_i)(\hat{G}_e + \hat{G}_i)^{-1} P_i, \quad \textbf{(4-84)}$$

(iii) the fields created by the active coating (**Fig. 4–28–c**)

$$\upsilon_e[0,0,\upsilon_c] = +\hat{G}_i(\hat{G}_e + \hat{G}_i)^{-1}\upsilon_c, \quad \textbf{(4-85)}$$

$$\upsilon_e[0,0,\upsilon_c] = -\hat{G}_i(\hat{G}_e + \hat{G}_i)^{-1}\upsilon_c, \quad \textbf{(4-86)}$$

$$p_i[0,0,\upsilon_c] = p_e[0,0,\upsilon_c] = \hat{G}_e\hat{G}_i(\hat{G}_e + \hat{G}_i)^{-1}\upsilon_c. \quad \textbf{(4-87)}$$

4.8.7. Solution of Some of the Sound Field Control Problems

According to the superposition principle, from partial solutions (4-81)-(4-87) we obtain general solutions (4–66) to the initial boundary-value problem

$$p_i[P_e,P_i,\upsilon_c] = p_i[P_e,0,0] + p_i[0,P_i,0] + p_i[0,0,\upsilon_c], \quad \textbf{(4-88)}$$

$$p_e[P_e, P_i, \upsilon_c] = p_e[P_e, 0, 0] + p_e[0, P_i, 0] + p_e[0, 0, \upsilon_c]\,, \tag{4-89}$$

$$\upsilon_i[P_e, P_i, \upsilon_c] = \upsilon_i[P_e, 0, 0] + \upsilon_i[0, P_i, 0] + \upsilon_i[0, 0, \upsilon_c]\,, \tag{4-90}$$

$$\upsilon_e[P_e, P_i, \upsilon_c] = \upsilon_e[P_e, 0, 0] + \upsilon_e[0, P_i, 0] + \upsilon_e[0, 0, \upsilon_c]\,. \tag{4-91}$$

Selecting a linear combination of partial solutions (4-88)-(4-91), it is possible to obtain the solutions to all the basic problems of the sound field control. For instance, the problem of sound insulation of the regions "i" and "e" from each other can be solved by satisfying at least one of the following conditions:

$$p_i = 0\,,\ p_e = 0\,,\ \upsilon_i = 0\,,\ \upsilon_e = 0 \tag{4-92}$$

for arbitrary $P_e, P_i \neq 0$. From (4-81)-(4-87) and (4-88)-(4-91), it is seen that, without an active coating (*i.e.*, for $\upsilon_c = 0$), this problem has no solution. However, with the use of an active coating, several solutions are possible in the form of variants of expressing υ_c in terms of p_i, p_e, υ_i, υ_e to provide for the fulfillment of one of the conditions (4-92). These solutions are not considered in this Section.

4.8.8. Synthesis of the Prescribed DNVP at the Surface S_e

Let us assume that, for $P_e = 0$ and $P_i = 0$, it is necessary to create the prescribed DNPV for υ_e at the surface S_e by using an active coating. The DNPV at the surface S_i is assumed to be free, *i.e.*, as it will come out. In other words, the DNPV at S_i is a by-product of the synthesis of DNPV at S_e. According to (4-85), the DNPV of υ_e appears at S_e with

$$\upsilon_c = +\hat{G}_i^{-1}(\hat{G}_e + \hat{G}_i)\upsilon_e\,. \tag{4-93}$$

In turn, according to (4–86), the DNPV at S_i will be

$$\upsilon_i = -\hat{G}_e\hat{G}_i^{-1}\upsilon_e\,. \tag{4-94}$$

It is easy to see that (4–93) and (4–94), in spite of the generality of the approach, lead us to the simplest, local in time and space, relationship between the controlled velocity υ_c and the measured velocities υ_e and υ_i (unlike boundary condition (4-79)); *i.e.*, we have

$$\upsilon_c(\mathbf{r},t) = \upsilon_e(\mathbf{r},t) - \upsilon_i(\mathbf{r},t)\,. \tag{4-95}$$

This algorithm (4-95) can be classified as {LS/LT/CA/F} (see **Fig. 1-3**) [94]. Thus, for the synthesis of the prescribed DNPV by the active coating with controlled thickness, even with a finite compliance of the support (surface S_i), data on operators $\hat{G}_e$ and $\hat{G}_i$ is unnecessary even when $\upsilon_i(\mathbf{r},t)$ really depends on both $\upsilon_c(\mathbf{r},t)$ and $\upsilon_e(\mathbf{r},t)$. However, here we implicitly assumed that the *system is stable.*

4.8.9. Stability of the DNVP Synthesis

The active coating, varying its thickness, in general, rests on both surfaces S_i and S_e. However, since the prescribed velocity υ_e should be provided at S_e, we assume that the support is the surface S_i. The operator

$$\hat{\gamma} = \hat{G}_e\hat{G}_i^{-1} \tag{4-96}$$

represents the functional expression of the feedback (or the dependence of vibrations υ_i of the support on the vibrations υ_c of the active coating thickness), which involves both the static and resonance compliance of the surface S_i. If the regions "i" and "e" are homogeneous liquid half-spaces divided by a plane active coating, we have $\hat{\gamma} = (\rho_e c_e)(\rho_c c_c)^{-1}$. Obviously, in the case of a fixed (stationary) surface S_i, we obtain the operator $\hat{\gamma} = \hat{0}$ and the system is certainly stable. The action of the operator is similar to that of the above operators (see (4-68), (4-72)), and its kernel has the form

$$\gamma(\mathbf{r},\mathbf{r}_0;t-t_0) = \int_{-\infty}^{+\infty} d\tau \iint_{S_i} G_e(\mathbf{r},\mathbf{R};t-\tau)G_i^{-1}(\mathbf{R},\mathbf{r}_0;\tau-t_0)dS_i(\mathbf{R})\,. \quad \mathbf{(4\text{-}97)}$$

The operator $\hat{\gamma} = \hat{G}_e\hat{G}_i^{-1}$, at least at some frequencies and in some regions of the surface S_i, can provide a positive feedback, which reduces the stability of the active system. With repeated passage through the feedback channel, a certain initial perturbation $(\delta\upsilon_c)_0$ of the velocity $\upsilon_c(\mathbf{r},t)$ of the coating thickness variation gives rise to the resulting perturbation $(\delta\upsilon_c)_\Sigma$ in the form of a functional series $(\delta\upsilon_c)_\Sigma = \sum_{n=1}^{\infty} \hat{\gamma}^n(\delta\upsilon_c) = \hat{\gamma}/(\hat{1}-\hat{\gamma})(\delta\upsilon_c)_0$.

As a sufficient condition for the stability of the system, we accept the following condition: the vibration $\upsilon_i(\mathbf{r},t)$ of the surface S_i that is induced by an arbitrary distribution of the velocity $\upsilon_c(\mathbf{r},t)$ of the active coating thickness variation at no point $\mathbf{r}$ and no instant t exceeds in absolute value the maximum magnitude $|\upsilon_c(\mathbf{r},t)|$ of the perturbation υ_c that induced this vibration; *i.e.*, $|\upsilon_i(\mathbf{r},t)| < \max_{\{\upsilon_c(\mathbf{r},t),\mathbf{r},t\}} |\upsilon_c(\mathbf{r},t)|$. The prescribed needleshaped distribution $|\upsilon_c(\mathbf{r},t)| = A_c \exp(i\omega t) I\left[a - |\mathbf{r}-\mathbf{R}|\right]$ (where $a \to 0$, $I(\xi)=1$ for $\xi < a$, and $I(\xi)=0$ for $\xi \geq a$) of monochromatic vibrations of the coating thickness at the frequency ω in an arbitrarily close vicinity a of the point $\mathbf{R}$ of the active coating produces the vibrations $\upsilon_i(\mathbf{r},t) = A_c \exp(i\omega t)\overline{\gamma}(\mathbf{r}_0,\mathbf{R},\omega)$ at a certain point $\mathbf{r}_0$ of the surface S_i. Here, $\overline{\gamma}$ is the Fourier transform $\overline{\gamma}(\mathbf{r}_0,\mathbf{R},t) = \int_{-\infty}^{+\infty}\gamma(\mathbf{r}_0,\mathbf{R},t)\exp(i\omega t)dt$ of the kernel γ of the operator $\hat{\gamma}$. Let us try to create at the point $\mathbf{r}_0$ of S_i the maximum vibration amplitude induced by the vibrations $\upsilon_c(\mathbf{r},t)$ of limited amplitude; *i.e.*, $|\upsilon_c| \leq A_c$. Then, creating the distribution $\upsilon_c(\mathbf{r},t) = A_c \exp[i\omega t - \arg\overline{\gamma}(\mathbf{r}_0,\mathbf{R},\omega)]$ over the whole active coating, we obtain the maximum possible value $|\upsilon_i(\mathbf{r},t)| = A_c \oiint_{S_i} |\overline{\gamma}(\mathbf{r}_0,\mathbf{r},\omega)|\, dS_i$ of vibration of the surface S_i at the point $\mathbf{r}_0$, this vibration being induced by the distribution υ_c of the limited amplitude $\upsilon_c(\mathbf{r},t) = A_c \exp[i\omega t - \arg\overline{\gamma}(\mathbf{r}_0,\mathbf{R},\omega)]$. Let us apply the operation described above to all the points $\mathbf{r}_0$ of the surface S_i. Now, if, for all frequencies $-\infty > \omega > +\infty$ and all points $\mathbf{r}_0$ on S_i, the condition

$$\max_{\{\mathbf{r}_0,\omega\}} \oiint_{S_i} |\overline{\gamma}(\mathbf{r}_0,\mathbf{r},\omega)| dS_i(\mathbf{r}) < 1 \quad \mathbf{(4\text{-}98)}$$

is satisfied, where maximum is taken over all possible variations of $\mathbf{r}_0$ and ω, the system is stable. If, at least for one frequency ω one point $\mathbf{r}_0$ of the surface S_i, condition (4-98) fails, the stability of the system is not guaranteed. Thus, if condition (4-98) is satisfied, the system is stable, and, for synthesizing the prescribed DNPV at S_e (according to (4-95)), we do not need full information about operators $\hat{G}_e$ and $\hat{G}_i$.

4.8.10. Suppression of the Fields of Radiation and Scattering of Region "i"

In this section, we consider the possibilities for suppressing the fields generated at the surface S_e by the sources in the region "i", as well as the scattered fields from the sources in the region "e". As an indicator of the absence of radiated and scattered fields of the region "i", we take, for instance, condition (4-59) of matching the normal velocities of the surface S_e and the projections of particle velocities onto the normal to S_e in the case of a passive matching of the region "i" with the region "e":

$$\upsilon_e[P_e, P_i, \upsilon_c] = V_e \quad \textbf{(4-99)}$$

Taking into account (4-82), (4-84) and (4-85), for the velocity mismatch $\upsilon_\varepsilon = \upsilon_e - V_e$, we obtain the expression

$$\upsilon_\varepsilon = (\widehat{G}_e + \widehat{G}_i)^{-1}(\widehat{H}_e\widehat{G}_i - \widehat{G}_e)V_e + (\widehat{G}_e + \widehat{G}_i)^{-1}(\widehat{1} + \widehat{H}_i)\widehat{G}_e V_i + (\widehat{G}_e + \widehat{G}_i)^{-1}\widehat{G}_i\upsilon_c \,. \quad \textbf{(4-100)}$$

4.8.11. Passive Matching of Region "i" with Region "e"

If $P_i = 0$ and $\upsilon_c = 0$, the passive matching (*i.e.*, $\upsilon_\varepsilon = 0$) of the region "i" with the region "e" can be achieved by satisfying the relationship

$$\widehat{H}_e\widehat{G}_i - \widehat{G}_e = 0\,, \quad \textbf{(4-101)}$$

where the internal Green operator

$$\widehat{G}_i = \widehat{\widehat{G}}_i = \widehat{H}_i^{-1}\widehat{G}_e \quad \textbf{(4-102)}$$

corresponds to the region "i", which is transparent for the waves from the region "e". The operator $\widehat{G}_i = \widehat{\widehat{G}}_i = \widehat{H}_i^{-1}\widehat{G}_e$ means the relationship of the form

$$p_i = -\widehat{\widehat{G}}_i\upsilon_i \quad \textbf{(4-103)}$$

between the pressure and normal velocity at the surface S_i. Note that such a passive matching does not exclude the radiation from "transparent" sources usually described by the right-hand part of the wave equation.

From (4-100), we also obtain the relationship

$$p_\varepsilon = \widehat{G}_e\upsilon_\varepsilon \quad \textbf{(4-104)}$$

between the pressure p_ε of mismatch (or scattering) and the normal velocity mismatch $\upsilon_\varepsilon = \upsilon_e - V_e$, which means that matching of the surface S_e to incident waves in pressure (4-60) is equivalent to matching in velocity (4-59).

4.8.12. Active Matching of Region "i" with Region "e"

If $\widehat{G}_i \neq \widehat{\widehat{G}}_i$ and $P_i \neq 0$, the condition $p_\varepsilon = \upsilon_\varepsilon = 0$ can be fulfilled only by including an active coating with nonzero $\upsilon_c \neq 0$ velocity of active coating thickness variation, which we represent as the sum of two components

$$\upsilon_c = \upsilon_{ci} + \upsilon_{ce}\,, \quad \textbf{(4–105)}$$

where the component

$$\upsilon_{ci} = -(\widehat{1} + \widehat{H}_i)\widehat{G}_e\widehat{G}_i^{-1}V_i \quad \textbf{(4–106)}$$

compensates for the influence of the sources located in the region "i" on υ_e and the component

$$\upsilon_{ce} = -\widehat{G}_i^{-1}\widehat{H}_e(\widehat{G}_i - \widehat{\widehat{G}}_i)V_e \quad \textbf{(4-107)}$$

compensates for the influence of the sources located in the region "e" on υ_e. Using (4-77)-(4-80), we express

$$V_e = \widehat{G}_e^{-1}(\widehat{1} + \widehat{H}_e)^{-1}(p_e - \widehat{G}_e\upsilon_e) \text{ and } V_i = \widehat{G}_i^{-1}(\widehat{1} + \widehat{H}_i)^{-1}(p_i + \widehat{G}_i\upsilon_i) \quad \textbf{(4-108)}$$

in terms of the total fields υ_i, p_e, υ_e, υ_i measured at the surfaces S_i and S_e. Taking into account (4-106)-(4-108), we obtain a rather simple relationship

$$\upsilon_c = -\hat{H}_e \hat{G}_e^{-1} p_i - \upsilon_i . \qquad \textbf{(4-109)}$$

From (4-109), according to (4-79), we obtain the expression (analogous to (4-103))

$$\upsilon_e = -\hat{H}_e \hat{G}_e^{-1} p_e , \qquad \textbf{(4-110)}$$

which means the transparency of the region "i" for the waves coming from the region "e". It should be noted that, in deriving (4-110), we assumed a nonzero wave source inside the region "i" and a nonzero component υ_{ci} for its compensation. Therefore, as opposed to the passive matching (4-102) of the region "i" to the region "e", the active coating compensates also for the radiation of "transparent" sources in region "i". In the adopted kinematic approach, algorithm (4-109) controlling the rate of variation of the active coating thickness solves the formulated problem, provided that the system is stable. Therefore, if condition (4-98) is satisfied, algorithm (4-109) solves the problem without any data on the operators $\hat{H}_i$, $\hat{G}_i$ (or on the structure of the region "i").

By interchanging the indices i and e in the expressions obtained above, we get similar results for the problem of suppressing the sound fields in the region "i". In conclusion, we note that the operators $\hat{H}_e$, $\hat{G}_e$, $\hat{H}_i$, $\hat{G}_i$ have the simplest form in the case when all vibrations are of the same frequency ω and the boundary-value problem allows separation of variables together with orthogonality of eigenfunctions at the surface S_e (or S_i). For instance, in an azimuth-symmetric boundary- value problem for a spherical surface S_e of radius R, we obtain the following matrix representation of the operators and acting on the corresponding polar modes of the quantities p_i, p_e, υ_i, υ_e, υ_c:

$$[\hat{G}_{i,e}, \hat{H}_{i,e}]_{n,m,\omega} = \int_{-\infty}^{+\infty} \exp(i\omega t)dt \int_0^{\pi} \sin(\vartheta)d\vartheta \int_0^{\pi} \sin(\vartheta_0)d\vartheta_0 P_n(\vartheta)P_m(\vartheta_0)\{G_{i,e}(\vartheta,\vartheta_0,t), H_{i,e}(\vartheta,\vartheta_0,t)\}$$

where $P_m(\vartheta)$ is the Legendre polynomial of degree m ($m = 0,1,2,...$). The matrix operators $[\hat{H}_{i,e}]_{n,m,\omega}$ and $[\hat{G}_{i,e}]_{n,m,\omega}$ for a homogeneous filling of the regions "i" and "e" have the diagonal form

$$[\hat{H}_{i,e}]_{n,m,\omega} = \delta_{n,m}(\hat{H}_{i,e})_{m,\omega}, \quad [\hat{G}_{i,e}]_{n,m,\omega} = \delta_{n,m}[\hat{G}_{i,e}]_{m,\omega},$$

where

$[\hat{H}_{i,e}]_{m,\omega} = [d\psi_m^{+-}(\xi)/d\xi]/[d\psi_m^{-+}(\xi)/d\xi]_{\xi=kR}$, $[\hat{G}_{i,e}]_{m,\omega} = i\omega\rho_{i,e}[\psi_m^{-+}(\xi)]/[d\psi_m^{-+}(\xi)/d\xi]_{\xi=kR}$, . is the spherical Hankel function of the wave propagating from the center $r = 0$ to infinity, $\psi_n^-(kr)$ is the spherical Hankel function of the wave propagating from infinity to the center, $\psi_n^-(kr) = \{\psi_n^+(kr)\}^*$, $k = \omega / c_{i,e}$, $\delta_{n,m}$ is the discrete delta-function. In addition, the matrices $[\hat{H}_{i,e}^{-1}]_{n,m,\omega}$ and $[\hat{G}_{i,e}^{-1}]_{n,m,\omega}$ inverse of $[\hat{H}_{i,e}]_{n,m,\omega}$, $[\hat{G}_{i,e}]_{n,m,\omega}$, are also diagonal. Thus, with the above-stated assumptions, the operators $[\hat{H}_{i,e}^{-1}]_{n,m,\omega}$, $[\hat{G}_{i,e}^{-1}]_{n,m,\omega}$, $[\hat{H}_{i,e}]_{n,m,\omega}$, $[\hat{G}_{i,e}]_{n,m,\omega}$ act on the independent polar harmonics of the quantities p_i, p_e, υ_i, υ_e, υ_c as algebraic multipliers. If a spherical surface S_e of radius R has neutral buoyancy ($\rho_i = \rho_e$) and is perfectly rigid ($c_i >> c_e$) with respect to the region "e", the vibrations of the undeformable sphere in the field of a plane incident wave completely coincide with the first polar harmonic, and the boundary-value problem is divided into independent components describing the scattering of each individual polar harmonic. For all the spherical modes, except for the first one, the active coating support will be fixed (ideal). As is known [84], the associated mass of a vibrating sphere of small wave dimension is exactly two times smaller than the mass of the surrounding liquid displaced by the sphere. Consequently, at low frequencies, the rest of the active coating on a rigid sphere of neutral buoyancy does not lead to the instability of the system. However, this, unlike condition (4–98), does not guarantee the stability of the system as a whole.

4.8.13. Shaping the Assigned Zeroes of Scattering

Below we consider briefly the formation of the predetermined zeroes in the angle diagram of radiation or (and) scattering of vibrating body without active coating, but with the help of multipole sources of unidirectional radiation (MSUR). Using the formula (4-104) $p_\varepsilon = \hat{G}_e \upsilon_\varepsilon$, obtained above, we can calculate the scattering acoustical field on the basis of measured values of normal oscillation velocity on the surface of the vibrating body.

We consider the following scenario (**Fig. 4-28**): several plane acoustical monochromatic (on the frequency ω) incident waves with known wave vectors $\mathbf{k}_1, \mathbf{k}_2, \ldots, \mathbf{k}_M$ are running to the passive protected body (with the external surface S and without any active coating and own radiation). We must suppress the scattering field in the far zone of the body (*i.e.* on the distance $R >> D_V^2 \lambda^{-1}$ from body) in the known directions $\tilde{\mathbf{k}}_1, \tilde{\mathbf{k}}_2, \ldots, \tilde{\mathbf{k}}_{\tilde{M}}$ (wave vectors of suppressed waves).

We do not know any vibroacoustical characteristics of the protected body but only know its shape. We measure the complex amplitudes E_m ($m = 1, 2, \ldots, M$) of acoustical pressure of incident wave by the receiving multipole sources of unidirectional radiation (MSUR) of multipolarity order $N > 1$ (**Fig. 4-28-a**, see also the section 2.3.). The receiving MSURs are centered on points $\mathbf{r}_n$ ($n = 1, 2, \ldots, M$). Output signals u_n (from n-th receiving MSUR) of complex amplitude E_m of sound pressure of m-th incident wave are given by known receiving (in sound pressure) directivity pattern $F(\mathbf{q}_n, \mathbf{k}_m, \omega)$ of n-th receiving MSUR (in amplitude, $\mathbf{q}_n$ –axis vector of n-th MSUR, $|\mathbf{q}_n| = 1$), *i.e.*:

$$u_n = F(\mathbf{q}_n, \mathbf{k}_m, \omega) E_m, \tag{4-111}$$

where $\mathbf{q}_n = -\mathbf{k}_n / |\mathbf{k}_n|$. Further we can formulate the DNOV

$$\upsilon_{\varepsilon m}(\mathbf{r}, \omega) = \upsilon_m(\mathbf{r}, \omega) - \upsilon(\mathbf{r}, \omega)$$

mismatch to the m-th incident wave, where

$$\upsilon_m(\mathbf{r}, \omega) = (\Diamond(\mathbf{r}), \mathbf{k}_m) E_m / (c\rho), \tag{4-112}$$

$(\Diamond(\mathbf{r}), \mathbf{k}_m)$ –scalar product of normal $\Diamond(\mathbf{r})$ (where $|\Diamond(\mathbf{r})| = 1$, $\mathbf{r} \in S$) vector to S, $\mathbf{k}_m$ –wave vector of m-th incident wave, $\upsilon(\mathbf{r}, \omega)$ –measured DNOV (**Fig. 4–29–b**) on S in the presence of incident waves. Mismatch sound pressure on S we take from the formula

$$p_{\varepsilon m} = \hat{G}_e \upsilon_{\varepsilon m} \quad (\mathbf{r} \in S) \tag{4-113}$$

(see (4-104)). Then we calculate sound scattering pressure field $p_{\varepsilon m}(\mathbf{R}, \omega)$ of m-th incident wave, caused by $\upsilon_{\varepsilon m} \neq 0$ in far zone ($R >> D_V^2 \lambda^{-1}$), using the known formula

$$p_{\varepsilon m}(\mathbf{R}, \omega) = -\frac{i\omega\rho}{4\pi} \oiint_S \left[\left(p_{\varepsilon m}(\mathbf{r}, \omega) / i\omega\rho \right) \frac{\partial}{\partial n} \left(\frac{\exp(-ik\Re)}{\Re} \right) - \frac{\exp(-ik\Re)}{\Re} \upsilon_{\varepsilon m}(\mathbf{r}, \omega) \right] dS(\mathbf{r}), \tag{4-114}$$

where $\Re = |\mathbf{R} - \mathbf{r}|$, $\mathbf{R}$ –coordinate of observation point in far zone $|\mathbf{R}| = R$. Knowing the scattering field (4-114) of m-th incident wave, we space the radiating MSURs centered on the points $\tilde{\mathbf{r}}_n$ ($n = 1, 2, \ldots, \tilde{M}$) and directivity patterns $\tilde{F}(\tilde{\mathbf{q}}_n,\tilde{\mathbf{k}}_m,\omega)/R$ (in amplitude of sound pressure, $\tilde{\mathbf{q}}_n$ –axis vector of n -th radiating MSUR). Total scattering field p_ε is equal to

$$p_\varepsilon(\mathbf{R},\omega)=\sum_{m=1}^{M} p_{\varepsilon m}(\mathbf{R},\omega). \quad \textbf{(4-115)}$$

We need to suppress the total scattering field (4-115) in the known directions $\tilde{\mathbf{k}}_1,\tilde{\mathbf{k}}_2,\ldots,\tilde{\mathbf{k}}_{\tilde{M}}$, so we space $\tilde{M}$ radiating MSURs (**Fig. 4-28-c**), centered on points $\tilde{\mathbf{r}}_n$ ($n=1,2,...,\tilde{M}$) and with axis vectors

$$\tilde{\mathbf{q}}_n=\tilde{\mathbf{k}}_n/\left|\tilde{\mathbf{k}}_n\right| \ (\left|\tilde{\mathbf{q}}_n\right|=1). \quad \textbf{(4-116)}$$

Amplitudes φ_n ($n=1,2,...,\tilde{M}$) for excitation of radiating (compensating) MSURs we take get from (4-115) and the following equations

$$\varphi_n[R\tilde{F}_n(\tilde{q}_n,\tilde{\mathbf{k}}_n,\omega)]_{R\to\infty}+[Rp_\varepsilon(\mathbf{k}_n,\omega)]_{R\to\infty}=0 \quad \textbf{(4-117)}$$

for each $n=1,2,...,\tilde{M}$. So using (4-112), (4-115), (4-117), we obtain zeros of scattering field in all directions $\tilde{\mathbf{k}}_1,\tilde{\mathbf{k}}_2,\ldots,\tilde{\mathbf{k}}_{\tilde{M}}$ (**Fig. 4-28-d**). This algorithm can be classified as {NS/NT/CA/F} (**see Fig. 1-3**). So the above active damping system makes the7 protected body invisible for incident waves from directions $\mathbf{k}_1,\mathbf{k}_2,\ldots,\mathbf{k}_M$, if the observers are spaced far from the body in directions $\tilde{\mathbf{k}}_1,\tilde{\mathbf{k}}_2,\ldots,\tilde{\mathbf{k}}_{\tilde{M}}$.

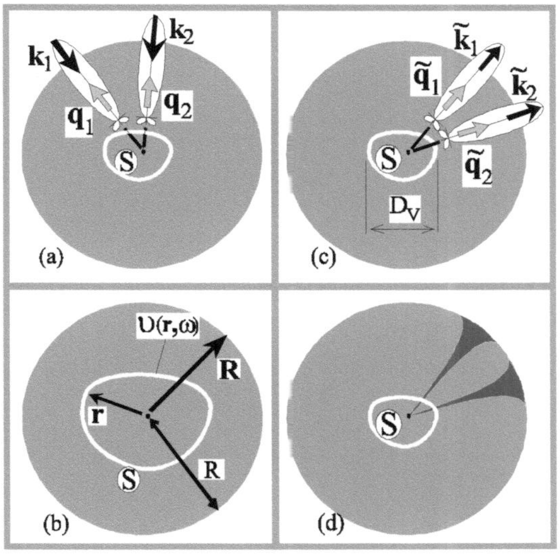

Figure 4–28: On the geometry of active damping system. Receiving MSURs with axis vectors $\mathbf{q}_1,\mathbf{q}_2$ against wave vectors $\mathbf{k}_1$, $\mathbf{k}_2$, of incident waves (a). Vector coordinate $\mathbf{r}\in S$ of surface S – and vector coordinate $\mathbf{R}$ of observation point in far zone, module of this vector R, measured DNOV $\upsilon(\mathbf{r},\omega)$ on S (b). Radiating MSURs with axis vectors $\tilde{\mathbf{q}}_1,\tilde{\mathbf{q}}_2$ along wave vectors $\mathbf{k}_1$, $\mathbf{k}_2$ of scattered waves to be damped, D_V –characteristic linear scale of body protected (c). Synthesized zeroes of scattering field in far zone (d). For the simplicity of a picture the directivity pattern of scattered field is taken isotropic.

We must note the special requirements for the elements of the above active damping system:

1) the spacing of receiving and radiating MSURs must ensure, that all points $\mathbf{r}\in S$ are in back-space relating to each MSUR. This condition excludes any action of S on the receiving MSURs and excludes any action of radiating MSURs on S.

2) each receiving MSUR with axis vector $\mathbf{q}_n$ is sensitive only to the incident wave with vector $\mathbf{k}_n$ and insensitive to other incident waves and the field of radiating MSURs

3) power directivity pattern of radiating MSURs must not intersect between each other.

4) to satisfy the conditions 1-4 we must have sufficiently large order $N > 1$ of multipolarity of MSURs and sufficiently rarefield spectrums of wave vectors $\mathbf{k}_1, \mathbf{k}_2, \ldots, \mathbf{k}_M$ and $\tilde{\mathbf{k}}_1, \tilde{\mathbf{k}}_2, \ldots, \tilde{\mathbf{k}}_{\tilde{M}}$.

Of course due to reciprocity principle the normalized directivity patterns in receiving and in radiation are identical, *i.e.* $F(\mathbf{q}_n, \mathbf{k}, \omega) / F(\mathbf{q}_n, -\mathbf{q}_n |\mathbf{k}|, \omega) = [\tilde{F}(\mathbf{q}_n, \mathbf{k}, \omega)]^* / [\tilde{F}(\mathbf{q}_n, \mathbf{q}_n |\mathbf{k}|, \omega)]^*$.

CHAPTER 5

Acoustical Sources of One-sided Radiation

Abstract: In monochromatic description (sections 5.1, 5.2), the results of the studies of the physical characteristics of unidirectional acoustic sources used in active sound control systems are presented. A discrete unidirectional source in the form of two phased monopoles (section 5.1) and a planar array of such unidirectional sources is considered (section 5.2.1). One-dimensional boundary-value problems with two (the two-point problem) and three (the three-point problem) controlled parallel planar boundaries between homogeneous media with arbitrary impedances are studied (sections 5.2.2-5.2.6). The boundaries (two or three) are subjected to the action of external forces. The case of the zero sum of external forces applied to the controlled boundaries corresponds to a supportless unidirectional source (SUS). It is shown that a unidirectional source can be created within the two-point boundary-value problem, whereas a supportless unidirectional source can be created within the three-point problem (sections 5.2.5, 5.2.6). Such parameters such as transparency, small size, absence of support, and broad frequency band can be achieved for a unidirectional source in the form of two piezoelectric layers with the same impedance and velocity of sound as those of the surrounding medium (5.2.7-5.2.9). The aspects of linearity of the transparent SUS and its application to active sound control problems are described (sections 5.2.10, 5.2.11). A spatially one-dimensional model of a plane active double layer between two homogeneous elastic half-spaces is studied analytically in temporal representation (section 5.3). The layer synthesizes a preset smooth trajectory of the controlled boundary between the media without any mechanical support. The outer layer of the coating is piezoelectric, and the inner layer is a polymer that is transparent for low-frequency sound and opaque for high-frequency sound because of dissipation. An algorithm for controlling the piezoelectric elements of the layer on the basis of signals from surface particle velocity sensors is proposed (section 5.3.6), and a method for measuring the particle velocity is developed simultaneously (section 5.3.9). Conditions of stability and efficiency of the synthesis are formulated (section 5.3.7). It is shown that the active layer thickness can be much smaller than the wavelength corresponding to the minimal time scale of the boundary trajectory to be formed. The accuracy of the trajectory synthesis depends on the accuracy of measuring, computing, and actuating elements of the system but does not depend on the vibroacoustic characteristics of the half-spaces separated by the active layer or on the presence of smooth waves in these half-spaces. For the synthesis to be efficient, the operating frequency band and the dynamic range of sensors and actuators should be many times greater than the frequency band and the dynamic range of the trajectory to be formed.

5.1. DISCRETE UNIDIRECTIONAL ACOUSTIC SOURCES

In some active sound control problems, it is necessary that the measurement of the field under control be independent of the radiation of control sources, which guarantees the stability of the active system. For this purpose, unidirectional sources (USs) (or unidirectional receivers, *i.e.*, vector microphones) are used. These devices are physical analogs of the virtual (fictitious) Huygens sources that are sometimes used to describe the processes of wave propagation and diffraction. The USs usually have the form of multielement sources characterized by a considerable relative difference in power fluxes emitted in two opposite directions. For non-one-dimensional sources, this property refers to the difference in integral (over the solid angle) powers emitted into the front and rear half-spaces. Ideally, USs provide a zero backward and a finite-power forward radiation. USs are useful in designing active sound control systems, which were considered, by G.D. Malyuzhinets in his classical works [15-19]. The possibility of creating such sources has long been known for waves of different physical nature, which is a consequence of the classical wave equation. This paper presents a detailed study of a number of properties of acoustic USs in a free space in the absence of dissipation, specifically: constraints on the dynamic range of emitted waves, acoustic transparency of the source, dependence of its characteristics on its wave dimensions, energy transfer processes in the USs, versions of using USs in active sound control, *etc.* Particular attention is paid to the possibility of supportless operation of USs. A three-dimensional acoustic US [100], [101] is usually considered as a combination of in-phase excitement and spatially collocated monopole and dipole (this object is sometimes is called a tripole) with a cardioid directional pattern, or as a combination of three monopoles or pulsating spheres with radius a placed at a distance h from one another (see **Fig. 5-1-a**). In this design, $a << h << \lambda$ and the ratio $V_M / V_D = 2kh$ of the complex amplitudes $+V_D$, $+V_M$, and $-V_D$ of the pulsation velocity is purely real, where $k = \omega / c = 2\pi / \lambda$, ω is the frequency, λ is the wavelength, and c is the velocity of sound in the medium. Below, we consider an alternative version of a three-dimensional US (**Fig. 5-1-b**, **Fig. 5-1-c**), which has a form of two monopoles separated by a

Vladimir V. Arabadzhi

distance $2h$ from one another. The monopoles oscillate with velocity amplitudes $+V\exp(+ikh)$ and $-V\exp(-ikh)$. A similar US in the form of two surfaces filled with phased monopoles was considered earlier by Fedoryuk [17]. Let us call this US a phased US. Its radiation pattern has the same cardioid shape as that of the tripole. At distance $r > \lambda >> h$ from the US, the emitted power flux density is

$$\Pi_H = |V|^2 \omega\rho a^4 k^3 h^2 r^{-2} (1 + \cos\vartheta)^2, \qquad \textbf{(5-1)}$$

where is the density of the medium and ϑ is the polar angle, $0 \le \vartheta < \pi$. Pattern (1) is anisotropic in space, *i.e.*, $\Pi_H(\pm\vartheta) \ne \Pi_H(\pi + \vartheta)$, unlike the acoustic power flux density

$$\Pi_D = |V|^2 \omega\rho a^4 k^3 h^2 r^{-2} \cos^2\vartheta, \qquad \textbf{(5-2)}$$

of the dipole (monopole) of the same design, where Due to the phase adjustment of the spatially separated (by a distance $2h$) monopoles, the US guarantees zero radiation in the backward direction $\vartheta = \pi$, however, at the expense of a decrease by a factor $\sim (kh)^{-1} >> 1$ of in the field emitted in the forward direction $\vartheta = 0$ with respect to the radiation produced by a monopole with the same excitation amplitude. The phase difference between the monopoles is a linear function of frequency, and, in the time representation, it is reduced to the delay $\tau = 2h/c$, which facilitates the design of a broadband US in spite of its small wave dimensions. The examples of the tripole and phased USs illustrate the non-uniqueness of the solution typical of many inverse problems, in which the same radiation pattern (5-1) can be produced by each of the two types (tripole or phased) of USs, which have quite different designs.

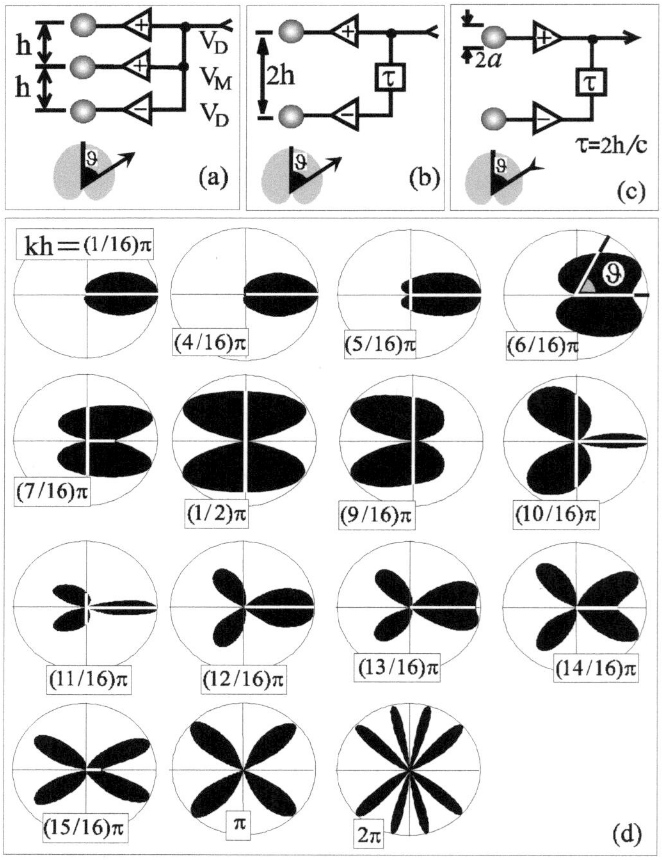

Figure 5-1: Tripole US (a) and phased US in the radiation (b) and reception (c). Power flux density (d) produced by a phased US (directivity patterns) normalized by its maximum, $\Pi_H(\vartheta)/\max[\Pi_H(\vartheta)]$, as a function of the polar angle $0 \le \vartheta < \pi$ for different values of kh.

5.1.1. Mechanical Support of a Discrete US

The requirement for the US to have no support is imposed, in particular, in hydroacoustic problems on vibrations of thin-walled steel shells in water. In these cases, steel objects that are stiff in air appear to be compliant when immersed in water. Therefore, based on the variety of materials existing in practice, we may state that there are actually no means to acoustically fix structures and their boundaries immersed in a liquid. Note that the question on the necessity of a mechanical support for USs was not considered by Malyuzhinets and his colleagues. A more general statement of the problem assumes that there exists a discrete system of active controlled elements that are in mechanical contact with the wave-guiding medium alone (a liquid, gaseous, or elastic one). Is this system of active elements capable of generating an acoustic field that propagates in one direction and does not propagate in the opposite direction? A simplified interpretation of the momentum conservation law suggests that it is impossible to combine unidirectionality of the source with the absence of the support. For the discrete US considered above, the support force is necessary, though its magnitude is small, because it is provided by the dipole scattering of the fields produced by neighboring monopoles (oscillating spheres) from each other. Obviously, we only consider sinusoidal forces of the first order in acoustic amplitude, rather than time-independent radiation pressure, which exerts an even smaller effect, quadratic in acoustic amplitude. The assumption that the centers of the oscillating spheres are movable yields the following estimate of the relative total (acting on both monopoles) supports reaction force: $|F_H|/|F_0| = a^2/(4h^2) << 1$, where F_H is the support reaction force and F_0 is the support reaction force in the following problem: a circular piston of radius a oscillates in an infinite planar stiff baffle by creating the same far-field amplitude as the US at the maximum of its radiation pattern.

5.1.2. Dynamic Range of the US: Linearity

Let a sphere of radius a oscillate with the velocity amplitude $V = V_0$, producing a certain acoustic field in the far-field zone. To produce the same field at the same distance, the phased US composed of two such spheres needs to have an oscillation amplitude $V = V_H$ of the spheres of the same US (with radius a) that is $|V_H|/|V_0| = (2kh)^{-1} << a$ times greater than the excitation amplitude of the monopole. For the boundary-value problem to remain linear, it is also necessary to satisfy the condition $|V_H|/\omega << a$.

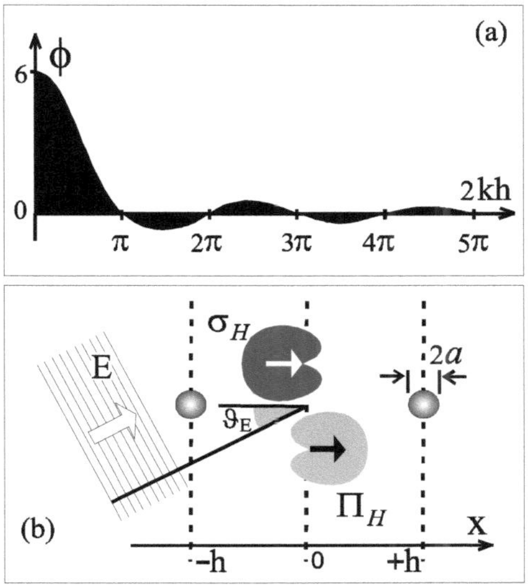

Figure 5-2: Unidirectionality (in terms of radiated power) of a phased US as a function (a) of frequency (or kh), and geometry (b) of the problem of absorption of an incident wave by a phased US.

5.1.3. Radiation Pattern of a Phased US

At small wave dimensions $kh << 1$, the cardioid power radiation pattern of the US is actually frequencyindependent (from $kh = 0$ through $(5/16)\pi$, see **Fig. 5-1-d**), which makes small-size USs broadbands. As the frequency increases, the pattern qualitatively changes when $(1/16) \le kh \le (7/16)\pi$ by acquiring new lobes and gradually ceasing to be unidirectional. When a US has dimensions of $kh = m\pi / 2$ ($m = 1, 2, ...$), its pattern coincides with the pattern of the simple dipole with a purely real amplitude ratio of its elements and kh equal to an integer multiple of $\pi / 2$. In the backward direction $\vartheta = \pi$, the pattern is always zero. **Fig. 5-1-a** represents $\varphi = \left[(\Pi_H)_F / (\Pi_H)_B\right] - 1$ versus US wave dimension kh (frequency), where $(\Pi_H)_F$ is the total power flux emitted by the US into the forward half-space and $(\Pi_H)_B$ is emitted into the backward half-space.

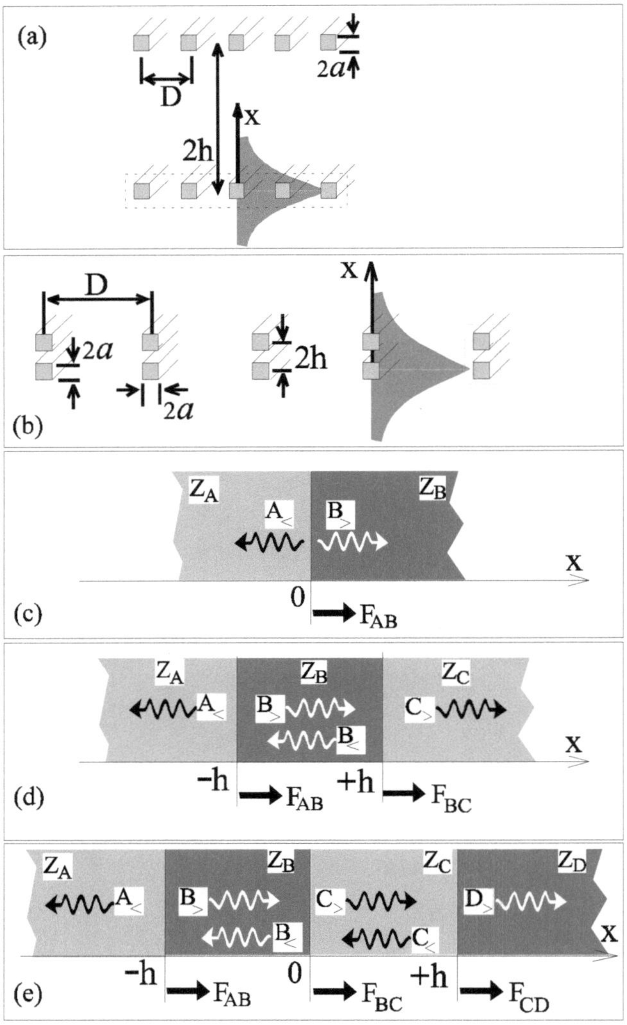

Figure 5-3: Versions of array-type USs (a), (b); forces and waves in the one- (c), two- (d), and three-point (e) boundary-value problems, respectively.

5.1.4. A Phased US under the Maximum (Resonance) Absorption Conditions

If the excitation amplitude and phase of the US are specially chosen in accordance with the incident wave, the incident wave will perform positive work on the US's monopoles and the US will absorb the incident wave energy [60], [67]. Let the velocity amplitudes of monopoles of the discrete US at the points $x = -h$ and $x = +h$ (see **Fig. 5-2-b**) be $V_1 = -V \exp(-ikh)$ and $V_2 = +V \exp(+ikh)$ (where $V = |V| \exp(i\psi_V)$, $\mathrm{Im}(\psi_V) = 0$) and the plane wave be incident at angle ϑ_E to the US axis and produce pressure with amplitudes of $E_1 = E \exp(+ikh \cos \vartheta_E)$ and $E_2 = E \exp(-ikh \cos \vartheta_E)$ (where $E = |E| \exp(i\psi_E)$, $\mathrm{Im}(\psi_E) = 0$) on the US monopoles. Under the maximum (resonance) absorption at frequency ω, the reactive component of the impedance of the device (drive) that excites the US compensates to the reactive component of the US impedance, while the active components of their impedances are identical. At a velocity magnitude of $|V| = (3/16)(1-\cos\vartheta_E)\omega^{-1}k^{-2}\rho^{-1}h^{-1}|E|$ and phase of $\psi_V = \psi_E - (\pi/2)$, the power W_H absorbed by the US is maximized and the absorption cross-section is $\sigma_H = 3\lambda^2(1+\cos\vartheta_E)^2 2^{-3}\pi^{-1}$. It is important to note that the US absorption pattern $\sigma_H(\vartheta_E)$ is opposite to its power flux density radiation pattern (5-1) (and also to the scattering pattern or to the power of the signal received by the US) in the far-field zone: the maximum of the absorption pattern coincides with zeros of the radiation, scattering, and reception patterns, *i.e.*, $W_H(\vartheta)/W_H(0) = \Pi_H(\vartheta-\pi)/\Pi_H(\pi)$. In other words, the absorbed power is maximized at a US orientation for which its received signal is zero. This is the qualitative difference between the phased pair of monopoles (*i.e.*, US) and dipoles (multipoles) with purely real ratios between complex oscillation amplitudes of their elements: in these sources, the absorbed power W_D, power flux density, and received signal power have the same dependence on the polar angle, *i.e.*, $W_D(\vartheta)/W_D(0) = \Pi_D(\vartheta)/\Pi_D(0)$.

The real parts $\mathrm{Re}\, Z_H$, $\mathrm{Re}\, Z_D$ and imaginary parts $\mathrm{Im}\, Z_H$, $\mathrm{Im}\, Z_D$ of impedances Z_H, Z_D of the US and the dipole of the same design, respectively, are related as $\mathrm{Re}\, Z_H = \mathrm{Re}\, Z_D = 2^5 3^{-1}\pi k^3 h^2 a^4 \omega\rho$, $\mathrm{Im}\, Z_H = \mathrm{Im}\, Z_D = 8\pi\rho\omega a^3$ and the maximum absorption cross sections of the US and the dipole are identical. Under the maximum absorption conditions, the US scattering cross section $\tilde{\sigma}_H$ is equal to the absorption cross section: $\tilde{\sigma}_H = \sigma_H$ [67], its power flux density of the scattered field coincides with its radiation power flux density, and the backscattering cross section (in directions $|\vartheta| \le \pi/2$) $\tilde{\sigma}_B$ and forward scattering cross section (in directions $\pi/2 \le |\vartheta| \le \pi$) $\tilde{\sigma}_F$ are related as $\tilde{\sigma}_B/\tilde{\sigma}_F = 1/7$ (see, *e.g.*, **Fig. 5-2-b**) and $\tilde{\sigma}_H = \tilde{\sigma}_B + \tilde{\sigma}_F$.

5.2. ONE-DIMENSIONAL UNIDIRECTIONAL SOURCES (MONOCHROMATIC REPRESENTATION)

Active sound control problems also employ continuous unidirectionally radiating (or unidirectionally receiving) surfaces (physical analogs of Huygens surfaces [1]) that can completely and abruptly separate spatial regions insonified or not these surfaces. Further, we consider the simplest version of a unidirectionally radiating surface: a planar, or one-dimensional, US. A planar US can be designed as a discrete array of USs, as described above, as well as a set of infinite planar homogeneous elastic layers of a finite thickness, whose boundaries control (for example, piezoelectric) forces, regular over the surface, are applied.

5.2.1. Array-Type USs

In this section, for simplicity, we consider equidistant planar US arrays composed of oscillating (along the normal to the array plane, see **Fig. 5-3**), parallel, square, $2a \times 2a$ rods with fixed centers. To provide transparency of the array required in active sound control problems [3], it is necessary to satisfy the condition $2a \ll D$. A one-dimensional US can be constructed from discrete elements using one of the following two approaches:

(a) One can compose a US from parallel fluctuating rods placed on a plane with a period D (a planar monopole radiator) and then form the one-dimensional US from two such planar radiators separated by a distance $2h$. The configuration of such a US (**Fig. 5-3-a**) is defined by the condition $2a \ll D \ll 2h \ll \lambda$. Single dipole scattering of sound from the rods produces a nonzero force F_H of the support reaction and the dimensionless reaction force $|F_H|/|F_0| = a^2 D^{-1} h^{-1}$, where F_0 is the force applied to a light rigid plane to generate a plane wave of the same amplitude as that of the output plane wave of the array-type US. The presence of discrete radiators in the array-type US produces a reactive (near, exponent shaded in **Fig. 5–3**) field, which remains significant within a certain distance from the array; this field does not allow the array to be placed directly on the surface of the protected bodies [17]. Under condition $h \gg D^2/\lambda$, the external (with respect to the array) near field consists of the uncompensated field of only one planar array. Let us describe the thickness of the region occupied by the reactive field by the dimensionless function $|P_r|/|P_H| = [D/(8h)]\exp[-(2\pi/D)x]$ of the decay of the reactive field with distance x from the array, where P_r is the amplitude of the reactive pressure field of the single-layer array of oscillating rods and P_H is the pressure amplitude of the plane wave unidirectionally radiated by the array-type US. Let us estimate the dynamic range overshoot as follows. Assume that an infinite, light, rigid plane, oscillating with a velocity amplitude $V = V_0$, creates a plane wave of the same amplitude. Then, for the array-type

US to unidirectionally radiate a plane wave of the same amplitude, spheres of the US must oscillate with a velocity amplitude $V = V_H$ by a factor of $|V_H|/|V_0| = D/(4akh) >> 1$ greater than V_0 under the condition that the field remains linear.

(b) One can compose a discrete phased US from two oscillating rods separated by a distance $2h$ and then place such USs on a plane with a uniform spacing D to form a one-dimensional US. The configuration of this US (**Fig. 5-3-b**) is determined by the condition $2a << 2h << D << \lambda$. The dimensionless support force defined similar to that of case (a) as $|F_H|/|F_0| = 1/2\pi$ is independent of geometrical parameters of the problem. Unlike the case considered above, near fields of closely spaced planar arrays partially compensate each other (like in forming the unidirectional radiation); as a result, we obtain the estimate $|P_r|/|P_H| = (\pi/4)\exp[-(2\pi/D)x]$ for the reactive field decay as a function of the distance x from the array. The dynamic range of the US is characterized by the same conditions as in case (a).

5.2.2. A Continuous Planar Unidirectional Source

While discrete tripole USs are physically feasible, a continuous planar US is unfeasible in principle, because particle velocities or bulk forces in a continuous medium are in this case unavoidably double-valued (when the monopole and dipole excitation regions are collocated). In this section, we consider the examples of one-, two-, and three-point boundary-value problems to investigate the possibility of creating a US and the consistency of its radiation in the absence of a mechanical support, *i.e.*, the design of a supportless US (or SUS).

5.2.3. The One-Point Boundary-Value Problem

Using the one-point boundary-value problem as an example (**Fig. 5-3-c**), we will define the terms *controlled wall* and *transparent force source* used below. Considering the field $u(x)$ of complex amplitudes of longitudinal particle displacement at the frequency ω in an infinite elastic rod of cross-sectional area S, whose parameters are distributed as follows: for $x \le 0$, the impedance, wave number, density, and elastic modulus are $Z_A = S\sqrt{\rho_A E_A}$, $k_A = \omega\sqrt{\rho_A / E_A}$, and E_A; for $x > 0$, they are, respectively, $Z_B = S\sqrt{\rho_B E_B}$, $k_B = \omega\sqrt{\rho_B / E_B}$, ρ_B, and E_B.

5.2.4. Transparent Force Source

We introduce a thin, rigid, light, planar wall perpendicular to the rod in its $x = 0$ section with a single degree of freedom, namely, the motion along the x axis. If such a wall is free, it is transparent and the incident waves are reflected only from the discontinuity in the step of the parameter of the medium. Let force F_{AB} be applied to the wall along the rod axis. This force must be independent of the kinematic characteristics of wall oscillations, for example, oscillations under the action of the incident wave with characteristic amplitude U_w and frequency ω_w. This means that the intrinsic impedance of the source, Z_F, complies with the condition $|Z_F| << \max\{|Z_A|,|Z_B|\}$. In the absence of incident waves, the field of longitudinal particle displacements in the rod has the form $u = u_A = A_< \exp(+ik_A x)$ for $x < 0$ and $u = u_B = B_> \exp(-ik_B x)$ for $x > 0$ and satisfies the boundary conditions $u_A = u_B$ and $(u_A)'_x E_A S - (u_B)'_x E_B S = F_{AB}$ at $x = 0$. As a result, we obtain the amplitudes of the displacement waves (the amplitudes of the displacement waves are identical, while the amplitudes of the stress waves are not): $A_< = B_> = -i\omega F_{AB}(Z_A + Z_B)^{-1}$. Because the source of force F_{AB} is transparent and the boundary value problem is linear, the total wave field is a superposition of waves $A_<$ and $B_>$ plus the incident field at F_{AB} and plus the field scattered from the boundary $x = 0$ at $Z_A \ne Z_B$.

5.2.5. The Two-Point Boundary-Value Problem

Let two controlled walls be placed in the sections $x = -h$ and $x = +h$ of the rod and the parameters be distributed as follows (**Fig. 5-3-d**): or $x < -h$, the rod's impedance, wave number, density, and elastic modulus are $Z_A = S\sqrt{\rho_A E_A}$, $k_A = \omega\sqrt{\rho_A / E_A}$, ρ_A, and E_A; for $-h \le x \le +h$, the impedance, wave number, density, and elastic modulus are

$Z_B = S\sqrt{\rho_B E_B}$, $k_B = \omega\sqrt{\rho_B / E_B}$, ρ_B, and E_B; and for $x > +h$, the impedance, wave number, density, and elastic modulus are $Z_C = S\sqrt{\rho_C E_C}$, $k_C = \omega\sqrt{\rho_C / E_C}$, ρ_C, and E_C, respectively. The wave field $u(x)$ of particle shift has the following forms (with magnitudes $A_<$, $B_<$, $B_>$, $C_>$): $u = u_A = A_< \exp(+ik_A x)$ for $x \le -h$; $u = u_B = B_< \exp(+ik_B x) + B_> \exp(-ik_B x)$ for $-h \le x < +h$; $u = u_C = C_> \exp(-ik_B x)$ for $x > +h$. For short, we will not thoroughly analyze the two-point problem and will note only a few things. In the two-point problem, it is possible to create a US (*i.e.*, choose such force amplitudes F_{AB} and F_{BC} that the external radiated field is unidirectional). The sum of the forces $F_{AB} + F_{BC} \ne 0$ for the unidirectional radiation is the same as the force in the one-point problem that generates the wave of the same amplitude. Therefore, in the two-point problem, unidirectional radiation of the US complies with the no-support condition ($F_{AB} + F_{BC} = 0$) only at frequencies of $2k_B h = \pi m$ ($m = 0,1,2,...$) at which the radiation is zero in both directions. Therefore, it is impossible to create a two-point SUS. One can easily show the inapplicability of the tripole ideology with $F_{AB} = +F_M + F_D$ and $F_{BC} = -F_M + F_D$, where F_M is the monopole component of the force that compresses and stretches the walls in the phase and F_D is the dipole component of the force that moves both walls leftwards or rightwards in phase. The point is that it is impossible to obtain nonzero unidirectional radiation when the phase shift between F_M and F_D is $m\pi$ ($m = 0,\pm1,\pm2,...$) even in the presence of a support. When the US radiates unidirectionally rightwards (or leftwards), the rear wall is fixed and the front wall oscillates with a displacement amplitude equal to the amplitude of the emitted wave. Therefore, unlike the discrete USs considered above, there is no dynamic range overshoot for displacement in the two-point problem. However, at $kh \ll 1$, an overshoot is observed in the forces applied to the walls: $\max\{|F_{AB}|,|F_{BC}|\} / \omega Z|A_<| = (2kh)^{-1} \gg 1$, $\max\{|F_{AB}|,|F_{BC}|\} / \omega Z|C_>| = (2kh)^{-1} \gg 1$ (at $Z = Z_A = Z_B = Z_C$ and $k = k_A = k_B = k_C$) *i.e.*, the forces that excite the US must be much greater than the forces that generate the same wave in the one point problem. On the internal surface of the front wall, the pressure and velocity are nonzero and shifted in phase by $\pi/2$, otherwise, the US would blow up. The external surface of the rear wall is energy neutral. The external surface of the front US wall emits the wave. In the presence of waves incident from outside, the field consists of the waves produced in the US radiation problem plus the field generated in the problem of scattering of the incident waves from the boundaries $x = -h$, $x = 0$, and $x = +h$ at $F_{AB} + F_{BC} = 0$.

5.2.6. The Three-Point Boundary-Value Problem

Let three controlled walls be present in the $x = -h$, $x = 0$, $x = +h$ sections of the rod (**Fig. 5-3-e**), and let three forces F_{AB}, F_{BC} and F_{CD} created by transparent sources be applied to these walls. At $x < -h$, the rod's impedance, wave number, density, and elastic modulus are $Z_A = S\sqrt{\rho_A E_A}$, $k_A = \omega\sqrt{\rho_A / E_A}$, ρ_A, and E_A; at $-h \le x < 0$ the rod's impedance, wave number, density, and elastic modulus are $Z_B = S\sqrt{\rho_B E_B}$, $k_B = \omega\sqrt{\rho_B / E_B}$, ρ_B, and E_B; at $-h \le x \le +h$ the rod's impedance, wave number, density, and elastic modulus are $Z_C = S\sqrt{\rho_C E_C}$, $k_C = \omega\sqrt{\rho_C / E_C}$, ρ_C, and E_C; at $+h < x < +\infty$ the rod's impedance, wave number, density, and elastic modulus are $Z_D = S\sqrt{\rho_D E_D}$, $k_D = \omega\sqrt{\rho_D / E_D}$, ρ_D, and E_D, respectively. Without waves incident from $x < -h$ and $x > +h$, the field of complex amplitudes $u(x)$ of the rod's longitudinal particle displacements has the form

$$u = u_A = A_< \exp(+ik_A x) \textbf{ for } x < -h, \tag{5-3}$$

$$u = u_B = B_< \exp(+ik_B x) + B_> \exp(-ik_B x) \textbf{ for } -h \le x < 0,$$

$$u = u_C = C_< \exp(+ik_C x) + C_> \exp(-ik_C x) \textbf{ for } -h \le x \le +h,$$

$$u = u_D = D_> \exp(-ik_D x) \textbf{ for } x > +h$$

and satisfies the boundary conditions

$$u_A = u_B,\ (u_A)'_x E_A S - (u_B)'_x E_B S = F_{AB} \text{ for } x = -h, \quad (5\text{-}4)$$

$$u_B = u_C,\ (u_B)'_x E_B S - (u_C)'_x E_C S = F_{BC} \text{ for } x = 0,$$

$$u_C = u_D,\ (u_C)'_x E_C S - (u_D)'_x E_D S = F_{CD} \text{ for } x = +h.$$

Substitution of (5-3) into (5-4) yields the system of six equations

$$+A_< \exp(-ik_A h) - B_< \exp(-ik_B h) - B_> \exp(+ik_B h) = 0, \quad (5\text{-}5)$$

$$+\omega Z_A A_< \exp(-ik_A h) - \omega Z_B B_< \exp(-ik_B h) + \omega Z_B B_> \exp(+k_B h) = -iF_{AB},$$

$$+B_> + B_< - C_> - C_< = 0,$$

$$-\omega Z_B B_> + \omega Z_B B_< + \omega Z_C C_> - \omega Z_C C_< = iF_{BC},$$

$$+C_> \exp(-ik_C h) + C_< \exp(+ik_C h) - D_> \exp(-ik_D h) = 0,$$

$$-\omega Z_C C_> \exp(-ik_C h) + \omega Z_C C_< \exp(+ik_C h) + \omega Z_D D_> \exp(-ik_D h) = -iF_{CD}.$$

Equations (5-5) give expressions for the desired complex amplitudes $A_<$, $A_>$, $B_<$, $B_>$, $C_<$, $C_>$, $D_>$ of the waves in the three-point problem in the form of linear functions of the forces F_{AB}, F_{BC}, F_{CD}:

$$A_< = A_{<AB} F_{AB} + A_{<BC} F_{BC} + A_{<CD} F_{CD}, \quad (5\text{-}6)$$

$$B_> = B_{AB>} F_{AB} + B_{BC>} F_{BC} + B_{CD>} F_{CD},$$

$$B_< = B_{<AB} F_{AB} + B_{<BC} F_{BC} + B_{<CD} F_{CD},$$

$$C_> = C_{AB>} F_{AB} + C_{BC>} F_{BC} + C_{CD>} F_{CD},$$

$$C_< = C_{<AB} F_{AB} + C_{<BC} F_{BC} + C_{<CD} F_{CD},$$

$$D_> = D_{AB>} F_{AB} + D_{BC>} F_{BC} + D_{CD>} F_{CD}.$$

Coefficients of the forces F_{AB}, F_{BC}, F_{CD} are defined as follows:

$$B_{AB>} = +(i/\omega)(Z_A - Z_B)^{-1} \alpha_{11} (\alpha_{12} - \beta_1^{-1} \alpha_{11}) \exp(+ik_B h), \quad (5\text{-}7)$$

$$B_{<AB} = +(i/\omega)(Z_A + Z_B)^{-1} \alpha_{12} (\alpha_{11} - \beta_1 \alpha_{12}) \exp(-ik_B h),$$

$$B_{BC>} = +(i/\omega) Z_C^{-1} (\beta_2 - 1)(\alpha_{12} - \beta_1^{-1} \alpha_{11})^{-1},$$

$$B_{<BC} = +(i/\omega) Z_C^{-1} (\beta_2 - 1)(\alpha_{11} - \beta_1 \alpha_{12})^{-1},$$

$$B_{CD>} = -(i/\omega)(Z_D + Z_C)^{-1}(\alpha_{12} - \beta_1^{-1}\alpha_{11})^{-1} 2\exp(-ik_C h),$$

$$B_{<CD} = -(i/\omega)(Z_D + Z_C)^{-1}(\alpha_{11} - \beta_1\alpha_{12})^{-1} 2\exp(-ik_C h),$$

$$C_{AB>} = -(i/\omega)(Z_A + Z_B)^{-1}(\alpha_{21} - \beta_2\alpha_{22})^{-1} 2\exp(-ik_B h),$$

$$C_{BC>} = +(i/\omega)Z_B^{-1}(\beta_1 - 1)(\alpha_{21} - \beta_2\alpha_{22})^{-1},$$

$$C_{CD>} = +(i/\omega)(Z_D + Z_C)^{-1}\alpha_{22}(\alpha_{21} - \beta_2\alpha_{22})^{-1}\exp(-ik_C h),$$

$$C_{<AB} = -(i/\omega)(Z_A + Z_B)^{-1}(\alpha_{22} - \beta_2^{-1}\alpha_{21})^{-1} 2\exp(-ik_B h),$$

$$C_{<BC} = +(i/\omega)Z_B^{-1}(\beta_1 - 1)(\alpha_{22} - \beta_2^{-1}\alpha_{21})^{-1},$$

$$C_{<CD} = +(i/\omega)(Z_D - Z_C)^{-1}\alpha_{21}(\alpha_{22} - \beta_2^{-1}\alpha_{21})^{-1},$$

and

$$A_{<AB} = \exp(+ik_A h)[B_{<AB}\exp(-ik_B h) + B_{AB>}\exp(+ik_B h)],$$

$$A_{<BC} = \exp(+ik_A h)[B_{<BC}\exp(-ik_B h) + B_{BC>}\exp(+ik_B h)],$$

$$A_{<CD} = \exp(+ik_A h)[B_{<CD}\exp(-ik_B h) + B_{CD>}\exp(+ik_B h)],$$

$$D_{AB>} = \exp(+ik_D h)[C_{<AB}\exp(+ik_C h) + C_{AB>}\exp(-ik_C h)],$$

$$D_{BC>} = \exp(+ik_D h)[C_{<BC}\exp(+ik_C h) + C_{BC>}\exp(-ik_C h)],$$

$$D_{CD>} = \exp(+ik_D h)[C_{<CD}\exp(+ik_C h) + C_{CD>}\exp(-ik_C h)],$$

or

$$A_{<AB} = +\frac{i\alpha_{12}\exp[+i(k_A - 2k_B)h]}{\omega(Z_A + Z_B)(\alpha_{11} - \beta_1\alpha_{12})} + \frac{i\alpha_{11}\exp[+i(k_A + k_B)h]}{\omega(Z_A - Z_B)(\alpha_{12} - \beta_1^{-1}\alpha_{11})}, \quad \textbf{(5-8)}$$

$$A_{<BC} = +\frac{i(\beta_2 - 1)\exp[+i(k_A - k_B)h]}{\omega Z_C(\alpha_{12} - \beta_1\alpha_{12})} + \frac{i(\beta_2 - 1)\exp[i(k_A + k_B)h]}{\omega Z_C(\alpha_{12} - \beta_2^{-1}\alpha_{11})},$$

$$A_{<CD} = -\frac{2i\exp[+i(k_A - k_C - k_B)h]}{\omega(Z_D + Z_C)(\alpha_{11} - \beta_1\alpha_{12})} + \frac{2i\exp[+i(k_A - k_C + k_B)h]}{\omega(Z_D + Z_C)(\alpha_{12} - \beta^{-1}\alpha_{11})},$$

$$D_{AB>} = +\frac{2i\exp[+i(k_D - k_B + k_C)h]}{\omega(Z_A + Z_B)(\alpha_{22} - \beta_2^{-1}\alpha_{21})} - \frac{2i\exp[+i(k_D - k_B - k_C)h]}{\omega(Z_A + Z_B)(\alpha_{21} - \beta_2\alpha_{22})},$$

$$D_{BC>} = +\frac{i(\beta_1 - 1)\exp[+i(k_D + k_C)h]}{\omega Z_B(\alpha_{22} - \beta_2^{-1}\alpha_{21})} + \frac{i(\beta_1 - 1)\exp[+i(k_D - k_C)h]}{\omega Z_B(\alpha_{21} - \beta_2\alpha_{22})},$$

$$D_{CD>} = +\frac{i\alpha_{12}\exp[+i(k_D + 2k_C)h]}{\omega(Z_D - Z_C)(\alpha_{22} - \beta_2^{-1}\alpha_{21})} + \frac{i\alpha_{22}\exp[+i(k_D - k_C)h]}{\omega(Z_D + Z_C)(\alpha_{21} - \beta_1\alpha_{22})},$$

where

$$\alpha_{11} = 1 + Z_B Z_C^{-1} - \beta_2(Z_B Z_C^{-1} - 1),\ \alpha_{12} = 1 - Z_B Z_C^{-1} + \beta_2(Z_B Z_C^{-1} + 1), \tag{5-9}$$

$$\alpha_{21} = 1 + Z_C Z_B^{-1} - \beta_1(Z_C Z_B^{-1} - 1),\ \alpha_{22} = 1 - Z_C Z_B^{-1} + \beta_1(Z_C Z_B^{-1} + 1),$$

$$\beta_1 = (Z_A - Z_B)(Z_A + Z_B)^{-1}\exp(-2ik_B h),\ \beta_2 = (Z_D - Z_C)(Z_D + Z_C)^{-1}\exp(-2ik_C h).$$

Using (5-8) and (5-9), we formulate the system of equations for the forces that generate the wave $A_<$ in the absence of the wave in the opposite direction and in the absence of a support:

$$A_< = A_{<AB}F_{AB} + A_{<BC}F_{BC} + A_{<CD}F_{CD}, \tag{5-10}$$

$$D_{AB>}F_{AB} + D_{BC>}F_{BC} + D_{CD>}F_{CD} = 0,$$

$$F_{AB} + F_{BC} + F_{CD} = 0.$$

Its solutions are the forces $F_{AB} = F_{<AB}$, $F_{BC} = F_{<BC}$, $F_{CD} = F_{<CD}$:

$$F_{<AB} = [(A_{<AB} - A_{<BC}) - \mu_>(A_{<BC} - A_{<CD})]^{-1}A_<, \tag{5-11}$$

$$F_{<CD} = [(A_{<AB} - A_{<BC})\mu_>^{-1} - (A_{<BC} - A_{<CD})]^{-1}A_<,$$

$$F_{<BC} = -F_{<AB} - F_{<CD},$$

where $\mu_> = (D_{AB>} - D_{BC>})/(D_{BC>} - D_{CD>})$.

The system of equations for the forces that generate the wave $D_>$ in the absence of the wave in the opposite direction and without a support has the form:

$$A_{<AB}F_{AB} + A_{<BC}F_{BC} + A_{<CD}F_{CD} = 0, \tag{5-12}$$

$$D_> = D_{AB>}F_{AB} + D_{BC>}F_{BC} + D_{CD>}F_{CD},$$

$$F_{AB} + F_{BC} + F_{CD} = 0.$$

Its solution are the forces $F_{AB} = F_{AB>}$, $F_{BC} = F_{BC>}$, $F_{CD} = F_{CD>}$:

$$F_{AB>} = [(D_{AB>} - D_{BC>}) - \mu_<(D_{BC>} - D_{CD>})]^{-1}D_>, \tag{5-13}$$

$$F_{CD>} = [(D_{AB>} - D_{BC>})\mu_<^{-1} - (D_{BC>} - D_{CD>})]^{-1}D_>,$$

$$F_{BC>} = -F_{AB>} - F_{CD>},$$

where $\mu_< = (D_{<AB} - D_{<BC}) / (D_{<BC} - D_{<CD})$. Thus, in this three-point boundary-value problem forces (5-11) and (5-13) define a supportless unidirectional source (SUS) of radiation at an arbitrary ratio between impedances and velocities of sound in media that contact the controlled boundaries.

5.2.7. Transparent SUS Boundaries

In the case of acoustically homogeneous [90] SUS layers,

$$Z_A = Z_B = Z_C = Z_D = Z\,,\; k_A = k_B = k_C = k_D = k\,, \tag{5-14}$$

we obtain the following coefficients:

$$A_{<CD} = B_{<CD} = B_{AB>} = C_{<CD} = C_{AB>} = D_{AB>} = \varphi \exp(-ikh)\,, \tag{5-15}$$

$$A_{<AB} = D_{CD>} = \varphi \exp(+ikh)\,,$$

$$A_{<BC} = B_{<BC} = C_{BC>} = D_{BC>} = \varphi\,,$$

$$B_{<AB} = B_{BC>} = B_{CD>} = C_{<AB} = C_{<BC} = C_{CD>} = 0\,,$$

where $\varphi = (2i\omega Z)^{-1}$ amplitudes of all the waves in the three-point boundary-value problem :

$$A_< = \varphi \exp(+ikh) F_{AB} + \varphi F_{BC} + \varphi \exp(-ikh) F_{CD} \tag{5-16}$$

$$B_> = \varphi \exp(-ikh) F_{AB}$$

$$B_< = \varphi F_{BC} + \varphi \exp(-ikh) F_{CD}$$

$$C_> = \varphi \exp(-ikh) F_{AB} + \varphi F_{BC}$$

$$C_< = \varphi \exp(-ikh) F_{CD}$$

$$D_> = \varphi \exp(-ikh) F_{AB} + \varphi F_{BC} + \varphi \exp(-ikh) F_{CD}$$

Then, we find the forces from Eqs. (5–16)

$$F_{AB} = F_{<AB} = -\varphi^{-1}[1 - \exp(+ikh)]^{-1}[1 - \exp(-2ikh)]^{-1} A_<$$

$$F_{BC} = F_{<BC} = -F_{<AB} - F_{<CD}\,,$$

$$F_{CD} = F_{<CD} = -\varphi^{-1}[1 - \exp(-ikh)]^{-1}[1 - \exp(+2ikh)]^{-1} A_<\,,$$

where $F_{<AB} = F_{<CD} \exp(+ikh)$, that provide supportless radiation of the wave with the amplitude $A_<$ leftwards; and the forces

$$F_{AB} = F_{AB>} = -\varphi^{-1}[1 - \exp(-ikh)]^{-1}[1 - \exp(+2ikh)]^{-1} D_> \tag{5-17}$$

$$F_{BC} = F_{BC>} = -F_{AB>} - F_{CD>}\,,$$

$$F_{CD} = F_{CD>} = -\varphi^{-1}[1-\exp(+ikh)]^{-1}[1-\exp(-2ikh)]^{-1}D_{>}, \quad \textbf{(5-18)}$$

(where $F_{CD>} = F_{AB>}\exp(+ikh)$), that provide supportless unidirectional radiation of the wave with the amplitude $D_{>}$ rightwards. It can easily be seen from Eqs. (5-17) and (5-18) that similar to all the US types considered above, the three-point SUS does not operate at the so-called "resonant" frequencies either; hence, it is necessary to impose the condition $kh \neq n\pi$ ($n = \pm 1, \pm 2, ...$). The conditions for the SUS to produce unidirectional radiation are provided by the force ratios

$$F_{<AB} / F_{<CD} = (\alpha_{>} - \alpha_{<}^{-1}) / (1 - \alpha_{>}^{-1}\alpha_{<}^{-1}), \; F_{CD>} / F_{AB>} = (\alpha_{>}^{-1} - \alpha_{<}) / (\alpha_{>}^{-1}\alpha_{<}^{-1} - 1),$$

where $\alpha_{<} = (A_{<AB} - A_{<BC}) / (A_{<BC} - A_{<CD})$, $\alpha_{>} = (D_{CD>} - D_{BC>}) / (D_{BC>} - D_{AB>})$. Next, we obtain rough estimates of the effect of small variations in media impedances and wave frequency on the violation of the unidirectional radiation conditions for the SUS. For this purpose, consider the case of a symmetric (with respect to x = 0) parameter distributions $Z_A = Z_D = Z_0$, $Z_B = Z_C = Z_0$, $k_A = k_D = k_0$, $k_B = k_C = k_1$, which means (see Eqs. (5-9)) that

$$\alpha_{11} = \alpha_{21} = 2, \; \beta_1 = \beta_2 = \beta, \; \alpha_{11} = \alpha_{22} = 2\beta, \; \beta = (Z_0 - Z_1)(Z_0 + Z_1)^{-1}\exp(-2ik_1 h). \quad \textbf{(5-19)}$$

In the case determined by Eqs. (5–19), we obtain a quantity $\alpha_{>} = \alpha_{<} = \alpha$ independent of k_0, *i.e.*

$$\alpha = \frac{(\eta-1)[\exp(-4ik_1 h) - \exp(-3ik_1 h)] + (\eta+1)[\exp(-ik_1 h) - 1]}{(\eta-1)\exp(-3ik_1 h) + (2+\eta-\eta^2)\exp(-2ik_1 h) - (\eta+1)\exp(-ik_1 h)}, \quad \textbf{(5-20)}$$

where $\eta = Z_0 / Z_1$ (under conditions (5-14), $\alpha = \exp(+ik_1 h)$ for any $k_1 h$). Then, the sensitivity of operating conditions of the thin ($k_1 h << 1$) SUS to deviations in media parameters can roughly be estimated by the quantities $|\partial\alpha / \partial\eta|$ and $|\partial\alpha / \partial(k_1 h)|$: (a) $\eta = 1$ (a transparent SUS), $|\partial\alpha / \partial\eta| \approx 1/4$, and $|\partial\alpha / \partial(k_1 h)| \approx 23/16$; (b) $\eta << 1$ (the medium inside the SUS is denser than the outside), $|\partial\alpha / \partial\eta| \approx 2k_1 h$, and $|\partial\alpha / \partial(k_1 h)| \approx \sqrt{4\eta^2 + 9(k_1 h)^2}$; and (c) $\eta >> 1$ (the medium outside the SUS is denser than inside), $|\partial\alpha / \partial\eta| = 2kh_1\eta^{-2}$, and $|\partial\alpha / \partial(k_1 h)| \approx 2/\eta$. As follows from the above estimates (a), (b), and (c), the unidirectional radiation conditions of the thin SUS can be little sensitive to variations in the field frequency and media impedances. Under the conditions (5-14), the coefficients (5-15), amplitudes (5-16), and forces (5-17), (5-18) could be obtained much easier as a superposition of solutions to (a) the problem of excitation of the field by a pair of forces $\pm F_{AB>}$ (or $\pm F_{<AB}$) applied to the walls $x = -h$ and $x = 0$ (with the free wall $x = +h$) and (b) the problem of excitation of the field by a pair of forces $\pm F_{BC>}$ (or $\pm F_{<BC}$) applied to the walls $x = 0$ and $x = +h$ (with the free wall $x = -h$). The superposition is valid for transparent force sources, but it would be invalid for devices that specify a kinematic variable (coordinate or velocity) of the controlled boundary.

5.2.8. Transparency of the US

Note that the equality $Z_A = Z_B = Z_C = Z_D$ of media impedances is not necessary to provide the transparency of the impedance boundaries with respect to incident waves. It has been shown (see (4-61) in section 4.8.2.) that, when the conditions $k_A h,\; k_D h << 1$ and $\omega h\rho_B,\; \omega h\rho_C << Z_A, Z_D$ are satisfied simultaneously, the three-point SUS is also transparent, however, excitation forces for it are determined by general expressions (5-11), (5-13), rather than by (5-17), (5-18). In addition, the force sources must also be transparent.

5.2.9. Piezoelectric Model of the SUS

In the three-point problem, unidirectional radiation was obtained for two pairs of mutually compensated forces without any external support, and this result sometimes evokes to be suspicions whether the momentum conservation law was satisfied in spite of the rigor in the statement of the problem. However, excitation of the US by two pairs of forces corresponds to rather a simple electromechanical problem, in which intervals between the plated boundaries $x=-h$, $x=0$, and $x=+h$ are filled with a longitudinally polarized piezoelectric (**Fig. 5-4**). The electric voltages (between the boundaries $x=-h$ and $x=0$ and between the boundaries $x=0$ and $x=+h$) are in this case determined by the force ratio $\bar{\varepsilon}_> / \bar{\bar{\varepsilon}}_> = -F_{AB>} / F_{CD>}$, and application of the electric voltages to the boundaries introduces almost no mechanical resistance.

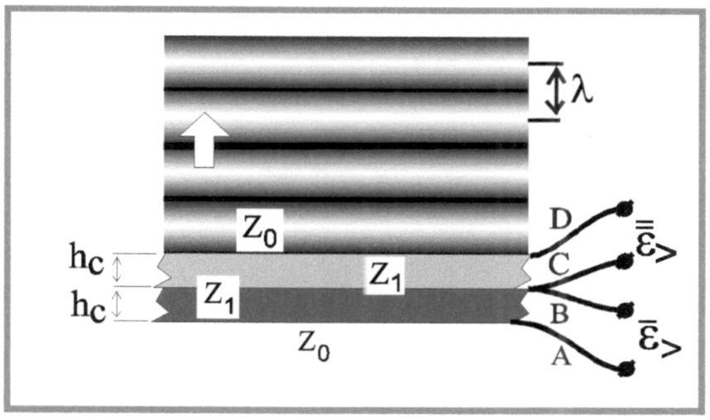

Figure 5–4: Piezoelectric Model of the SUS

5.2.10. The Dynamic Range Overshoot: Linearity of the Transparent SUS

Let us estimate the overshoot of the SUS displacement dynamic range as the ratio of the maximum of the SUS's wall displacement amplitudes to the displacement amplitude in the output wave (*i.e.*, the SUS's front wall displacement amplitude). Using (5–18), it can easily be shown that the inner wall of the impedance boundary $x=0$ has the greatest displacement amplitude ($U_{<BC} = B_< + B_>$, $U_{BC>} = C_< + C_>$, $U_{<CD} = 0$, $U_{AB>} = 0$), where $B_> = B_>(F_{<AB}, F_{<CD})$, $B_< = B_<(F_{<AB}, F_{<CD})$ or $C_> = C_>(F_{AB>}, F_{CD>})$, $C_< = C_<(F_{AB>}, F_{CD>})$, respectively. Equation (5-18) yields $U_{<BC} / U_{<AB} = (B_> + B_<) / A_< = [1-\exp(+ikh)]^{-1}$, $U_{BC>} / U_{CD>} = (C_> + C_<) / D_> = [1-\exp(-ikh)]^{-1}$ and overshoots at $kh \ll 1$, respectively, $U_{<BC} / U_{<AB} = -i / kh$ and $U_{BC>} / U_{CD>} = +i / kh$. Taking into account the usual condition $|U_{<BC}|, |U_{<AB}| \ll 1/k$ for the waves to be linear, we obtain a much stronger constraint on the displacement amplitude in the output wave: $|U_{<BC}|, |U_{<AB}| \ll 1/(k^2 h)$. Also typical of SUSs with small wave dimensions $kh \ll 1$ is the force dynamic range overshoot $|F_{<CD}| / |\omega Z A_<| = |F_{AB>}| / |\omega Z D_>| = (kh)^{-2} \gg 1$ relative to the one-point boundary-value problem. Naturally, the acoustic field inside the SUS, which is strong at $kh \ll 1$, is reactive; *i.e.*, its velocity and pressure are mutually orthogonal in time. In the presence of incident waves, the total wave field is a superposition of the waves considered above in the SUS excitation problem plus the field produced by scattering (if conditions (5-14) are violated) of the incident waves from the boundaries $x=-h$, $x=0$, $x=+h$ of the SUS at $F_{AB} = F_{BC} = F_{CD} = 0$. Unlike the discrete SUS models consisting of monopole arrays considered above, the two-layer model of the one-dimensional SUS with arbitrary layer materials produces reactive acoustic field only inside the SUS. Outside the three-point SUS, reactive (near) fields are absent in the spatially one-dimensional problem. Therefore, such an SUS can be placed directly on the protected surface without any modification of the latter (in the spatially one-dimensional problem).

5.2.11. Application of the SUS to Active Sound Control Problems

Let us consider the operation of the two-layer (three-point and supportless) transparent phased SUS in some of the active sound control problems. **Fig. 5-5** gives a one-dimensional representation of the protected body (interval $|x| \le D$) and two SUSs $2h$ thick placed at distance D from the body. Let the rear wall $x=-h$ be connected mechanically to an elastic rod with arbitrary parameter distributions and length. Then, in the absence of waves incident from $x<-h$ and $x>+h$, the velocities of the walls $x=-h$, $x=0$, $x=+h$ and the displacement field will remain the same as those in the original problem with the infinite rod, including the cases of free and fixed wall $x=-h$. Let us focus on four scenarios of SUS operation.

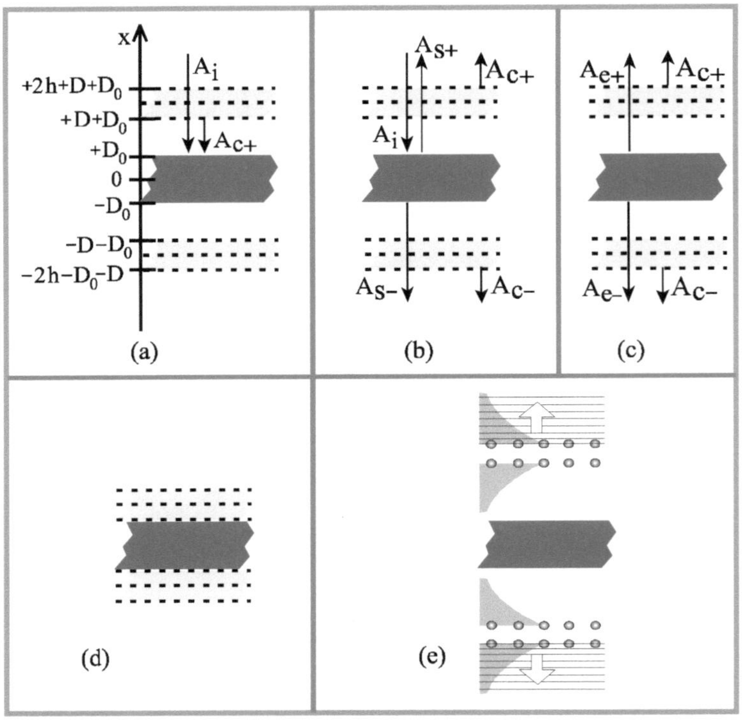

Figure 5–5: One-dimensional SUS (a–d) and an array-type US (e) in active sound control problems.

5.2.11.1. Absorption of the Incident Wave (a Blackbody)

Walls $x=+D+D_0$, $x=+h+D+D_0$, $x=+2h+D+D_0$ are transparent and, in the presence of a wave incident from $x=+h$, the SUS field obtained in the boundary-value problem without incident waves is simply added to the field of the incident waves (**Fig. 5-5-a**). Specifically, in the problem of absorption of the incident wave, the SUS field with amplitude $A_{C+}=-A_i$ compensates the incident field with the amplitude A_i on the surface of the body. In this process, the SUS itself absorbs the energy of the incident wave, while the protected body does not feel the incident wave. The front wall of the SUS is at rest, its rear wall absorbs the energy, and the internal wall executes energy-neutral oscillations (*i.e.*, with the pressure and velocity phases shifted on the surfaces by $(\pi/2)\pm m\pi$ $(m=1,2,...)$ with the displacement amplitude much higher than that of the front wall.

5.2.11.2. Absorption of the Scattered Waves (a Transparent Body)

In the one-dimensional problem of suppressing the scattered field produced by an incident wave, the SUS renders the protected body invisible to an external observer, in no way changing the scattering characteristics of the protected body or the boundary conditions on its surface $x=\pm D_0$ (**Fig. 5-5-b**). In other words, under the action of the incident wave with the amplitude A_i, the body executes the same oscillations as those in the absence of the SUS [15-17]. In particular, on the surface $x=2h+D+D_0$, the field of the wave reflected from the body with amplitude A_{S+} compensates the SUS with amplitude $A_{C+}=-A_{S+}$, thereby absorbing the energy of the reflected wave. At the same time, on the surface $x=2h-D-D_0$, the field of the scattered wave with amplitude $A_{S-}-A_i$ compensates the SUS with amplitude $A_{C-}=-(A_{S-}-A_i)$ while absorbing the scattered wave (where A_{S-} and A_{S+} are the amplitudes of the waves refracted and reflected by the body).

5.2.11.3. Absorption of the Waves Radiated by the Body

In the one-dimensional sound control problem, two SUSs compensate the radiated field. The SUSs make the protected body nonradiating (*i.e.*, the velocities of the boundaries $x=\pm 2h\pm D\pm D_0$ are zero) to an external observer while keeping its boundary conditions on its surface $x = \pm D_0$ unchanged (**Fig. 5-5-c**). Under the action of internal force sources, the body executes the same oscillations as those in the absence of SUSs [15-17]. Specifically, on the surface $x = 2h + D + D_0$, the field of the wave with amplitude $-A_{e+}$ radiated by the body compensates the SUS with amplitude $A_{C+} = -A_{e+}$ by absorbing the energy of the radiated wave. At the same time, on the surface $x = -2h - D - D_0$, the field of the wave with amplitude A_{e-} radiated by the body compensates the SUS with amplitude $A_{C-} = -A_{e-}$ by absorbing the energy of the radiated wave. Since the direction of SUS radiation in the problem of suppressing radiation is the same as that in the scattering problem, these modes can be combined. The distance D between the SUS and the protected body can be arbitrary, in particular, zero (**Fig. 5-5-d**), which corresponds to an active coating of the body. This situation is impossible for the SUS in the form of a discrete array of monopoles, which is characterized by an intense reactive field symmetric about the array center (**Fig. 5-5-e**).

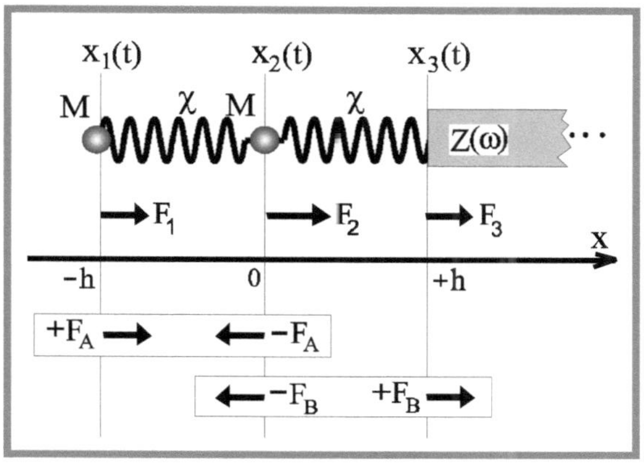

Figure 5-6: Simplest discrete model of a supportless unidirectional pressure source.

5.2.12. A Supportless Unidirectional Pressure Source

The three-point SUS considered above can be regarded as a supportless unidirectional pressure source at frequency $\omega \neq 0$. There is a simple, mechanical, oscillatory (but not of wave nature), discrete analog of this source. Indeed, consider two springs *A* and *B* characterized by stiffness χ and two point masses M connected to a load with an impedance $Z(\omega)$, as shown in **Fig. 5–6**. Applied to the ends of the springs with coordinates of $x_1(t)$, $x_2(t)$, $x_3(t)$ are the pairs of external forces with amplitudes of F_1, F_2, F_3, that are assumed independent of oscillations. These forces are in turn composed of two pairs of forces $\pm F_A$ and $\pm F_B$: $F_1 = +F_A$, $F_2 = -F_A - F_B$, $F_3 = +F_B$. Assume that, at $F_1 = F_2 = F_3 = 0$, *i.e.*, at rest, the coordinates of the spring ends are $x_1(0) = -h$, $x_2(0) = 0$, $x_3(0) = +h$. Complex amplitudes $X_1(\omega)$, $X_2(\omega)$, $X_3(\omega)$, respectively, of deviations $\tilde{x}_1(t) = x_1(t) - x_1(0)$, $\tilde{x}_2(t) = x_2(t) - x_2(0)$, $\tilde{x}_3(t) = x_3(t) - x_3(0)$ from the equilibrium states satisfy the system of equations $-M\omega^2 X_1 = +F_A + f_A$ and $iZ(\omega)X_3 = +F_B + f_B$, where $f_A = \chi(X_1 - X_2)$ and $f_B = \chi(X_2 - X_3)$ are the complex amplitudes of elastic forces of springs A and B, respectively. Assuming that the condition $X_1 = 0$ corresponding to the mass at the point $x = -h$ being at rest is satisfied, we obtain the forces $F_A = \chi[i\omega Z(\omega) - \chi]\omega^{-2}M^{-1}X_3$ $F_B = [1 - (\omega^2 M / \chi)]F_A$ that provide the specified complex amplitude $X_3 \neq 0$ of the load oscillations and amplitude $F = i\omega Z(\omega)X_3$ of the force applied to it.

Of course, in this mode of operation, the mass at point $x = -h$ may be arbitrary or totally absent, while the system provides zero pressure force on point $x = -h$ at a given nonzero pressure force F on the mechanical load and at a zero sum $F_1 + F_2 + F_3 = 0$ of external forces (*i.e.*, in the absence of a mechanical support) that act upon the system. The reaction force of the load, which is equal to $-F_B$, has already been taken into account in the force balance, and it is due to it that the absence of a support is provided for this mechanical source of unidirectional pressure at nonzero oscillation frequencies $\omega \neq 0$. In the unidirectional pressure mode, the ratio of active forces F_A and F_B is independent of the load impedance $Z(\omega)$ and, as the frequency decreases, it is necessary that F_A and F_B grow without limit. Note that pairs $\pm F_A$ and $\pm F_B$ can be implemented with the help of two electric solenoids with movable magnetic cores and light-weight flexible feed wires.

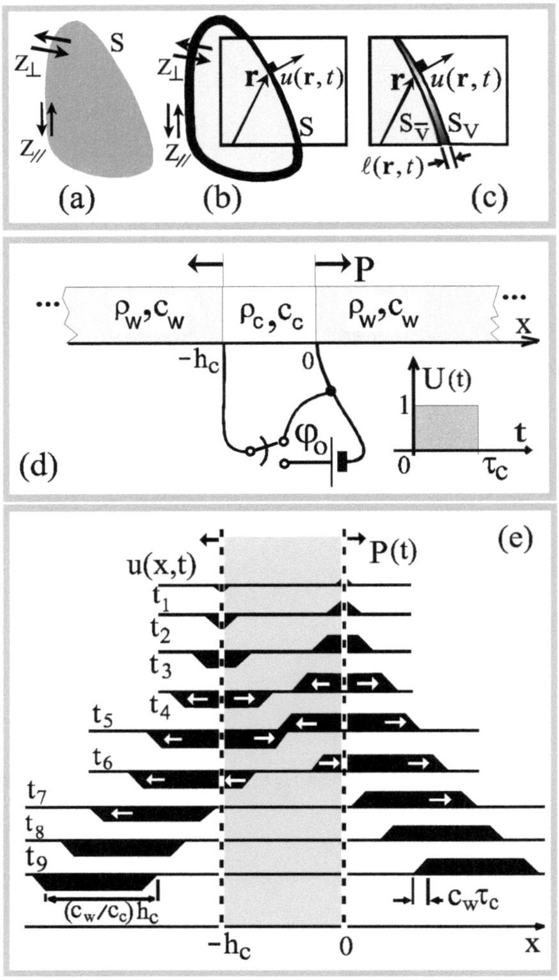

Figure 5-7: Geometry of a three-dimensional problem of controlling the DNPV $u(\mathbf{r},t)$ (a) on the surface S of a homogeneous elastic body and (b) on a shell surface; $Z_{\perp}$ and $Z_{//}$ are the impedances for waves of normal and tangential stresses of the shell and $\ell(\mathbf{r},t)$ is the thickness of the controlled layer; piezoelectric layer $-h_c < x \le 0$ between uniform compressible half-spaces $x < -h_c$ and $x > 0$, and the graph $\phi(t)$ of electric voltage, applied to the layer $-h_c < x \le 0$; instantaneous distributions of the particle displacements $u(x,t)$ in the medium at the instants $0 < t_1 < t_2 < t_3 < t_4 < h_c/(2c_c)$, $h_c/(2c_c) < t_5 < t_6 < h_c/c_c$, $h_c/c_c < t_7 < t_8 < t_9$ under the effect of one voltage pulse $\phi(t)$ of duration $\tau_c << h_c/c_c$.

5.3. A CONTINUOUS PLANAR UNIDIRECTIONAL SOURCE (TEMPORAL REPRESENTATION)

Many problems of controlling vibroacoustic fields are reduced to the formation of a preset space–time distribution of normal particle displacements $u_{\otimes}(\mathbf{r},t)$ or velocities (DNPV) $\{u_{\otimes}(\mathbf{r},t)\}'_t$ at the boundary S between two elastic media (or, for instance, at the outer surface S of a closed shell in a liquid, **Fig. 5-7-b**). One of these media is inside S and has arbitrary vibroacoustic properties. The external medium is supposed to be infinite, homogeneous, and isotropic with a density ρ_w and a sound speed c_w. The zero DNPV corresponds to the solution to the problem of soundproofing and suppression of radiation. An arbitrary nonzero DNPV may be, for example, the solution to the problem of the formation of a preset radiation field at the boundary S or the problem of matching the distribution to incident waves (suppression of scattering) [15-19]. The instrument commonly used for solving such problems is an active piezoelectric layer [66] (**Fig. 5-7-c**) of a controlled thickness $\ell(\mathbf{r},t)$ separating the two media and lying between the outer surface S_V and the inner surface $S_{\overline{V}} = S$. The prescribed DNPV $u_{\otimes}(\mathbf{r},t)$ should be created on the outer surface S_V. The solution of the aforementioned problems, as a rule, is complicated by the fact that $u_{\otimes}(\mathbf{r},t)$ should be formed in real time, *i.e.*, by knowing only the past and current values of the prescribed DNPV and the degree of its smoothness in space and time. For this purpose, exhaustive and periodically updated information on the vibroacoustic characteristics of the boundaries $S_{\overline{V}}$ and S_V separated by the layer is needed. In the simplest case, it may be assumed that the surface $S_{\overline{V}}$ is immobile, *i.e.*, that this surface is in contact with a stationary mechanical support. In particular, this suggests that the impedance $Z_{\perp}$ for waves of normal stresses of surface S_V (in the absence of contact with the external medium) is much greater than the impedance of the external medium, which simplifies the solution (section 4.4.) of noise control problems for, *e.g.*, steel shells in air, when $|Z_{\perp}| >> \rho_w c_w$. However, for a steel shell in water, $|Z_{\perp}| \sim \rho_w c_w$ or $|Z_{\perp}| << \rho_w c_w$ (**Fig. 5-7-b**); in this case, the surface $S_{\overline{V}} = S$ cannot play the role of an acoustically rigid mechanical support for the active layer. n addition, under the condition of neutral floatability of the shell in the liquid, even the perfect rigidity of the shell does not provide a sufficient support for the active layer. Such a perfectly rigid shell should oscillate under the action of the active layer as a monolithic body whose mass is limited by the condition of neutral buoyancy. In the general case, it is necessary to determine an integral impedance operator (section 4.4.) of the surfaces $S_{\overline{V}}$ and S_V, because only in this case is it possible to rigorously formulate the problem of stability and efficiency of the active system. In the cases of practical interest, the volume of required information is so large that the process of learning (or updating the information) for an adaptive control system [21] in many cases lags behind the natural drift of parameters of the boundary-value problem under the effect of changes in temperature, hydrostatic pressure, ageing of materials, and so on. In addition, the linear operator modeling normal-to-surface vibrations of a closed shell disregard the factor of "nonextensibility" of its walls, when the impedance $Z_{//}$ of waves of tangential stresses on the shell surface is much greater in absolute value than the impedance of waves of normal stresses; *i.e.*, $|Z_{//}| >> |Z_{\perp}|$. This makes the fundamental difference between vibrations of a shell and vibrations of a homogeneous elastic body (**Fig. 5-7-a**), where $|Z_{//}| \sim |Z_{\perp}|$. The smallness of changes in the perimeters of the shell compared to its normal deformations imparts nonlinear properties to the model system and severely complicates the control of its vibrations. As a result, the sound-field control system of interest cannot be based on the interaction with the shell.

5.3.1. Characteristic Scales

A fundamental feature of the approach presented in this paper is the absence of any requirements imposed on the rigidity of the surface $S_{\overline{V}}$ (support) or on any information about its vibroacoustic characteristics [90], [98]. It is required on the surface S_V and, to form a prescribed DNPV $u_{\otimes}(\mathbf{r},t)$, the spectral power of which is mainly concentrated in the frequency range

$$\omega_{\min} \le |\omega| \le \omega_{\max} \qquad \textbf{(5-21)}$$

The DNPV $u_{\otimes}(\mathbf{r},t)$ is characterized by the minimal $\tau_{\min} = \pi / \omega_{\max}$ and maximal $\tau_{\max} = \pi / \omega_{\min}$ time scales and by the displacement amplitude $u_{\otimes} \sim A_w$. The thickness $\ell(\mathbf{r},t)$ of the active layer should be much smaller than the wavelength corresponding to the upper boundary of frequency range (5-21) and much greater than the displacement amplitude A_w; *i.e.*,

$$A_w << \ell(\mathbf{r},t) < 2\pi c_w / \omega_{\max} \qquad \textbf{(5-22)}$$

To form the DNPV $u_{\otimes}(\mathbf{r},t)$ on S_V, we use a periodic stepped change of space distributions $u(\mathbf{r},t_n) = u_n(\mathbf{r})$ of normal displacements of surface S_V at the instants $t = t_n = nT$ ($n = 1, 2, \ldots$) with the period

$$T << 2\pi / \omega_{\max} \tag{5-23}$$

The jump like change of distributions $u_n(\mathbf{r})$ takes place as a result of impact-control actions of duration

$$\tau_c << T \tag{5-24}$$

Hence, it is necessary to remove the restriction on possible radiation in the range $|\omega| > \omega_{\max}$. Thus, we try to minimize the deviation $u - u_{\otimes}$ of the surface S from the trajectory prescribed in the interval $(-\infty, t)$ (but unknown beforehand) in frequency range (5-21); *i.e.* $\int\limits_{|\omega|\in[\omega_{\min},\omega_{\max}]} |\tilde{u} - \tilde{u}_{\otimes}|^2 d\omega \to \min$, where $\tilde{u} = \int\limits_{-\infty}^{+\infty} u(\mathbf{r},t)\exp(-i\omega t)dt$ and $\tilde{u}_{\otimes} = \int\limits_{-\infty}^{+\infty} u_{\otimes}(\mathbf{r},t)\exp(-i\omega t)dt$ are the spectra of the actual and prescribed trajectories. Beyond range (5–21), it is sufficient to require that the vibration power be bounded, $\int\limits_{|\omega|\notin[\omega_{\min},\omega_{\max}]} |\tilde{u} - \tilde{u}_{\otimes}|^2 d\omega < \infty$ to provide for the stability of the system. For the one-dimensional case considered below, this means the formation of a prescribed trajectory of displacement $u(0,t) = u_{\otimes}(t)$ of a certain plane boundary, equilibrium position of which corresponds to the point $x = 0$. The time derivative $u_t'(0,t)$ of the displacement represents the particle velocity, whose spectrum should be made close to the spectrum of the function $[u_{\otimes}]_t'$ in frequency range (5-21), while outside this range, it is only restricted by the finiteness of the displacement.

5.3.2. The Boundary-Value Problem

We consider a one-dimensional problem involving the displacements $u(x,t)$ of particles in an elastic medium ($-\infty < x < +\infty$) and assume that these displacements are described by the equation $u_{tt}'' = c^2 u_{xx}''$, where c is the sound speed in the medium. A homogeneous medium with density ρ_w and sound speed c_w corresponds to the domains $x < -h_c$ and $x > 0$. The interval $x \in [-h_c, 0]$ (**Fig. 5-7-d**) is separated by metalized surfaces, electrically independent of one another, and filled with longitudinally polarized piezoelectric with density ρ_c, sound speed c_c, and Young's modulus E_c. We assume that the boundaries $x = 0$, $x = -h_c$ are soundtransparent; *i.e.*,

$$|\rho_w c_w - \rho_c c_c| / (\rho_c c_c) << 1. \tag{5-25}$$

5.3.3. A Rectangular Electric Pulse

If, in the absence of incident waves, a constant electric voltage ϕ_0 is applied to one piezoelectric layer (for instance, to $x \in [-h_c, 0]$) of thickness h_c, then the surface (along the plane $x = 0$) energy density $W_{mech} = E_c h_c^{-1} \psi^2(\infty)/2$ of mechanical deformation $\psi(t)$ and the surface density of electrostatic energy $W_{e\ell} = \varepsilon_0 \varepsilon_c h_c^{-1} \phi_0^2 / 2$ are connected by the electromechanical coupling coefficient $\eta = W_{mech} / W_{e\ell}$ characterizing this piezoelectric material (typically, $\eta \sim 0.2 - 0.7$), where ε_c is the relative dielectric permittivity of piezoelectric material, ε_0 is the dielectric permittivity of vacuum and $\psi(\infty)$ is the value of static (compression–tension) deformation of the layer $x \in [-h_c, 0]$ of thickness h_c: due to $t \to \infty$ there is no difference: we have any medium in half spaces $x < -h_c$ and $x > 0$ or we have vacuum. For $\psi(\infty)$, we obtain the expression $\psi(\infty) = (\varepsilon_0 \varepsilon_c \eta / E_c)^{1/2} \phi_0$, which is independent of the layer thickness. With fixed layer $x \in [-h_c, 0]$ boundaries, we obtain the relationship $P(\infty) = \mu_c \phi_0$ between the piezoelectric pressure $P(\infty)$ and the applied voltage ϕ_0, where $\mu_c = h_c^{-1} (\varepsilon_0 \varepsilon_c \eta E_c)^{1/2}$. Now, we assume that, during the time interval $t \in [0, \tau_c]$ and in the absence of incident waves, the electric voltage $\varphi(t) = \varphi_0 U(t)$, where $U(t) = 1$ at $t \in [0, \tau_c]$ and $U(t) = 0$ at $t \notin [0, \tau_c]$ (**Fig. 5-7-d**). Due to the symmetry of the boundary-value problem, the displacements $u(0,t)$ and $u(-h_c,t)$ of the boundaries $x=0$ and $x=-h_c$ are connected by the relation $u(0,t)=-u(-h_c,t)=\psi(t)$ at $\psi(0)=0$. If, within the time τ_c of action of the electric pulse, the displacement $\varphi(t)$ of the boundaries of the segment $x\in[-h_c,0]$ is much smaller than the static limit, *i.e.*, $|\psi(\tau_c)|/|\psi(\infty)|<<1$, then the action of all piezoelectric layers is equivalent to pressure (compression or tension). This pressure is instantaneously (with the light velocity) and uniformly distributed over the segment $x\in[-h_c,0]$ and is equal to zero outside it. According to the Euler equation for a continuous medium, the particles are subjected to a force proportional to the gradient of this pressure. Therefore, the boundary-value problem is reduced to a simultaneous impact action of pressure $P(t)=\mu_c\varphi(t)$ on the boundary $x=-h_c$ and pressure $-P(t)=\mu_c\varphi(t)$ on the boundary $x=0$. In this case, the boundary displacement $\varphi(t)$ for

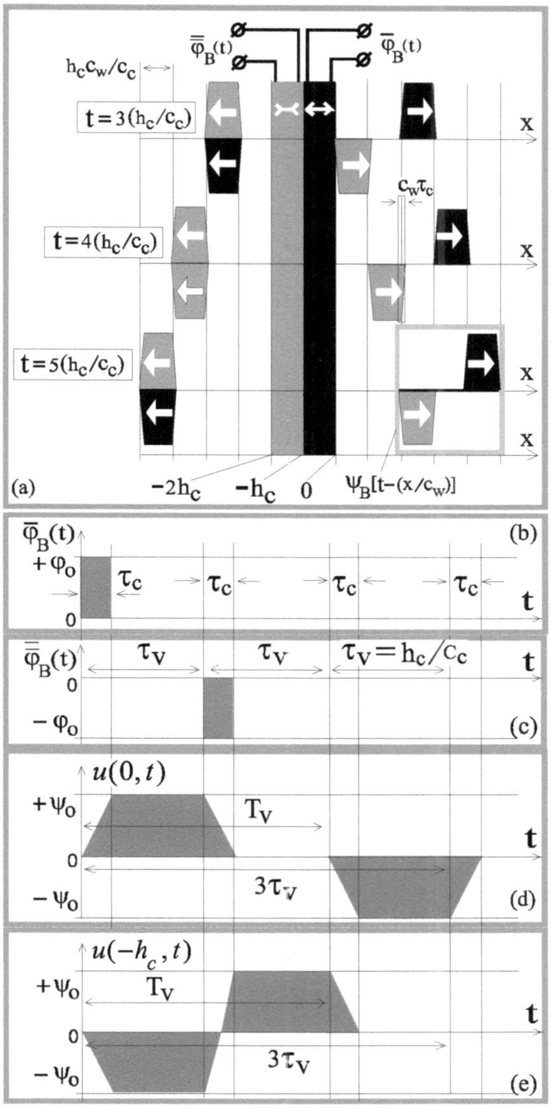

Figure 5-8: Instantaneous spatial distributions of particle displacements in the medium under the effect of piezoelectric layers excited by two electric pulses $\overline{\varphi}_B(t)$ and $\overline{\overline{\varphi}}_B(t)$ (grey graphs are produces by $\overline{\varphi}_B(t)$, black graphs are produced by $\overline{\overline{\varphi}}_B(t)$) (a); voltage pulse $\overline{\varphi}_B(t)$ (b), applied to layer $x\in[-h_c,0]$; voltage pulse $\overline{\overline{\varphi}}_B(t)$ (c), applied to layer $x\in[-2h_c,-h_c]$; temporal distribution $u(0,t)$ of displacement of face boundary $x=0$ (d); temporal distrbution $u(-h_c,0)$ of displacement of inner boundary $x=-h_c$.

$0 \le t \le \tau_c$ is described by the integral $\psi(t) = (z_w/2)^{-1} \int_0^t P(\xi)d\xi$ which yields the "plasticity" of the boundaries $x = -h_c$ and $x = 0$ with respect to short pulses (impacts) of local pressure: after termination of the pulse $P(t)$, the displacement of every boundary under the effect of the pulse persists until the arrival of waves from the other boundary. Below, we assume that the duration of the electric pulses satisfies the condition

$$\tau_c << h_c / c_c . \tag{5-26}$$

Transient processes (and proper time scales) do not manifest themselves in this system because of condition (5-25) of the transparency of the boundaries $x = 0$, $x = -h_c$, and $x = -2h_c$. **Fig. 5-7-e** shows the spatial distributions $u(x,t)$ of particle displacements in the medium under the effect of a single short (see relation (5-26)) pulse of electric voltage $\varphi(t)$ within the interval $x \in [-h_c, 0]$ at various instants of time.

5.3.4. A Huygens Source

In the section concerning the scales, the formation of the trajectory $u(0,t) = u_\otimes(t)$ of the boundary $x = 0$ was assumed to occur within the time $\sim T << 2\pi/\omega_{max}$ in the absence of interaction of the active layer with the underlying surface (support) or with the boundary $x = -h_r - 2h_c$. Therefore, below, we will try to find the shortest time combination of the voltages $b\overline{\varphi}(t)$ and $b\overline{\overline{\varphi}}(t)$ (**Fig. 5-8-b, Fig. 5-8-c**) apply to the piezoelectric layers $x \in [-h_c, 0]$ and $x \in [-2h_c, -h_c]$ that provides a zero field $u(0,t) = 0$ in the domain $x < -2h_c$ and a nonzero field $u(x,t) \neq 0$ in the domain $x > 0$ for any coefficients b. We confine the class of desired voltages $\overline{\varphi}(t)$ and $\overline{\overline{\varphi}}(t)$ to rectangular pulses

$$\overline{\varphi}_B(t) = \overline{b}\varphi_0 U(t - \overline{\tau}), \quad \overline{\overline{\varphi}}_B(t) = \overline{\overline{b}}\varphi_0 U(t - \overline{\overline{\tau}}) \tag{5-27}$$

of duration τ_c satisfying the condition (5-26). Now, it is easy to acertain that the above-mentioned goals can be achieved only for $\overline{\overline{\tau}} - \overline{\tau} = h_c / c_c = \tau_V$, as is shown in **Fig. 5-8-a**. Trying to provide for $u = 0$ in the domain $x < -2h_c$, we obtain a unique combination of delays $\overline{\tau}$, $\overline{\overline{\tau}}$ and amplitudes $\overline{b}$, $\overline{\overline{b}}$, which, in turn, produces a bipolar rectangular wave of displacements propagating rightward in the domain $x > 0$: $u(x,t) = \psi_B[t - (x/c_w)]$. Here, $\psi_B(t)$ is characterized by the following properties: (a) $\psi_B(t) = 0$ for $t < 0$; (b) $\psi_B(t) = const > 0$ for $0 < t \le \tau_V$; (c) $\psi_B(t) = const < 0$ for $2\tau_V < t \le 3\tau_V$; (d) $\psi_B(t) = 0$ for $t > \tau_V$, where $\tau_V = h_c / c_c$. The function $\psi_B(t)$ also satisfies the integral relations inside the finite interval, $\int_0^{\tau_V} \psi_B(t)dt = +\psi_0\tau_V$, $\int_{2\tau_V}^{3\tau_V} \psi_B(t)dt = -\psi_0\tau_V$ and, outside this interval, it equals zero. The functions $\overline{\varphi}_B(t)$ and $\overline{\overline{\varphi}}_B(t)$ satisfy the relation $\int_0^{\tau_c} \overline{\varphi}_B(t)dt = -\int_{\tau_V}^{\tau_V+\tau_c} \overline{\overline{\varphi}}_B(t)dt = \varphi_0\tau_c$, where $\varphi_0 = 2\rho_c c_c \tau_c^{-1} \mu_c^{-1} \psi_0$. The system described above represents a pulsed version of a classical Huygens wave source [90] characterized by one-sided radiation with the sound energy concentrated at the jumps of the function $\psi_B(t)$. The equality $\int_{-\infty}^{+\infty} \psi_B(t)dt = \int_0^{3\tau_V} \psi_B(t)dt = 0$ is a consequence of the momentum conservation law for the medium of wave propagation. Note that the wave radiated only to the right (a wavelet [102]) $\psi_B[t - (x/c_w)]$ of minimal duration $3\tau_V$ necessarily has a pause τ_V between the maxima of different polarities. It is important to note that, in the effect of one-sided radiation, a role of fundamental significance is played by the wave deformations of the layers $x \in [-h_c, 0]$ and $x \in [-2h_c, -h_c]$. For example, in the case of $\rho_c c_c >> \rho_w c_w$ (unlike relations (5-25)), the layers $x \in [-h_c, 0]$ and $x \in [-2h_c, -h_c]$ are perfectly rigid bodies, into which no waves penetrate from the external medium ($x < -2h_c, x > 0$). Such bodies (layers) have the given thickness $d(t) = |u(-2h_c, t) - u(h_c, t)|$ and $\bar{d}(t) = |u(-h_c, t) - u(0, t)|$ and a common boundary $u(-h_c, t)$. In this case, none of the combinations of the functions $d(t)$ and $\bar{d}(t)$ can provide the desired effect of one-sided radiation.

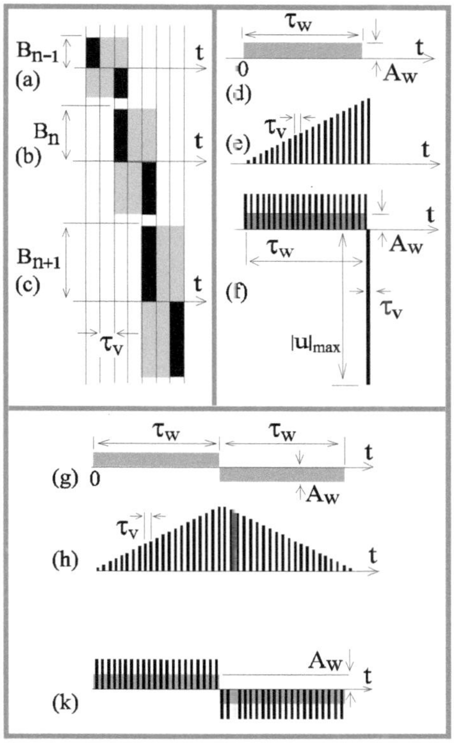

Figure 5-9: Examples of synthesis of the prescribed displacement trajectory of the boundary $x = 0$ by means of controlling the amplitudes B_n of sequence (5–30) of base pulses at zero noise and $\Im = 1$:

(a), (b), (c) successive superposition in time of the base pulses $\psi_B(t)$ with the amplitudes B_{n-1}, B_n, B_{n+1};

(d) the desired displacement $u_{\otimes}(t)$ with a nonzero mean value;

(e) amplitudes B_n of sequence (5–30) of the base pulses $\psi_B(t)$ that approximates the prescribed displacement $u_{\otimes}(t)$ of the boundary $x = 0$;

(f) the resulting displacement u(0, t) of the boundary $x = 0$;

(g) the desired displacement $u_{\otimes}(t)$ with a zero mean value;

(h) amplitudes B_n of sequence (5–30) of the base pulses $\psi_B(t)$ that approximates the prescribed displacement $u_{\otimes}(t)$ of the boundary $x = 0$;

(k) the resulting displacement $u(x, t)$ of the boundary $x = 0$.

5.3.5. Base Pulses

As was shown above, in the absence of incident waves, the displacement $u(0,t)=\psi_B(t)$ of the boundary $x=0$ is a result of the action of the voltage pulses $\overline{\varphi}_B(t)$ and $\overline{\overline{\varphi}}_B(t)$ connected by the relation

$$\overline{\overline{\varphi}}_B(t)=-\overline{\varphi}_B(t-\tau_V) \tag{5-28}$$

and applied to the layers $x\in[-h_c,0]$ and $x\in[-2h_c,-h_c]$. Now, we represent the current voltages

$$\overline{\varphi}(t)=\sum_n B_n\overline{\varphi}_B(t-nT_V)\,,\quad \overline{\overline{\varphi}}(t)=\sum_n B_n\overline{\overline{\varphi}}_B(t-nT_V) \tag{5-29}$$

across the layers $x\in[-h_c,0]$ and $x\in[-2h_c,-h_c]$ as a sum of base voltage, pulses $\overline{\varphi}_B(t)$ and $\overline{\overline{\varphi}}_B(t)$, where $T_V=2\tau_V$ is the pulse repetition period. The current displacement of the controlled boundary x = 0,

$$\psi(t)=u(0,t)=\sum_n B_n\psi_B(t-nT_V) \tag{5-30}$$

is represented as a sum of base pulses of displacement $\psi_B(t)$. Note that the duration of base pulse is $T_{\psi B}=3\tau_V+\tau_c=(3/2)T_V+\tau_c$. The coefficients B_n of expansions (5-29), (5-30) (or the amplitudes of pulses $\overline{\varphi}_B(t)$ and $\overline{\overline{\varphi}}_B(t)$) are determined by the control algorithm.

5.3.6. The Algorithm for Controlling

The purpose of the algorithm for controlling the piezoelectric layers is the synthesis of the prescribed trajectory $u_\otimes(t)$ of displacement of the boundary $x=0$ on the basis of a sequence of bipolar antisymmetric pulses $\psi_B(t)$ with the left third part of every subsequent pulse being superimposed on the right third part of every preceding pulse. The novelty of the proposed approach is determined by the fact that, usually, the prescribed trajectory is approximated by a sequence of nonoverlapping pulses with nonzero mean value of every pulse (for example, by a sequence of delta-pulses), whereas, in the algorithm described below, intersecting bipolar base pulses (**Fig. 5-8-a**) with zero mean are used to synthesize the prescribed trajectory.

The algorithm averages the difference $u(0,t)-u_\otimes(t)$ between the desired $u_\otimes(t)$ and actual $u(0,t)$ displacements of the boundary $x=0$ over the time interval $t\in[T_Vn-T,T_Vn]$ and tends to compensate for this error signal on the average within the interval $t\in[nT_V,(n+1)T_V]$. For this purpose, at the instant $t=nT_V$, the algorithm begins to generate on the interval $t\in[nT_V,nT_V+T_\psi]$ the base pulse $\psi_B(t-nT_V)$ of displacement of the boundary $x=0$ with the amplitude

$$B_n=(\psi_0T/2)^{-1}\int_{T_Vn-T}^{T_Vn}\widehat{F}^{\varepsilon}[u_\oplus-u_\otimes]dt \tag{5-31}$$

where is the operator $\widehat{F}$ of action of one low frequency differentiating network (see below), ε is a positive integer, $u_\oplus(t)$ is the measured displacement of the boundary $x=0$, $u_\otimes(t)$ is the prescribed trajectory of the boundary $x=0$, $\psi_0=\tau_V^{-1}\int_0^{\tau_V}\psi_B(t)dt$, and $T=\Im T_{\psi B}$ ($\Im=1,2,3,\ldots$) is the duration of the interval of averaging. The expression (5-31) means the controlling algorithm, which can be classified as {LS/LT/CA/F} (see **Fig. 1-3**).

5.3.7. Stability of the System

Let us formulate the stability condition for a damping system to prevent the unlimited growth of the impact amplitude. The origin of instability may be both internal (connected with the compensation for the system's own inevitable random errors) and external (connected with the special features of the synthesized trajectory $u_{\otimes}(t)$ preset from outside).

(a) Internal source of instability. We assume that, for a zero preset displacement $u_{\otimes}(t)=0$, one "wrong" base pulse of displacement $\psi_B(t)$ with the amplitude $\xi_1 \neq 0$ accidentally appears at the instant $t=0$; then, algorithm (5-31) tends to provide for $u(0,t)=0$. Sequence of electric pulses (5-29), which serves to compensate for the disturbance of the form $\xi_1\psi_B(t)$ according to algorithm (5-31), has the amplitude distribution

$$B_n(\xi_1)=\xi_1(\psi_0\Im/2)^{-1}(-1)^n\Im^{-n} \quad \text{(5-32)}$$

where $\Im=1,2,3,...$ From (5-32) it follows that, for

$$\Im \geq 1 \quad \text{(5-33)}$$

the conditions $\lim\limits_{n\to\infty}|B_n|=0$ and $\lim\limits_{n\to\infty}\sum_{k=1}^{n}|B_n|<\infty$ are satisfied the system is stable. The value of the averaging interval being a multiple of $T_{\psi B}$ is needed for the most efficient suppression of the oscillating component in the error signal $u_{\oplus}-u_{\otimes}$.

(b) External source of instability. Assume that noise in the system is absent, $\Im=1$, and $u_{\otimes}(t)\neq 0$. The time average value of every base pulse $\psi_B(t-nT_V)$ equals zero. Then, for the synthesis of a constant displacement A_w (**Fig. 5-9-d**), during the characteristic maximum interval $\tau_{max}=\pi/\omega_{min}$ of sign constancy of the function $u_{\otimes}(t)$, the amplitude of base pulses B_n should increase linearly from $A_w\psi_0^{-1}$ at $t=0$ to

$$(B_n)_{max}=(1+\tau_{max}\tau_V^{-1})A_w\psi_0^{-1} \quad \text{(5-34)}$$

at $t=\tau_{max}$ (see **Fig. 5-9-e**). In addition, after the termination of algorithm (5-31), (*i.e.*, at $B_n=0$ and $t>\tau_{max}$), a powerful reverse displacement (**Fig. 5–9–f**) of the boundary with the amplitude

$$|u(0,\tau_{max}+\tau_V)|_{max}=|B_n|_{max}\psi_0=(1-\tau_{max}\tau_V^{-1})A_w >> A_w \quad \text{(5-35)}$$

is inevitable. The large amplitude of reverse displacement is caused by the necessity of obtaining the zero value of the integral $\int_0^{\tau_{max}+\tau_V}u(0,t)dt=0$, where the function $u(0,t)$ consists of the pulses $\psi_B(t-nT_V)$ with zero mean in time. The presence of a nonzero constant component in the signal $u_{\otimes}(t)$ is equivalent to the tendency $\tau_{max}\to\infty$ and, correspondingly, $|u(0,\tau_{max}+\tau_V)|_{max}\to\infty$ and $|B_n|_{max}\to\infty$, which implies instability of the system. **Figs. 5-9-d, e, f** illustrate the incorrectness of the synthesis of a function with nonzero mean using the function combination $\psi_B(t-nT_V)$ (**Fig. 5-9-d**) with a zero time-averaged value. **Figs. 5-9-d,e,f** also show the process of synthesis of the trajectory $u_{\otimes}(t)$ with a zero mean value by algorithm (5-31). In this case, the result of synthesis (**Fig. 5-9-k**) and the sequence of amplitudes B_n (**Fig. 5-9-h**) of base pulses are fully adequate to the stated problem. The prescribed rectangular trajectories $u_{\otimes}$ shown in **Fig. 5-9** are not smooth (as was supposed above) but, all the more, they demonstrate the efficiency of algorithm (5-31). For $\tau_V << \tau_{\min}$, the amplitude of the n-th base pulse may be approximately evaluated as $B_n = (\psi_0 \tau_V / 2)^{-1} \int_0^{\tau_V n} u_{\otimes}(t)dt$ which yields an unlimited increase in B_n with $n \to \infty$. Therefore, for stability of the system (or, for the finiteness of the quantities $|B_n| < \infty$ (and $|u(0,t)| < \infty$), it is necessary to exclude a constant component from the signal $u_{\oplus}(t) - u_{\otimes}(t)$. For this purpose, it is sufficient at the input of algorithm (5-31) to insert

$$\varepsilon \geq 1 \qquad \textbf{(5-36)}$$

series-connected differentiating RC networks with a time constant $\tau_D >> \tau_{\max}$, which are described by the operator $\hat{F}$ (see (5-31)). Such networks with large time constants suppress the Fourier components in the error signal $u_{\oplus}(t) - u_{\otimes}(t)$ in a narrow frequency band $|\omega| < 2\pi / \tau_D$ near the zero frequency but do not distort the signal at the frequencies $2\pi / \tau_D << |\omega| < \omega_{\min}$. As was mentioned above (see relation (5-21)), we consider the trajectories $u_{\otimes}(t)$ with a zero power at zero frequency. Estimates (5-34) and (5-35) of the quantities $|B_n|_{\max}$ and $|u(0, \tau_{\max} + \tau_V)|_{\max}$ imply stronger requirements on the dynamic range of compression (expansion) of the piezoelectric material. However, if, instead of one homogeneous piezoelectric layer of thickness h_c, we use an echelon (**Fig. 5-10-b**) of $N_c >> 1$ layers that have a thickness h_c / N_c and opposite polarizations and are electrically connected in parallel, then, their static expansion (compression) under the action of the same voltage will be N_c times greater. This is possible because the absolute value of linear expansion (compression) of a piezoelectric is proportional to the applied electric voltage but does not depend on the layer thickness. However, in this multilayer piezoelectric system, the recharging current is greater by a factor of N_c.

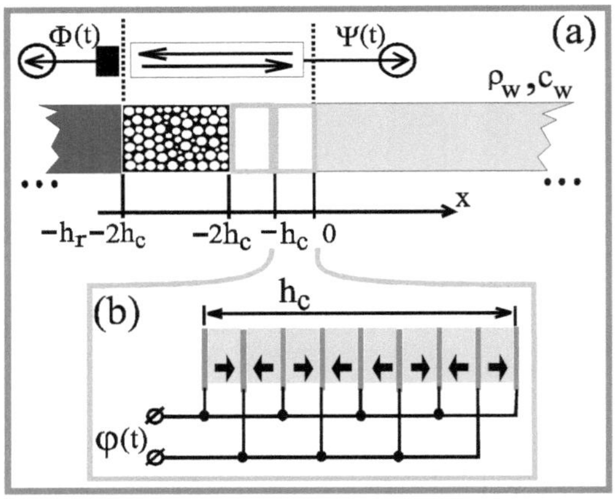

Figure 5-10: Modification of the boundary-value problem with piezoelectric layers $x \in [-h_c, 0]$ and $x \in [-2h_c, -h_c]$ (a) in the form of an echelon of thin layers with opposite polarization and metalized boundaries electrically connected in parallel (b).

5.3.8. The Efficiency of Control

If conditions (5-33) and (5-36) determining the stability of the system are satisfied, the efficiency of the system is characterized by the closeness of the measured coordinate $u_{\oplus}(t)$ of the displacement of the boundary $x = 0$ to the prescribed trajectory $u_{\otimes}(t)$. We take into account such disturbing factors as the following: (i) random noise (an additive factor) with the relative value $\delta_1 << 1$ in measuring and actuating units of the system and (ii) hardware distortions (a multiplicative dynamic factor) with the relative value $\delta_2 << 1$ in measuring and actuating units.

(i) *Random noise.* We consider a sequence of "improper" (to be suppressed by the algorithm) pulses $\xi_n \psi_B(t - nT_V)$ with the random amplitudes ξ_n, where $-\infty < n < +\infty$. We assume that the amplitudes ξ_n have a zero mean value $\langle \xi_n \rangle = 0$ and are uncorrelated; *i.e.*, $\langle \xi_n \xi_m \rangle = \chi^2 \delta_{nm}$, where $\chi = const > 0$, $\delta_{nm} = 0$ for $n \neq m$, and $\delta_{nm} = 1$ for $n = m$. Every n-th pulse beginning at the instant $t = nT_V$ generates a sequence $\Sigma_{m=n}^{\infty} B_m(\xi_n)\psi_B(t - mT_V)$ of compensating base pulses with the coefficients $B_m(\xi_n) = \xi_n(\psi_0 \Im / 2)^{-1}(-1)^m \Im^{-m}$ (see (5-32)). Then, we obtain the estimate of total average noise power $\langle \psi^2(\tilde{t}) \rangle = (1 - \Im^{-2})^{-1}\chi^2$ of the trajectory $u(0,t)$ at the instant $\tilde{t}$ due to the random errors ξ_n of amplitudes of the base pulses started at the instants $-\infty < t < \tilde{t}$ without taking into account the averaging over the interval $\tilde{t} - T < t < \tilde{t}$. With allowance for the interval T of time averaging in algorithm (5-31), the relative noise error δ_1 of the trajectory synthesis is $\delta_1 \leq (1 - \Im^{-2})^{-1/2}\delta_{\oplus}$, where $\delta_{\oplus} = \chi / A_w << 1$ is the total relative error of measuring and actuating units of the system (see the next section).

(j) (ii) *Hardware distortions.* With an increase in T (or $\Im$), the noise error δ_1 decreases, but the relative inertial dynamic distortion $\delta_2 = T / \tau_{\min}$ of the trajectory $u_{\otimes}(t)$ grows. If both δ_1 and δ_2 are small, then the total relative error δ_Σ of the synthesis can be represented as the sum $\delta_\Sigma = \delta_1 + \delta_2$. An increase in T leads to a decrease in δ_1 and an increase in δ_2. Then, there is a certain value of T for which $(\delta_\Sigma)_T = 0$ and δ_Σ is minimal.

5.3.9. Measurement of the Coordinate of the Controlled Boundary

To explain the measurement procedures we consider the modified one-dimensional problem. A homogeneous medium with density ρ_w and sound speed c_w corresponds to the domains $x < -h_r - h_c$ and $x > 0$. The intervals $x \in [-h_c, 0]$ and $x \in [-2h_c, -h_c]$ (**Fig. 5-10-a**) are separated by metalized surfaces, electrically independent of one another, and filled with longitudinally polarized piezoelectric with density ρ_c, sound speed c_c, and Young's modulus E_c. The domain $-h_r - h_c < x < -h_c$ is filled with a polymer material (for example, rubber with density $\rho = \rho_r$ and sound speed $c = c_r$) transparent at low frequencies (5-21) and opaque at high frequencies $|\omega| \geq 2\pi / T_V$ due to dissipation characterized by the space attenuation factor $\sim \exp[-\alpha_r(\omega)x]$ (where $\alpha_r = \alpha_1 |\omega|^{\alpha_2}$ and $\alpha_1, \alpha_2 = const \geq 0$). We assume that the boundaries $x = 0$, $x = -h_c$, $x = -2h_c$ are soundtransparent; *i.e.*,

$$|\rho_r c_r - \rho_c c_c| / (\rho_c c_c) << 1, \quad |\rho_w c_w - \rho_c c_c| / (\rho_c c_c) << 1, \quad \alpha_r(\omega_h) << \omega_h / c_r \tag{5-37}$$

where $\omega_h = \pi / T >> \omega_{\max}$. Some polymer piezoelectric materials have impedances close to the impedance of water [26], [27]. The impedances of various types of rubber may also vary over wide limits. The points $x = -h_r - h_c$ and $x = 0$ are the one-dimensional analogs of surfaces and S_V, and the quantity $(2h_c + h_r)$ is the one-dimensional analog of the active layer thickness $\ell(\mathbf{r}, t)$. We represent the measured coordinate $u_{\oplus}(t)$ of the boundary $x = 0$ in the form $u_{\oplus}(t) = \Psi(t) + \int_0^t d\zeta \int_0^\zeta \Phi(\xi) d\xi$, where $\Phi(t) \sim M\{u(-2h_c - h_r, t\}''_{tt}$ is the signal of an inertial accelerometer with a mass M (**Fig. 5-10-a**) and with a sensing element located at the point $x = -2h_c - h_r$; $\Psi(t) \sim [u(0,t) - u(-2h_c, -h_r, t)]$ is the signal of the sensor measuring the distance between the boundaries

$x = -2h_c - h_r$ and $x = 0$ (for instance, a miniature optical interferometer or electric capacitive sensor). A rather large mass of the inertial body of the accelerometer provides for its sensitivity to a relatively weak slow signal $\Phi(t)$. A powerful component of the signal $u_{\oplus}(t)$ at technological frequencies cannot be represented adequately by the accelerometer because of the propagation of sound waves in its inertial body. However, the source of high-frequency radiation is positioned at the point $x = -2h_c$ at a distance h_r from the accelerometer and, owing to the high-frequency attenuation factor

$$\delta_A = \exp[-\alpha h_r /(\tau_V \pi)] \quad \textbf{(5-38)}$$

influences the signal $\Phi(t)$ much less than $u_{\otimes}(t)$. Condition (5-37) is compatible with smallness (5-22) of the active layer thickness $\ell = h_r + 2h_c << c_w \tau_{\min}$ only when $h_r >> c_r \tau_V$. It is easy to see that neither the inertial accelerometer nor the optical interferometer taken separately can provide the measurement of displacement of the boundary $x = 0$ relative to the inertial reference system in such a wide frequency band and in the absence of a mechanical support.

5.4. SUMMARY

The SUS considered above exhibits unique physical properties, a number of which had been studied earlier [100], [101]. The novelty of this work primarily consists in a study of the possibility of supportless unidirectional radiation. It is shown that, at arbitrary parameters of the media in the layers, unidirectional radiation is possible in the two- and three-point boundary-value problems. In the three-point problem (unlike the two-point problem), unidirectional radiation can also be implemented in a supportless system. It is found that the supportless US, like the mechanical unidirectional pressure source, is an essentially oscillatory device, because its supportless operation is impossible at zero frequency. The spurious contradiction with the momentum conservation law relies on the static representation, which accounts only for the elastic stress in the medium, but neglects the inertia effects. At the same time, it is in the wave motion that elasticity and inertia are inseparable. It was also found that, at frequencies for which $2kh$ is an integer multiple of π, the forward and backward US radiation are zero and the forward and backward power fluxes integrated over the solid angle are identical. An estimate of the maximum absorption cross section of a discrete phased US is obtained, and the direction of maximum absorption is shown to correspond to zero in the receive and transmit directional patterns. It is shown that transparency of an array-type US is limited by the dynamic range of the waves, while transparency of a continuous one-dimensional US is independent of limitations imposed on the dynamic range only by the small wave dimensions of the US. The sharp boundary (with thickness h and wave dimension $kh << 1$) between the insonified and not insonified half-spaces formed by the one-dimensional US requires a high degree of reactivity of forces that excite the US. This situation is similar to the superdirectivity problem in array synthesis. The solutions obtained for three-point problems can also be applied to dispersive media, for which the wave number and impedance are intricate functions of frequency. The results can easily be formulated for compressible media (unlike elastic media considered above), in which the acoustic field is described by the potential. The following observations should be noted. If a sufficiently thin, perfectly rigid (with respect to the surrounding medium) double layer $|x| \le h$ oscillates without changing its shape, it may be transparent (under certain conditions [66]) to incident waves, while the boundaries $x = -h$, $x = 0$, $x = +h$ of this layer are completely nontransparent (and no waves are observed inside it). In the radiation mode, the thickness of this layer oscillates with its center of mass remaining at rest. Therefore, no matter how small the layer thickness is $kh << 1$, to provide the unidirectional radiation, at least partial transparency of the boundaries $x = -h$, $x = 0$ (for radiation leftwards) or $x = 0$, $x = +h$ (for radiation rightwards), rather than of the layer $|x| \le h$ as a whole, is necessary. The poorer the transparency of the boundaries, the higher the necessary excitation force amplitudes. Only a transparent SUS (and a transparent source on the right-hand side of the wave equation) simply adds its field to the field of any other source without affecting its dynamics. The case of transparent boundaries has not earlier been considered for piezoelectrics. The choice of acoustically active materials was limited only to piezoceramics and magnetostrictive alloys that could not be matched with water. Therefore, the proposed design was not important from the practical viewpoint. However, current polymeric technologies are capable of synthesizing piezoelectric materials [26], [27] with a wide and continuous impedance spectrum and of matching the piezoelectric, in particular, to water. Then we investigated the temporal representation of a one-dimensional system consisting of piezoelectric layers $x \in [-h_c, 0]$, $x \in [-2h_c, -h_c]$ and a layer of a high-frequency absorbing polymer in a homogeneous elastic isotropic medium occupying the regions $x < -2h_c - h_r$. The boundaries $x = 0$, $x = -h_c$, $x = -2h_c$ were assumed to be transparent to sound. The minimal-duration configuration of voltage pulses $\overline{\varphi}_B(t)$ and $\overline{\overline{\varphi}}_B(t)$ applied to the layers $x \in [-h_c, 0]$ and $x \in [-2h_c, -h_c]$ is determined, for which the field at the left (for $x < -2h_c - h_r$) is absent and, at the right (for $x > 0$), a wave (wavelet) of particle displacements $u(x,t) = \psi_B[t - (x / c_w)]$ propagates in the medium with an amplitude proportional to the amplitude of the electric pulses $\overline{\varphi}_B(t)$ and $\overline{\overline{\varphi}}_B(t)$. The function $\psi_B(t)$ has the duration $3h_c / c_c$ and consists of two rectangular pulses of different polarity and the same duration h_c / c_c separated by a pause of the same length h_c / c_c. The momentum conservation law for the medium in the absence of the field at the left determines the zero mean value of this wave. A wavelet approximation of the prescribed smooth trajectory $u_\otimes(t)$ of the boundary $x = 0$ by bipolar pulses $B_n \psi_B(t - nT_V)$ of displacement with a repetition period $T_V = 2h_c / c_c$ ($n = 1,2,...$), which are produced by the corresponding sequence of electric pulses $B_n \overline{\varphi}(t - nT_V)$ and $B_n \overline{\overline{\varphi}}(t - nT_V)$, is considered. Every pulse of the sequence is partially superimposed on the preceding one. A control algorithm determining the amplitudes B_n of the pulses $B_n \psi_B(t - nT_V)$ on the basis of the error signal measurement within the preceding time interval T_V is formulated. It is shown that, for stability of the synthesis, it is necessary to do the following: (a) to average the error signal over the interval T_V that is a multiple of the duration $3h_c / c_c$ of the function $\psi_B(t)$ (to suppress the oscillatory component) and (b) to eliminate the zero frequency from the error signal. The accuracy of the trajectory synthesis is evaluated. A technique for measuring the displacement of the boundary $x = 0$ with respect to an inertial reference system in a wide frequency range in the absence of mechanical support is suggested. The system of synthesis of a prescribed trajectory of a controlled boundary is a version [90] of a Huygens source [91]; it has the form of a thin continuous active coating on the protected surface and is considered in the temporal (pulse) representation.

REFERENCES

[1] N.N. Rozanov, "Invisibility: "pro" and "contra", *Priroda*, no. 6, pp. 3–10, 2008 (in Russian).

[2] M. Kerker, "Invisible bodies", *Journal of the Optical Society of America*, vol. 65, pp. 376–379, 1975.

[3] U. Leonhard, "Notes on conformal invisibility devices", *New Journal of Physics*, no. 8, pp. 118–123, 2006.

[4] H. Zhu, R. Raymany, K.A. Stelson, Active control of acoustic reflection, absorption and transmission using thin panel speakers, *Journal of the Acoustical Society of America*, vol. 2, pp. 852-870, 2003.

[5] D. Shurig, J. Mock Justice B., S. Cummer, J. Pendry, A. Starr, D. Smith, "Metamaterial electromagnetic cloak at microwave frequencies", *Science*, vol. 314, pp. 977–980, 2006.

[6] S. A. Cummer, B. I. Popa, D. Schurig, D. R. Smith, J. Pendry, M. Rahm and A. Starr, "Scattering theory derivation of a 3D acoustic cloaking shell", *Physical Review Letters,* vol. 100, pp. 02430, 2008

[7] J. B. Pendry, D. Schurig, D. R. Smith, "Controlling electromagnetic fields", *Science,* vol. 312, pp.17802, 2006.

[8] Pendry J B and J. Li "An acoustic metafluid: realizing a broadband acoustic cloak" *New Journal of Physics,* vol. 10, pp. 115032, 2008.

[9] S.A. Cummer, B.I. Popa, D. Shuring, D.R. Smith, J. Pendry, M. Rahm, A. Starr, "Scattering theory derivation of a 3D acoustic cloaking shell", *Physical Review Letters*, vol. 100, pp. 024301, 2008.

[10] H. Chen, C.T. Chan, "Acoustic cloaking in three dimensions using acoustic metamaterials", *Appied Physics Letters,* vol. 91, pp. 183518, 2007.

[11] Yu.I. Bobrovnitskii, "A nonscattering coating for a cylinder", *Acoustical Physics*, vol. 54, no. 6, pp. 758–768, 2008.

[12] I. Lasiecka, B. Morton, *Control Problems in Industry*. Birkhauser, London UK, 1995.

[13] G. Avalos, I. Lasiecka, "Exact controllability of structural acoustic interactions", *Journal de Matematiques Pure et Applique*, vol. 82, pp. 1074–1073, 2003.

[14] I. Lasiecka, Triggiani. *Research Monograph: Deterministic Control Theory for InfiniteDimensional Systems*. Cambridge University Press: Cambridge, UK, 1999.

[15] G. D. Malyuzhinets, On one theorem on analytical functions and its extension to the wave potentials, *Proceedings of 3rd All-Union Symposium on Wave Diffraction* (Moscow: Nauka), pp. 113–116, 1964

[16] G.D. Malyuzhinets, Nonstationary diffraction problems for a wave equation with finite right parts. *Proceedings of Acoustical Institute*, no. 15, pp. 124–139, 1971 (in Russian).

[17] M.V. Fedoryuk, "On the works by Malyuzhinets in the theory of wave potentials", *Proceedings of Acoustical Institute*, no. 15, pp. 169-171, 1971 (in Russian).

[18] A.A. Mazanikov, V.V. Tyutekin, "Investigation of active autonomic sound absorption system in multimode waveguide", *Proceedings of the 9th International Congress of Acoustics*, pp. 260, 1977.

[19] *Active methods of acoustic field suppression (based on Huygens surfaces).* Ed. by V.V. Tyutekin, TsNII Rumb, Moscow, 1982 (in Russian).

[20] David A.B. Miller, "Perfect cloaking", *Optics Express*, vol. 14, no. 25, pp. 12457–12466, 2006.

[21] B. Widrow, J.M Mc. Cool, "A comparison of adaptive algorithms based on the methods of steepest descent and random search", *IEEE Transaction: Antennas and Propagation.*, vol. 24, no. 5, pp. 615-637, 1976.

[22] T.R. Howarth, V.K. Varadan, X. Bao, V.V. Varadan, "Piezocomposite coating for active underwater sound reduction", *Journal of the Acoustical Society of America*, no. 2, pp. 823–831, 1992.

[23] A. Premount, *Mechatronics, Dynamics of Electromechanical and Peizoelectric Systems*. Springer: New York, NY USA, 2006.

[24] J.-F. Deu, W. Larbi, R. Ohayon, "Piezoelectric structural acoustic problems: symmetric variational formulations and finite element result", *Computer Methods in Applied Mechanics and Engineering*, vol. 197, pp. 1715–1724, 2008.

[25] R. Gentilman, D. Fiore, R. Torri, J. Glynn, "Piezocomposite smart panels for active control of underwater vibration and noise", *Proceedings of 5th SPIE International Symposium on Smart Structures and Materials*, pp.1–7, 1998.

[26] G. A. Lushcheykin, "New polymer containing piezoelectric materials", *Physics of Solid State*, vol.48, no.6, pp. 1023–1024, 2006.

[27] H. Kawai, "The piezoelectricity of polyvinylidene ftoride", *Japanese Journal of Applied Physics*, no. 8, pp. 975–976, 1969.

[28] C.R. Fuller, S.J. Elliott, P.A. Nelson, *Active Control of Vibration*. Academic Press: London, UK, 1996.

[29] C.H. Hansen, S.D. Snyder, *Active Control of Noise and Vibration.* E&FN: London, UK, 1997.

[30] E. Friot, C. Bordier, "Real-time active suppression of scattered acoustic radiation", Journal of Sound and Vibration, no. 3, pp. 563–580, 2004.

Vladimir V. Arabadzhi

[31] S. Uosukainen, "Active sound scatterers based on JMC method", *Journal of Sound Vibration.* Vol. 267, pp. 979–1005, 2003.

[32] S.J. Elliott, P. Joseph, P.A. Nelson, M.E. Johnson, "Active output absorption in the active control sound", *Journal of Acoustical Society of America*, no. 5, pp. 2501–2512, 1993

[33] P.A. Nelson, S.J. Elliott, *Active control of Sound.* Academic Press: London, UK, 1993.

[34] A. Premount, *Vibration Control of Active Structures, an Introduction.* Kluwer Academic Publishers, New York, NY, USA, 2002.

[35] M. Furstoss, D. Thenail, M.A. Galland, "Surface impedance control for absorption: direct and hybrid passive-active strategies", *Journal of Sound and Vibration*, no. 6, pp. 219–236, 2003.

[36] A.A. Mal'tsev, R.O. Maslennikov, V.V. Cherepennikov A.V. Khoryaev, "Adaptive active noise and vibration control systems", *Acoustical Physics*, vol. 51, no. 2, pp. 195–209, 2005.

[37] V.T. Ermolaev, A.A. Mal'tsev, K.V. Rodyushkin, "Statistical characteristics of the AIC and MDL criteria in the problem of estimating the number of multivariate signals in the case of a short sample", *Radiophysics and Quantum Electronics*, vol. 44, no. 12, pp. 977–983, 2001.

[38] A.A. Mal'tsev, S.V. Zimina, "Exact results of statistical analysis of multichannel adaptive systems with continuous gradient algorithms", *Radiophysics and Quantum Electronics*, vol. 42, no. 9, pp. 805–810, 1999.

[39] A.A. Mal'tsev, and S.V. Zimina, "Influence of fluctuations on the characteristics of adaptive antenna arrays", *Radiophysics and Quantum Electronics*, vol. 43, no. 1, pp. 76–84, 2005.

[40] A.A. Mal'tsev, "Improving the LMS adaptive filter", *Active Sound & Vibration Control News*, 1997, vol. 4. no. pp. 4, 1997.

[41] A.A. Belyakov, A.A. Mal'tsev, "An adaptive system for active cancellation of the acoustic field with an auxiliary identification channel", *Radiophysics and Quantum Electronics*, vol. 38, no. 3, 1995.

[42] A.A. Mal'tsev, R.O. Maslennikov, A.V. Khoryaev, V.V. Cherepennikov, "Experimental study of a system for automatic control of the boundary condition in a waveguide with adaptive tuning algorithm", *Radiophysics and Quantum Electronics*, vol. 13, no. 3, pp. 241–247, 1988.

[43] Yu.I. Bobrovnitski, "A new solution to the problem of an acoustically transparent body", *Acoust. Physics*, vol.50, no. 6, pp. 647–651, 2004.

[44] V. P. Dragunov, I. G. Neizvestnyi, and V. A. Gridchin, *Fundamentals of Nanoelectronics.* NGTU, Novosibirsk, 2000, (in Russian).

[45] P. Fay *et al.*, "Monolithically integrated high-speed MSM/HEMT and PIN/HEMT photoreceivers", *IEEE Transactions Microwave Theory and Techniques*, vol. 50, 2002.

[46] D.I. Voskresenskii and E.V. Ovchinnikova, "Characteristics of scanning antennas of super-short pulses on the base of spectral analysis", *Antenny*, vol. 46, no. 3, pp. 17–26, 2000.

[47] A.O. Kas'yanov and V.A. Obukhovets, "Intellectual radioelectronic coatings. Temporal position and problems. Review", *Antenny*, vol. 50, no. 4, pp. 4–11, 2001.

[48] A.O. Kas'yanov and V.A. Obukhovets, "Reflective arrays as a microwave components of intellectual coating", *Antenny*, vol. 50, no. 4, pp. 12–19, 2001.

[49] A.I. Semenikhin, "Impedance model of black states of cylindrical intellectual coating on the base of twist-effect", *Antenny,* vol. 50, no. 4, pp. 20–26, 2001.

[50] V.V. Arabadzhi, "On the cancellation of low frequency waves in laboratory tanks", *Fizika Atmosferi i Okeana*, vol. 28, no. 12, pp. 1205–1212, 1992.

[51] S.N. Pzhevkin, *Course of lectures on the theory of sound.* Published by Moscow University, 1960 [in russian].

[52] R.H. Lyon, "Noise reduction of rectangular enclosures with one flexible walls", *Journal of Acoustical Society of America*, vol. 35, pp. 1791–1796, 1963.

[53] K.V. Horoshenkov, and K. Sakagami, "A method to calculate the acoustic response of a thin, baffled, simply supported poroelastic plate", *Journal of Acoustical Society of America*, vol. 110, pp. 904–917, 2001.

[54] M. Cuesta, P. Cobo, "Active control of the exhaust noise radiated by an enclosed generator", *Applied Acoustics*, vol. 61, pp. 83–94, 2000.

[55] M. Cuesta, P. Cobo, "Optimization of an active control system to reduce the exhaust noise radiated by a small generator", *Applied Acoustics*, vol. 62, pp. 513–526, 2001.

[56] Lixi Huang, "Modal analysis of a drum like silencer", *Journal of Acoustical Society of America*, vol. 112, pp. 2015-2025, 2002.

[57] Lixi Huang, "Experiment studies of a drum like silencer", *Journal of Acoustical Society of America*, vol. 112, pp. 2026–2035, 2002.

[58] V.V. Arabadzhi, "On the Space-Time Local Active Control", *Journal of Low Frequency Noise and Vibration and Active Control*, V. 16, no. 2, pp. 89-108, 1997.

[59] V.V. Arabadzhi, "On the space-time local active wave control", *Proceedings of 10th International Conference on A daptive Structures and Technologies*, p. 626-633, 1999.

[60] V.V. Arabadzhi, "Local emitters in the regime of maximum absorption", *Journal of Low Frequency Noise and Vibration and Active Control*, vol. 18, no. 3, pp. 129–147, 1999.

[61] V.V. Arabadzhi, "Runing cyclical wave-bolt", in Proceedings of 25th ISMA-25 Noise and Vibration Engineering Conference, vol.1, pp. 1–6, 2000.

[62] V.V. Arabadzhi, "Cyclic wave gate", *Radiophysics and Quantum Electronics*, vol. 43, no. 7, pp. 593–599, 2000.

[63] V.V. Arabadzhi, "Absorption of long waves by nonresonant parametric microstructures", *Radiophysics and Quantum Electronics*, vol. 44, pp. 249–261, 2001.

[64] V.V. Arabadzhi, "Cyclical wave bolt", *Journal of Low Frequency Noise and Vibration and Active Control*, vol. 20, pp. 239–254, 2001.

[65] V.V. Arabadzhi, "Nonreflecting switching microstructure", *Journal of Communications Technology and Electronics*, vol. 50, no. 5, pp. 561–573, 2005.

[66] V.V. Arabadzhi, "Suppression of the sound of a vibrating body by monopoles attached to its surface", *Acoustical Physics*, vol. 57, no. 5, pp. 505-512, 2006.

[67] V.V. Arabadzhi, "Absorption of long waves by linear structures", *Acoustical Physics*, vol. 56, no. 6, pp. 996-1003, 2010.

[68] A.D. Lapin, "Sound absorption by monopole-dipole resonators in multimode waveguide", *Acoustical Physics*, vol. 51. No. 3, pp. 428–430, 2005.

[69] N.G. Kanev, M.A. Mironov, "Dipole resonant silencer on the edge of narrow tube", *Acoustical Physics*, vol. 52, no. 3. pp. 335–339, 2006.

[70] E.V. Korotayev, V.V. Tyutekin, "Experimental investigation of the active damping system of planar shape", *Acoustical Physics*, vol. 46, no. 1, pp. 84–88, 2000.

[71] H. Paul, P. Fisher, "Absorption of light by dipole", *Soviet Physics*, vol. 141, no. 2, pp.375–381, 1983.

[72] Sergei A. Shelkunoff, Harald T. Friis, *Antennas (theory and practice)*. New York. John Wiley& Sons, Inc. London. Chapman & Hall, Limited 1952.

[74] R..S. Hansen, "Connections between antennas as scatterers and emitters", *IEEE*, vol. 77, no. 5, pp. 30–34, 1989.

[73] M. Jessel, G.A. Mangiante, "Active sound absorbers in an air ducts", Journal of Sound and Vibration, no. 3, pp. 383-390, 1972.

[75] H.J. Riblet, "Note on the Maximum Directivity of Antenna", *Proceedings of Institute of Radio Engeneers*, vol. 36, pp. 620–624, May, 1948.

[76] Y.V.Gulyaev, A.N.Lagarkov, S.A.Nikitov, "Methamaterials: fundamental investigations and perspectives of application", *Proceedings of Russian Academy of Sciences*, vol. 78, no. 5, pp. 438–457, 2008 (in Russian).

[77] L. N. Zakhar'ev and A. A. Lemansky, *Wave scattering from black bodies.* Sovetskoe Radio, Moscow, 1972 (in Russian).

[78] S.P. Efimov, "Absolutely black body in the theory of diffraction", *Radiotekhnika i Elektroika*, vol. 23, pp. 7–12, 1978 (in Russian).

[79] E. Skudrzyk, *The foundations of acoustics.* Springer-Verlag, Wien New York, 1971.

[80] Yu.I. Bobrovnitskii, "Impedance theory of sound scattering and absorption: A constrained best absorber and the efficiency bounds of passive scatterers and absorbers", *Acoustical Physics*, vol. 53, no. 1, pp. 100-104, February, 2007.

[81] Yu.I. Bobrovnitskii, "Impedance theory of sound absorption: The best absorber and the black body", *Acoustical Physics*, vol. 52, no. 6, pp. 638–647, December, 2006.

[82] A.N. Tikhonov and A.A. Samarskii, *Equations of mathematical physics.* 4th ed. Nauka, Moscow, 1972; Pergamon, Oxford, 1964.

[83] L.D. Landau, E.M. Lifshitz, *Fluid mechanics, secondeEdition: volume 6 (course of theoretical physics).* Butterworth-Heinemann; 2 edition, January 15, 1987.

[84] L.F. Lependin *Acoustics.* Moscow, Visshaya shkola, 1978.

[85] V.V. Nikol'skii, *The electromagnetic field theory*. Vysshaya Shkola, Moscow, 1961 (in Russian).

[86] M.I. Kontorovich, M.I. Astrakhan, V.P. Akimov, and G.A. Fersman, *Electrodynamics of grid structures.* Radio i Svyaz', Moscow, 1987 (in Russian).

[87] S.B. Cohn, "Analysis of the metal-strip delay structure for micro-wave lences", *Applied Physics J.*, vol. 20, pp. 257–262, 1949.

[88] Y. Michel, R. Paushard, P. Vidal, "Le radant: nouven procede de balayage electronique antennas". L'onde electrique, vol. 59, no. 12, pp. 87–94, 1970.

[89] Yu.I. Bobrovnitskii, "Active control of sound in a room: method of global impedance matching", *Acoust. Physics*, no. 6, pp. 731–737, 2003.

[90] V.V. Arabadzhi, "Active Control of Normal Particle Velocity at a Boundary between Two Media", *Acoust. Physics*, vol. 51, No. 2, pp. 139–146, 2005.

[91] V.V. Arabadzhi, "Supportless unidirectional acoustic sources", *Acoust. Physics*, vol. 55, no. 1, pp. 120–131, 2005.

[92] V.V. Arabadzhi, "Conversion of an acoustically hard body into an acoustically transparent one in an initial-value problem", *Acoust. Physics*, vol.54, no. 6, pp. 749–757, 2008

[93] M.C. Junger, D. Feit, Sound, *Structure, and Their Interaction*. ASA Publications: New York, NY USA, 1993.

[94] V. Arabadzhi, "The Method of Active Suppression of Acoustical Noises of Radiating Objects", USSR Patent 1489460, April, 1986.

[95] E. Friot, C. Bordier, "Real-time active suppression of scattered acoustic radiation", *Journ. of Sound and Vibration*, no. 3, pp. 563–580, 2004.

[96] E. Friot, R. Guillermin, M. Winninger, "Active control of scattered acoustic radiation: a real-time implementation for a three-dimensional object", *Acustica Acta*, no. 2, pp. 278–283. 2006.

[97] A.L. Virovlyanskii, *Ray Theory of Long-Range Sound Propagation in the Ocean*. IPF RAN, Nizhni Novgorod, 2006 (in Russian).

[98] V.V. Arabadzhi, "Algorithm for active suppression of radiation and scattering fields by some physical bodies in liquids", *Algorithms*, 2009, vol. 2, no. 1, pp. 361–397, January 2009. [Online] Available: www.mdpi.com. [Accessed January 8, 2009].

[99] Danielle Moreau, Ben Cazzolato, Anthony Zander, Cornelis Petersen, "A review of virtual Sensing Algorithms for Active Noise Control", *Algorithms*, no.1, pp. 69–99, November 2008. [Online] Available: www.mdpi.com. [Accessed Oct. 29, 2008].

[100] M. Jessel, "Secondary sources and their energy transfer", *Acousical Letters,* vol. 9, no. 4, pp. 174-179, 1981.

[101] G. Mangiante and J. Vian, "Application du principe de Huygens aux absorbers acoustiques actifs", *Acoustica*, vol. 32, no. 3, pp. 175–182, 1977.

[102] I. Daubechies, *Ten Lectures on Wavelets.* Society for Industrial and Applied Mathematics, Philadelphia, 1992.

Index

Vladimir V. Arabadzhi

G

H

I

L

M

N

O

P

R

S

T

V

U

W

www.ingramcontent.com/pod-product-compliance
Lightning Source LLC
Chambersburg PA
CBHW041705210326
41598CB00007B/544
* 9 7 8 1 6 0 8 0 5 3 6 9 8 *